中法工程师学历教育系列教材

工业科学

(第1卷)

Industrial Science

(Volume 1)

(英文版)

Guillaume Merle, Vincent Crespel, Hervé Riou, Jiming Ma, Ping Xu 著

本书得到北京航空航天大学“研究生英文教材出版基金”资助

科学出版社

北京

内 容 简 介

本书主要介绍与现代复杂工业系统设计分析相关的基础知识。第 1 章介绍系统分析方法和理论，从系统工程角度介绍复杂工业系统的分析基础；第 2 章介绍经典的线性控制理论；第 3～6 章分别介绍机械机构的基础理论、机械机构的描述方法、运动学/动力学分析方法等；第 7 章介绍工业系统中的数字控制；第 8 章介绍时序控制的基础理论。

本书可作为工科学校高年级本科生的专业基础课教材，也可以作为工业工程专业研究生的基础课教材，还可供具有工业工程背景的专业领域工程师参考。

图书在版编目(CIP)数据

工业科学=Industrial science：英文/(法)付小尧(Merle, G.)等著.
—北京：科学出版社, 2013
中法工程师学历教育系列教材
ISBN 978-7-03-038429-4

I. ①工… Ⅱ. ①付… Ⅲ. ①工业-教材-英文 Ⅳ. ①T

中国版本图书馆 CIP 数据核字 (2013) 第 197179 号

丛书策划：匡 敏 余 江
责任编辑：余 江 李岚峰／责任校对：胡小洁
责任印制：闫 磊／封面设计：迷底书装

科学出版社 出版
北京东黄城根北街 16 号
邮政编码：100717
http://www.sciencep.com

北京文林印务有限公司印刷
科学出版社发行 各地新华书店经销
*
2013 年 5 月第 一 版 开本: 720×1000 B5
2013 年 5 月第一次印刷 印张:19
字数: 304 000

定价: 49.00 元

北京航空航天大学中法工程师学院
工程师教材融合编委会

丛 书 序

我国《国家中长期教育改革和发展规划纲要（2010–2020 年）》明确提出，要"适应国家经济社会对外开放的要求，培养大批具有国际视野、通晓国际规则、能够参与国际事务和国际竞争的国际化人才"，为此教育部于 2010 年启动了"卓越工程师教育培养计划"，并把培养国际化工程人才作为我国高等工程教育改革发展的战略重点之一。通过与国际高水平大学开展人才培养合作，借鉴国外先进经验，引入国外优质教育资源并结合自身优势，面向国家发展战略需求，建立植根于本土的工程师学历教育体系，是培养具有国际竞争力工程师人才的重要途径，也是贯彻落实"人才强国"战略、提升我国国际竞争力的重要举措。

采用精英培养模式的法国工程师学历教育对法国乃至世界经济、社会发展起到了重要的推进作用，许多工程师院校在世界范围内享有盛誉。为此，近年来我国许多大学对这种培养模式进行了深入研究，并成立了多家中法合作的工程师培养机构。这些具有国际化教育目标与理念的办学机构与项目，已经成为我国高等工程教育的重要组成部分，取得的成功经验深刻影响着我国高等工程教育改革与创新进程。

作为我国教育部批准的第一家中法教育合作培养通用工程师人才的教育机构，北京航空航天大学中法工程师学院于 2005 年由北京航空航天大学与法国中央理工大学集团合作建立，在创立和实施我国的国际通用工程师学历教育过程中，通过借鉴法国工程师培养理念，引进国外优质教育资源，结合北京航空航天大学自身优势，建立了卓越工程师培养本-硕统筹课程体系，赢得了国内外教育界、工业界的广泛认同与赞誉，并通过了法国工程师职衔委员会（CTI）和欧洲工程教育 EUR-ACE 体系的认证，成为迄今为止国内唯一一家具有在本土颁发法国和欧洲工程师文凭资质的办学机构，培养出来的毕业生得到了用人单位的普遍欢迎和高度评价。

为把探索实践过程中取得的成功经验和优质课程资源与国内外高校分享，我们在北京市教委和科学出版社的支持下，组织出版了这套《中法工程师学历教育系列教材》，其中包括由法国著名预科教师和法国工程师学院一线教师领衔编写的法文版、英文版和中文版的预科数学、物理、工业科学教材，以及适合工程师培养阶段的专业教材。本套教材可作为中法合作办学单位的预科和专业教材，也可作为其他相关专业的参考教材。

希望本套教材能为我国卓越工程师的教育培养作出贡献！

熊璋

北京航空航天大学中法工程师学院院长

2013 年 5 月

Preface

During the 19th and 20th centuries, technology was used to test new ideas. The technological solutions were the result of the imagination of genius men and women and were used by a few people and designed to answer problems of some specialists. Nowadays, the technological solutions are created by industries, result from scientific analysis and logical developments, and are designed to be used by everybody to solve the complex problems of their daily life. This evolution represents a real change in the paradigm of the creation of innovative solutions to address the future problems of mankind. Industries can now do this thanks to the training of excellent engineers and creators, based on a rational discipline called "industrial science".

Recently, Chinese and French scientists engaged a conversation on the possibility of publishing works on industrial science at the highest level to put at the disposal of the scientific community a synthesis of the modern methods of analysis and design of technological systems. The result of this conversation is this series of books (the present book is for the first year of University) that we have the pleasure to present. It realizes the synthesis of the scientific methods used to analyze, model, validate and design complex industrial systems. It is as complete as possible for the present time, according to the present degree of development of industrial science.

This book has been written by Chinese and French scientific colleagues, which creates the richness of this project. It tries to be as pedagogic as possible, to help anyone understand the mathematical and physical developments of industrial science and their technological applications, which is the fundamental point of this discipline. It represents a work which does not seem to have been undertaken by different cross-cultural views at this level anywhere else in the world.

Apart from this book, the Chinese and French colleagues have made a real investigation to understand when and how industrial science could be taught to train future engineers. Based on the knowledge of the French system of engineering schools, they decided to introduce industrial science as a lecture given to students from the early first year of their higher education until the last year.

This choice was motivated by the aim to give them some interdisciplinary competences (in mechanics of systems, automation engineering, hydraulic science, system control, computer science, applied mathematics, etc.) as soon as possible in their training. Moreover, they wanted to help the students to develop a strong technology culture through the study of innovative industrial systems, which takes time, as soon as possible.

The necessity to set up some experimental activities on real industrial systems seemed rapidly clear, as practical works can provide the students with some competences in the understanding of the industrial complexity of a real system. This practical works also give the students possibilities to make some interesting measurements and to validate their models. As a consequence, it was decided to create, at the École Centrale de Pékin (ECPk), an industrial science laboratory with real industrial systems. Such a laboratory is the first one created in China. It contains, for the present time, a boat autopilot, a power-assisted steering system, a packaging machine and an electrical assisted bike, and new investments have already been decided to develop this laboratory.

Based on the analysis of the Chinese and French colleagues and on the feedback of the students, the creation of this laboratory was an excellent choice as industrial science really helps to train future engineers by giving them a contact with the reality of industrial systems. Besides, the tasks that they have to carry out during their professional career are in relation with the design of new innovative systems, the measurement of performances, the prediction of behaviors, the modification of some technological parts or control laws, etc., and each of these tasks can be carried out in the industrial science laboratory that was created.

Today, the industrial science laboratory plays a major role in the training of the students. All the lectures are illustrated by some experimental activities and some concepts are first identified during the experimental activities. All the new scientific concepts are in relation with a problem defined on a real industrial system. The students can manipulate and test some innovative technologies coming from the industrial world. They can understand how these technologies work, what they contain, and how to pilot them. They can now easily describe a complex system and they are able to make some clear relations between the theoretical lectures and their immediate application in the industry or in real life. Based on these observations, they can imagine some new technologies with better performances. To sum up, they begin to become real engineers during their early higher education.

The Chinese and French colleagues have also worked on the best way to use the experimental activities to help the students in their future career. After some pedagogical and cooperative work, they have decided to train the students in having a global approach of complexity. The daily life of an engineer is indeed concerned with the modeling of complex problems, which quite never have trivial solutions. This is exactly what the students are being asked to do: model a complex industrial system with a global approach (by beginning with a global description of the problem and by making a zoom, if necessary, on some details), give one possible solution based on rational choices, and be able to justify this solution and to predict its validity limitation. Today, every student performs all these steps on every industrial system in the industrial science laboratory.

We hope that the reader of this book will be able to see that these pedagogical works have greatly influenced the redaction of its content. We have indeed tried to present a synthesis of all our works and benefic discussions in this field. One direct impact is the selected illustrations: they come from the devices that are used in the industrial science laboratory of ECPk.

Of course, we remain at the disposal of anyone to present this laboratory and explain how it is used in the engineers training.

Be one life a dream, and be this dream a reality thanks to industrial science.

Contents

Chapter 1

Systems Study

In this chapter, we will first define what a system is, and then analyze the characteristics and functions of systems.

1.1 General Definitions

1.1.1 Need

Definition 1 (Need)

A **need** is a necessity or a desire expressed by a customer. A need can be explicit or implicit.

Definition 2 (Product)

A **product** is what will be provided to a customer to meet one of his needs.

1.1.2 System

Definition 3 (System)

A **system** is a structured association of interrelated elements which constitutes an entity and has one or many functions.

For instance:

- the *nervous system* is a natural system in which the elements are organs and tissues, and the interactions between these elements are performed by means of electrical and chemical signals which allow communication between the human brain and the different parts of the human body;

- the *solar system* is a natural system in which the elements are the Sun and the planets of its planetary system, and the interactions between these elements are caused by the gravitational force, which makes these planets orbit the Sun;

- a *system of equations* is a modeling system in which the elements are equations, and the interactions between these elements are caused by the simultaneous presence of some variables in more than one equation;

- etc.

It is necessary to avoid the confusion between a **system** and a **set**. The knowledge of all the elements or components of a set is sufficient to know the whole set. However, the additional knowledge of the **interactions** which exist between these elements is necessary to understand the functioning of a system. For instance, a bicycle results from the structured association of two wheels, a frame, a chain, a handlebar, and lots of other components, but the set of all these components does not suffice to design a bicycle: interactions between these components are necessary.

Systems are used in a given **context**, which is the environment in which the system is used. This context can be defined by means of:

- the type of environment (marine, domestic, etc.);

- the application field (aeronautical, automotive, medical, etc.); and

- the type of customer (private, professional, young, old, etc.).

Technico-economical constraints can also be taken into account when studying a system, amongst which:

- its lifetime;

- its cost;

- the produced quantity (depending on whether the system is a unitary system, or a system with a small, medium or high series production);
- its reliability (ability of a system to perform its required function under stated conditions for a specific period of time); and
- its maintainability (ability of a system to be maintained in a state in which it can perform its required function).

Systems can be divided into two categories:

- **natural systems**, which have not been created by mankind and exist in Nature (e.g. the solar system, the nervous system, etc.); and
- **artificial systems**, which have been created by mankind to perform a given function (e.g. a bicycle, a nuclear power plant, etc.).

The content of this book will exclusively be dedicated to artificial systems. Amongst these, three types of artificial systems can be identified according to how much human beings are implicated:

- **non-mechanized systems** (also called basic systems, or manual systems), for which all the energy is provided by the user who commands the system: a bicycle is an example of non-mechanized system;
- **mechanized systems**, for which an external *power supply* allows the user to perform tasks that his own capacities do not allow him to perform: a motorcycle or an electrical drill are examples of mechanized systems; and
- **automated systems**, for which tasks are performed in a relatively autonomous way; such systems are driven and controlled by an external structure, the energy is provided by an external power supply, and the user has only a supervision function: a drink vending machine or a washing machine are examples of automated systems.

The remainder of this chapter, and this book as a whole, will focus on automated systems.

1.1.3 Function

From the point of view of the user, a system is not a set of interrelated components, but a *services generator* which aims at satisfying the user by meeting his needs. For instance, a bicycle driver does not consider his bicycle as a set of interrelated components, but as a mean of transport. He will hence be satisfied if his bicycle:

- is cheap;
- allows him to move without too many efforts;
- can go fast without too many efforts;
- has a nice design (shape, colors, etc.);
- etc.

All these aspects characterize the *global function* of the system.

Definition 4 (Global Function of a System)

The **global function** of a system represents the original and main use of the system. If the system does not perform the function for which it was designed, there is no reason for it to exist.

Definition 5 (Work Material)

The **work material** is the part of the environment of the system which is modified by the system. It can be a product (material), an energy, and/or a piece of information which is modified by means of an added value by the system.

Definition 6 (Added Value)

The **added value** of a system is the difference between the final state and the initial state of the work material which has been modified by the system. This basically is what the system provides to the initial work material.

For every artificial system, the global function of the system consists in providing an *added value* to a *work material* between an *initial* and a *final* expected state to meet a given *need* in a given *environment.*

Since it is this transformation of the work material by the system which will meet the user's need, we will consider that there is an equivalence between a system and a product in the remainder of this chapter.

This added value can be of any type such as, for example:

- a conservation over the time (of the mass, of the energy, etc.);
- a motion (its related distance, or even its related velocity or acceleration);
- a modification of the structure of the system;
- etc.

The global function of a system can be completed by some performance criteria which can be, amongst others:

- the range of variation of its inputs and outputs;
- the reaction rapidity to a variation of one particular input;
- its reliability;
- its initial cost;
- its energy consumption;
- its stability, and its ability to resist to disturbances;
- mechanical, electrical, optical, etc. data which are related to the system;
- its recyclability;
- etc.

1.2 Positioning of the Studies

Since the behavior of automated systems is too complex to be described easily, some graphical tools have been developed to allow to represent such systems in a simplified way, from a specific point of view. These points of view can be:

- the **structure** of the system, which describes how the system physically is, as an association of interrelated components: such an analysis is called **structural analysis**; and
- the **functions** which are performed by the system: such an analysis is called **functional analysis**.

These two types of analyses – structural and functional – will not be used at the same time during the life cycle of a product:

- the functional approach will be used during the *design* of a product, since its structure is yet unknown; and
- the structural approach will be used as soon as the structure of the product is known, for its *optimization* and *analysis*.

1.3 Functional Analysis Tools

Functional analysis aims at determining the functions which are performed by a product. These functions can be of two types:

- **service functions**, which represent the global function of the system and also the constraints which exist between the system and its environment, and which are determined by means of an **external functional analysis**; and
- **technical functions**, which represent the functions which are performed by the different sub-systems or components of the system, and which are determined by means of an **internal functional analysis**.

1.3.1 External Functional Analysis

External functional analysis aims at determining the service functions of a system.

Definition 7 (Service Function)

A **service function** is a function which is expected from a product, or performed by this product, in order to meet a user's need.

Service functions represent the interactions which exist between the system and its environment. To be able to determine them, the environment needs to be determined first thanks to the definition of a *boundary* for the system.

Definition 8 (Boundary of a System)

The **boundary** of a system is a real or fictive limit which separates the considered components into two groups: the components which are inside the boundary belong to the system, whereas the components which are outside the boundary belong to the environment of the system.

The determination of this boundary hence is mandatory to be able to model the system, as it defines the limits of the study. As soon as this boundary has been defined, it is possible to identify:

- the elements of the environment;
- the relations which exist between these elements and the product; and finally
- the inputs and outputs of the system.

As said in section 1.1.3, a system modifies a work material which belongs to its environment. A system hence has interactions with the elements of its environment, and these interactions are modeled by *service functions*. These service functions are identified by means of a tool named the **inter-actors diagram** (which is also called octopus diagram because of its shape). This tool allows to:

- **identify** the elements of the environment; and
- **determine** the interactions which exist between the system and the elements of its environment, which are service functions.

An example of an inter-actors diagram is given in Figure 1.1. The product is in an ellipse at the center of the diagram, surrounded by other ellipses which correspond to the elements of the environment. The interactions between the system and the elements of its environment are represented by arcs, which represent the service functions of the product.

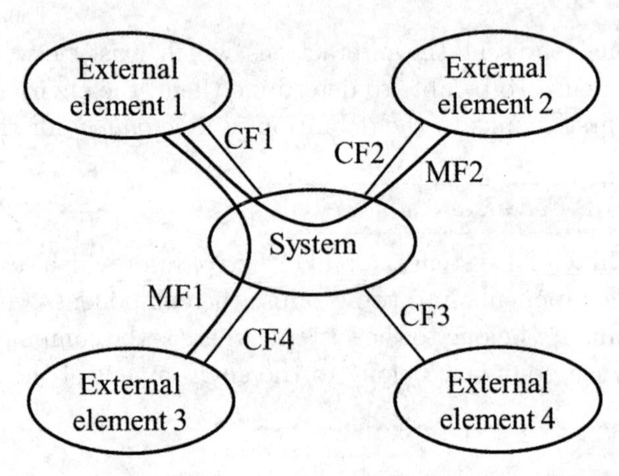

Figure 1.1 an inter-actors diagram

There are two types of service functions:

- the functions which correspond to interactions between the system and **a single** element of its environment are called **constraint functions** (CF): they express a constraint that the system must satisfy with respect to its environment (e.g. an adaptation to a specific type of energy, or safety constraints); and
- the functions which correspond to interactions between the system and **two or more** elements of its environment are called **main functions** (MF): they allow to meet some need; generally, there is only one main function.

The electrical power-assisted steering system of the Renault Twingo (Figure 1.2) is presented to introduce this inter-actors diagram clearly.

In order to change the direction of a vehicle, the driver turns the steering wheel which is connected to the direction column with a rack and pinion system: the translation of the rack turns the front wheels of the car. Because of the friction

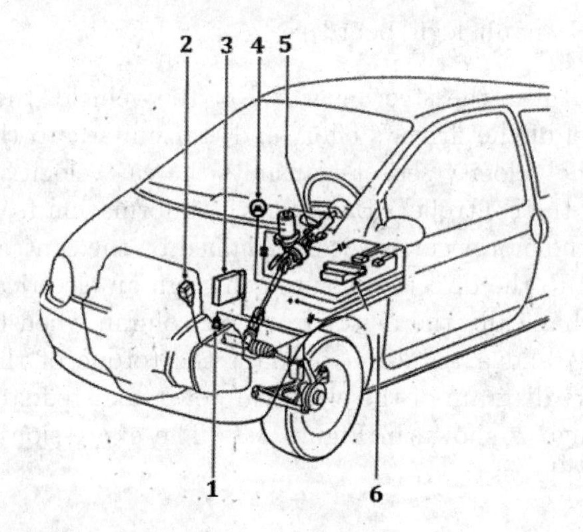

Figure 1.2 the electrical power-assisted steering system of the Renault Twingo

between the wheels and the road, this operation can sometimes be difficult: one solution is to insert a mechanical assistance in order to help the driver and limit the efforts on the steering wheel. This system is called "steering system" and the Renault Twingo was one of the first cars to have an electrical technology (the former technology was hydraulic), which is nowadays more and more common.

The electrical power-assisted steering system of the Renault Twingo is composed of a normal steering system with the addition of an electrical assistance provided by a motor. The goal of this system is to assist the driver to rotate the direction wheels of the car at low velocity with a level of assistance which decreases when the car velocity increases. The main components which appear in Figure 1.2 are as follows:

1. Velocity sensor;

2. Diagnosis socket;

3. Engine sensor;

4. Indicator lamp of the assistance;

5. Torsion sensor;

6. Controller (Renault logic board).

When the driver turns the steering wheel of the vehicle, the torsion sensor evaluates the level of the driver's effort and transmits it to the controller; at the same time, the velocity sensor transmits the car velocity and the engine rotation speed to the controller. Based on the information from all these sensors, the level of the motor current is determined by the controller. The motor is then connected to the direction column through an electrical clutch (which can separate mechanically the motor from the column when the assistance is no longer needed) and a reducer (to adapt the torque of the motor). The partial inter-actors diagram of the electrical power-assisted steering system of the Renault Twingo is shown in Figure 1.3. The expressions for the service functions are as follows:

- MF1: deliver a mechanical assistance during the rotation of the driving wheel;
- CF1: use the available energy (embedded 12 V/70 Ah battery);
- CF2: be robust to the external environment (sun, variation of temperature, vibrations of the car, etc.).

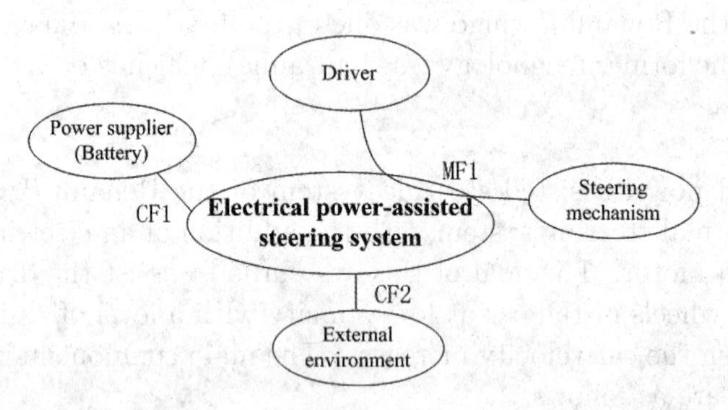

Figure 1.3 partial inter-actors diagram of the electrical power-assisted steering system of the Renault Twingo

1.3.2 Internal Functional Analysis

Internal functional analysis aims at determining the technical functions of a system.

Definition 9 (Technical Function)

A **technical function** represents an action which is **internal** to the product, i.e. between parts of the system. It is an interaction between components of the product defined by the designer in order to ensure one or many service functions.

The **FAST** (Function Analysis System Technic) diagram is a graphical functional description tool under an arborescent form. It allows to decompose the service functions of a system (which characterize the interactions between the system and its environment) into technical functions (which are internal to the product) in a logic manner, and from a functional point of view. The FAST diagram thus allows to obtain the description of each technological solution in front of the technical function(s) that it performs, as illustrated in Figure 1.4.

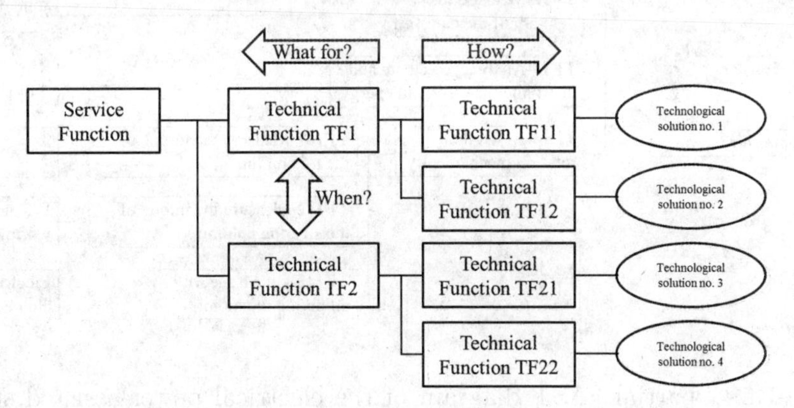

Figure 1.4 graphical expression of a FAST diagram

The construction and analysis of a FAST diagram are based on an interrogative approach. From a given function, the FAST diagram allows to answer the three following questions:

- **What** does this function have to be performed **for**? (the answer is obtained by going through the FAST diagram horizontally, **to the left**, and thus to a more global function);
- **How** does this function have to be performed? (the answer is obtained by going through the FAST diagram horizontally, **to the right**, and thus to more precise functions);

- **When** does this function have to be performed? (the answer is obtained by going through the FAST diagram vertically, upwards or downwards). There is absolutely no time meaning in this questioning, but just a functional point of view which means that the global function will be achieved when all the related technical functions have been achieved.

Thus, in the case of the electrical power-assisted steering system of the Renault Twingo, the FAST diagram of the function "Deliver a mechanical assistance during the driving wheel rotation" (denoted as MF1 in Figure 1.3) is shown in Figure 1.5.

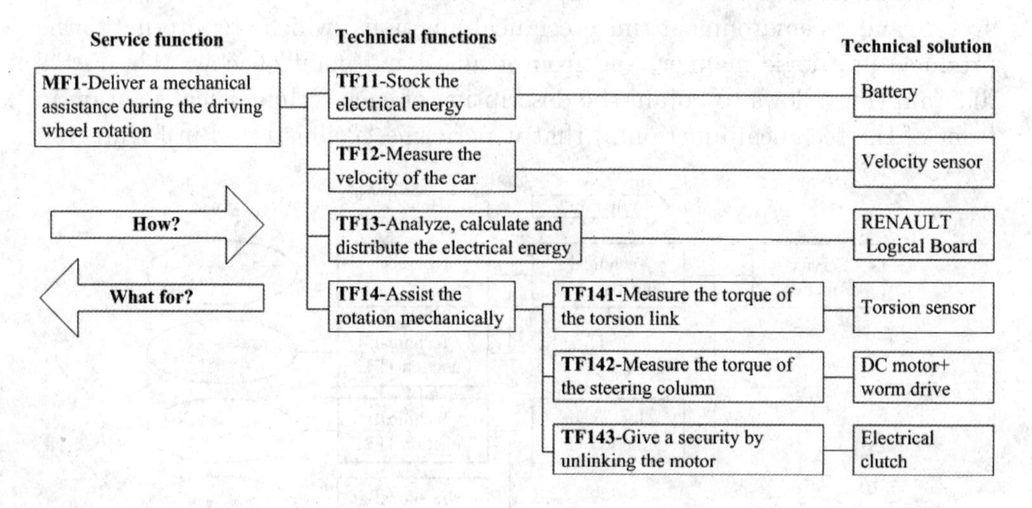

Figure 1.5 partial FAST diagram of the electrical power-assisted steering system of the Renault Twingo

In this example, the main function MF1 is decomposed into four technical functions:

- TF11: stock the electrical energy;
- TF12: measure the velocity of the car;
- TF13: analyze, calculate and distribute the electrical energy;
- TF14: assist the rotation mechanically.

In order to precise the organization of the mechanical assistance zone, we chose to divide the last technical function TF14 into three sub-functions. The corresponding technological solutions of these technical functions are shown in the right column of the FAST diagram.

By reviewing this FAST diagram horizontally to the right, we can answer the question "how does one function have to be performed?". For example, if we want to know how to perform the technical function TF11 "stock the electrical energy", by reviewing the FAST diagram horizontally to the right, we can find that the technological solution is the battery.

By reviewing this FAST diagram horizontally to the left, we can answer the question "what does this function have to be performed for?". For example, if we want to know the purpose of the technological solution "velocity sensor", we can go through the FAST diagram to the left, and we can determine that it is to satisfy the technical function TF12 "measure the velocity of the car" and the main function "deliver a mechanical assistance during the driving wheel rotation (MF1)" .

As a service function is performed by many technical functions, a service function will be *validated* if and only if all the technical functions which contribute to it are validated.

1.3.3 Structuring of the Functional Chains

As said in section 1.1.2, an automated system needs few or no human intervention. An automated system is composed of:

- an **operative system**, which is the part of the automated system that directly modifies the work material. It contains all the components which provide the added value by modifying the work material. From an informational point of view, it sends reports to the command system and receives orders from it to modulate or stop the transformation of the work material. The required energy is often high (e.g. 380 V for electrical energy, 250 bar for hydraulic energy, etc.).

- a **command system**, which ensures the coordination of the tasks that are necessary to perform the expected process, the supervision of the operative system (by means of orders), and the exchange of information with the user or other automated systems. The required energy is low (e.g. 5 V for electrical energy, 10 bar for hydraulic energy, etc.).

- a **control panel**, which allows to transmit the instructions of the user to the system and to follow the evolutions of the system.

The structure of an automated system can be described by the scheme in Figure 1.6.

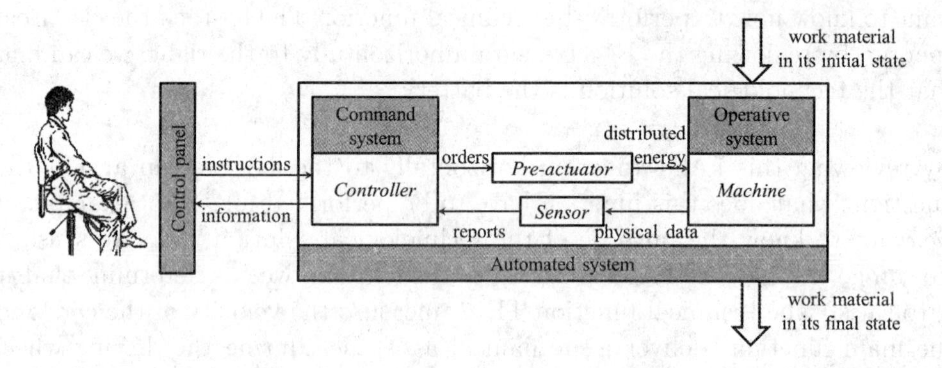

Figure 1.6 structure of an automated system

This decomposition can be refined in order to specify the architecture of both the command and the operative system. Indeed, an automated system can be decomposed into a command and an operative system, but it can also be decomposed into two so-called **functional chains**: an *information functional chain* and an *energy functional chain*:

- the **information functional chain** is the set and organization of the components which allow to acquire, treat and transmit the information; it is composed of:
  - **sensors**, which measure the state of physical data and transmit related information to the command system; and
  - the **command system**, which treats the information and transmits them to the operative system.
- the **energy functional chain** is the set and organization of the components which allow to supply, distribute, convert and transmit the energy; it is composed of:
  - **pre-actuators**, which distribute the available energy to the actuators at a level demanded by the command system; the most common

pre-actuators are relays and contactors (Figure 1.7) for electrical actuators, and control valves (Figure 1.8) for hydraulic and pneumatic actuators (fluidic technology).

Figure 1.7 a contactor

Figure 1.8 an hydraulic control valve

– **actuators**, which convert the energy provided by the pre-actuator – that cannot be used directly by the effectors – into mechanical energy; the most common actuators are electrical motors (Figure 1.9) and hydraulic or pneumatic cylinders (Figure 1.10).

Figure 1.9 an electrical motor

Figure 1.10 a pneumatic cylinder

– **power transmitters**, which adapt the mechanical energy produced by the actuators for it to have the characteristics (in terms of speed, torque, etc.) required by the effectors; the most common power transmitters are gear speed reducers (Figure 1.11).

Figure 1.11 a gear speed reducer

- **effectors**, which convert the mechanical energy provided by power transmitters into the work which is required to modify the work material; conveyors, the terminal fingers of robots (Figure 1.12) or production tools such as drills (Figure 1.13) are examples of effectors.

Figure 1.12 a mechanical arm

Figure 1.13 a drill

The organization of both functional chains is illustrated in Figure 1.14.

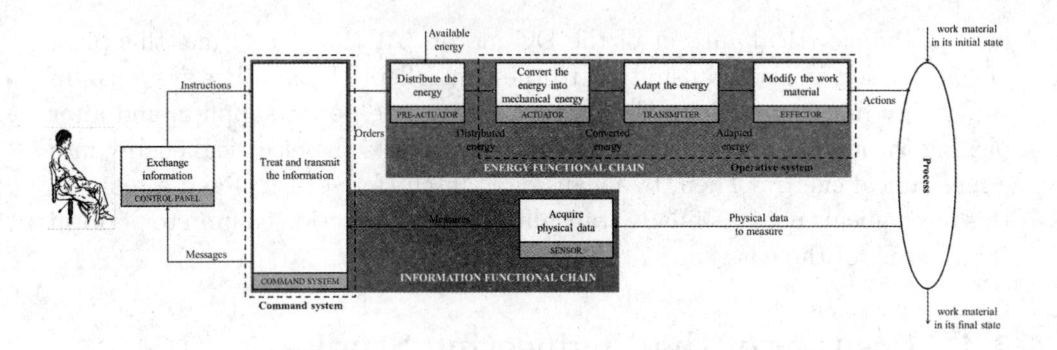

Figure 1.14 structure of an automated system

For instance, the structure of the electrical power-assisted steering system of the Renault Twingo is shown in Figure 1.15. The upper rectangle shows the **information functional chain** and the lower rectangle shows the **energy functional chain**. The rectangle on the right presents the main function of the system, which transforms the initial work material "torque on the driving wheel" into the final work material "torque on the direction column".

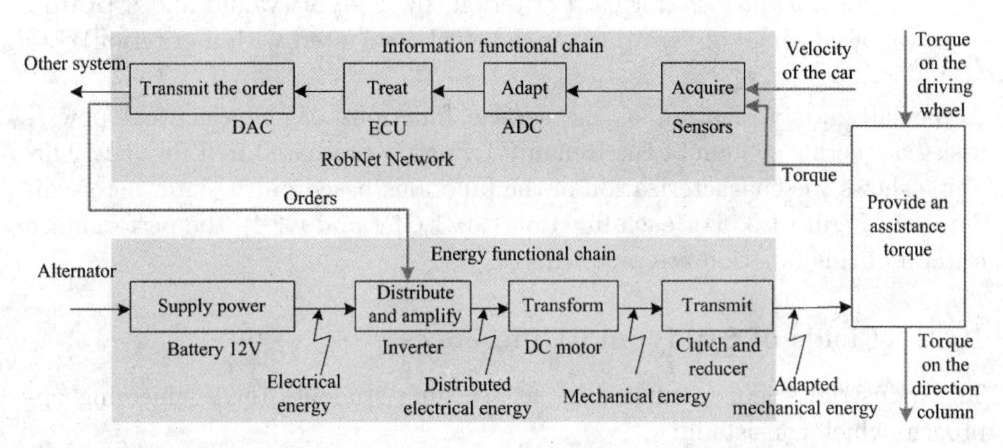

Figure 1.15 structure of the electrical power-assisted steering system of the Renault Twingo

In the information functional chain, the sensors acquire the velocity of the car and the torque provided by the user on the driving wheel, and transform them into an electrical signal. Then the ADC (Analog to Digital Converter) transforms the analog electrical signal into a numeric piece of information. Then the ECU (Electrical Center Unit) controller of the system calculates the level of the electrical current of the DC motor. At the same time, this piece of information is also transmitted to the controllers of the other systems. In the energy functional chain, the battery acts as the power supplier, and after passing through the inverter, the DC motor converts the electrical energy into a mechanical energy. Then, by means of a clutch (for security) and a reducer, this mechanical energy is finally transmitted to the direction column to perform the steering of the car.

1.4 Position of the Engineering Studies

1.4.1 Specifications

The **specifications** of a system are a contractual document in which a need is expressed in terms of service functions. These service functions are characterized, i.e.:

- they are *qualified* by means of appreciation **criteria** (measurable physical data); and
- they are *quantified*, for each criterion, by a **level** (value) and a **bound** (allowed range of variation for the level associated with a criterion).

A partial characterization of the service functions of the electrical power-assisted steering system of the Renault Twingo is presented in Table 1.1. This Table shows the characterization of the functions based on the inter-actors diagram in Figure 1.3. For each function (MF1,CF1 and CF2), the performance criteria of this function are precised.

1.4.2 Fields of Study and Differences

The industrial science course aims at teaching students the engineering approach, which consists in:

- performing functional, structural, and behavioral analyses on a pluritechnological system;
- validating the expected performances of a system by evaluating the differences between its specifications and the experimental results;
- proposing and validating multiple models for a system from tests, by evaluating the difference between the measured performances and the calculated or simulated performances;

Table 1.1 performance criteria of the technical functions in figure 1.3

Fct	Criterion	Level
MF1	Difference due to the twist of the torsion bar	at most 8°
	Response time to recover the angle after stopping the rotation	at most 1 s
	Maximal over-value to a rotation demand	0%
	Rotation of the steering wheel	±707°
	Rotation of the left wheel	between −39° and 30°
	Rotation of the right wheel	between −30° and 39°
	Torque with assistance	at most 9 N.m
	Torque without assistance	at most 24 N.m
	Velocity of the car	at most 150 km/h
	Maximal torque below 74 km/h	regular diminution
	Maximal torque above 74 km/h	minimal value
CF1	Battery voltage	at most 12 V
	Battery charge	at most 50 Ah
CF2	Materials	normalization controls

- predicting the performances of a system from models, by evaluating the differences between the calculated or simulated performances and the expected performances; and

- analyzing these differences and proposing solutions to improve the performances of the system.

Three different fields of study of a system can hence be considered:

- the **customer field**, which corresponds to the system that the customer expects to meet his needs; the specifications of the system allow to define its **expected performances**;

- the **laboratory field**, which corresponds to the real system; the performances of the system can be determined by means of sensors: they are the **measured performances** of the system; and

- the **simulation field**, which corresponds to the simulation of the system by means of models; these models provide the **simulated performances** of the system.

These three fields and the differences between the results that they provide are illustrated in Figure 1.16.

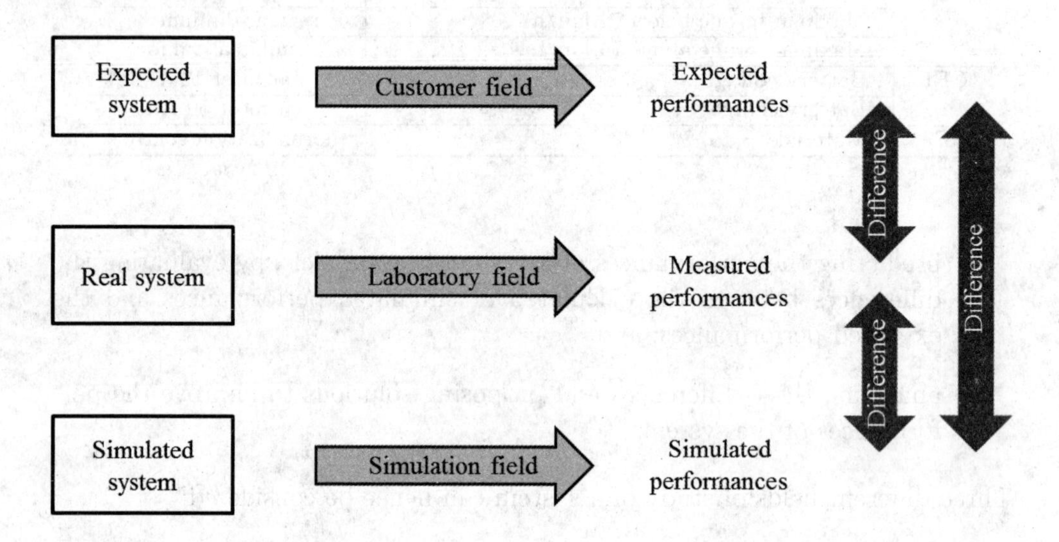

Figure 1.16 the different fields of study, and the differences between their results

Chapter 2

Linear Continuous-Time Time-Invariant Systems

The main problems which arise during the implementation of physical and industrial systems are related to the control of these systems, i.e. the way to sollicite them to get the expected behavior. In this chapter, we will see how to model linear continuous-time time-invariant systems to determine their temporal and frequential behavior.

2.1 History of Automation Control

The history of automation control can be divided into at least three main periods.

2.1.1 First Period: from the Antiquity until the Middle of the 19th Century

The Antiquity

During this "prehistory" of automation control, genial inventors designed automated systems in a purely intuitive way. As early as 250 BC, Greek inventors designed various systems such as the automatic water clock (clepsydra), the oil lamp, and the wine dosing machine.

The *clepsydra* (from the ancient Greek words $\kappa\lambda\epsilon\pi\tau\epsilon\iota\nu$, *kleptein*, "steal", and $\upsilon\delta\omega\rho$, *udor*, "water") was invented in Alexandria (in Egypt) during the 3rd century BC by the Greek mathematician Ctesibius. Ctesibius' clepsydra is depicted in Figure 2.1. A statue carrying an index is placed on a floater which goes up in a tank while the water slowly flows into the tank from the upper hole. At that time, the daytime period was divided into 12 hours whose duration depended on the season (these hours being longer in summer than in winter). This division of time is the reason why graduations are drawn on the time drum which is pointed by the index of the statue. The time drum hence only had to be turned for the duration of hours to correspond to the period considered. It can also be noticed that there is a floater in the regulation tank (in the upper left of Figure 2.1) which allows to regulate the flow of the water which fills the main tank of the clepsydra: indeed, for the clepsydra to work correctly, the water level in the regulation tank needs to remain constant, so that the water flow in the main tank is constant and that the clepsydra can be used as a clock. The behavior of the second floater can thus be described as follows:

- if the water level rises in the regulation tank, the floater goes up and reduces the incoming water flow; and
- if the water level drops, the floater goes down and the incoming water flow increases.

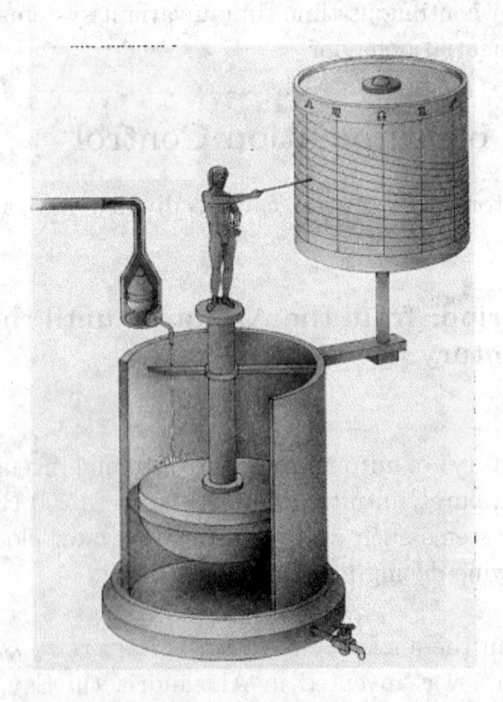

Figure 2.1 Ctesibius' clepsydra

The Greek engineer Philo of Byzantium used a similar principle at the end of the 3rd century BC when he designed a constant level oil lamp in his compendium of Mechanics, which is depicted in Figure 2.2. The pineapple-shaped vase contains oil which flows into many lamps through many tubes. These tubes allow to fill the lamps until the oil level reaches the tip of the tubes and hence obstructs them. The flowing then stops until the fire of the lamps has consumed enough oil to uncover the tip of the tubes.

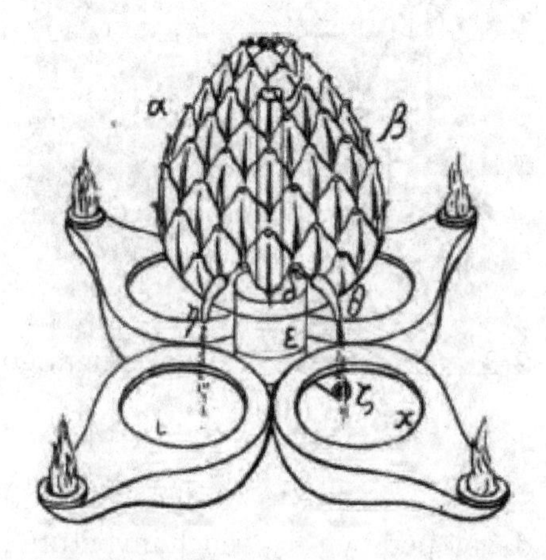

Figure 2.2 the constant level oil lamp of Philo of Byzantium

The 18th Century

Many centuries later, the French physicist René-Antoine Ferchault de Réaumur (1683-1757), the Scottish engineer James Watt (1736-1819) and his regulator, and the French inventor Joseph-Marie Jacquard (1752-1793) and his mechanical loom with punched cards made automation control progress.

Watt's centrifugal governor aims at controlling the speed of a steam turbine. The governor is connected to a throttle valve that regulates the steam flow supplying the turbine. As the speed of the turbine increases, the central spindle of the governor rotates at a faster rate and the kinetic energy of the balls increases. This allows the two masses on lever arms to move outwards and upwards against gravity. If the motion goes far enough, it causes the lever arms to pull down on a thrust bearing, which moves a beam linkage, which reduces the aperture of the throttle valve. The steam flow entering the turbine is thus reduced, and the speed of the turbine is controlled, preventing over-speeding. This behavior is illustrated in Figure 2.3. Watt's centrifugal governors were used on every steam machine, and hence on old locomotives. Nowadays, they are still used on some plane motors and in nuclear power plants as a final non-electronic switch.

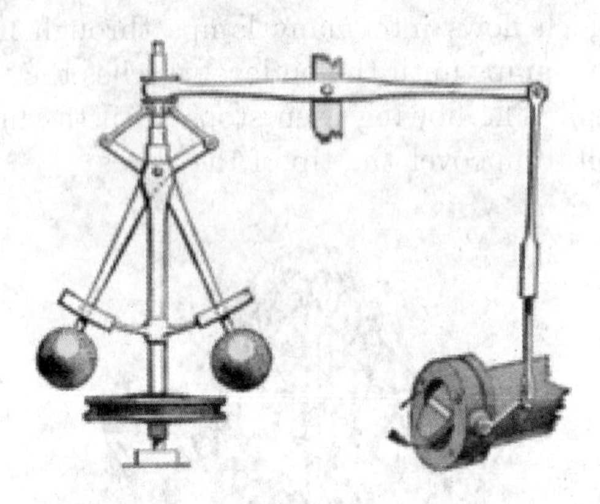

Figure 2.3 James Watt's centrifugal governor

Jacquard's mechanical loom, which was designed in 1801, combines different techniques which were developed by the French inventors Basile Bouchon, Jean-Baptiste Falcon and Jacques de Vaucanson (1709-1782). Each position of its punched cards corresponds to a hook which can either be raised or stopped depending on whether the hold is punched out of the card or the card is solid. The hook raises or lowers the harness, which carries and guides the warp thread so that the weft will either lie above or below it. The sequence of raised and lowered threads is what creates the pattern. Thanks to Jacquard's looms, a single worker could manipulate the loom whereas many of them used to be necessary. They are still used nowadays to weave complex patterns. The original Jacquard's loom is presented in Figure 2.4, and a recent Jacquard's loom currently used in the UK is presented in Figure 2.5.

2.1.2 Second Period: from the Middle of the 19th Century until the Middle of the 20th Century

The second period of automation control, which began during the 19th century, is characterized by the feedback control theory and by the applications of the Boolean algebra. The first works on control theory were carried out by the Scottish physicist Maxwell (in 1868), by the English mathematician Routh, who gave his name to an algebraic stability criterion (in 1872), and by the German mathematician Hurwitz (in 1890).

Figure 2.4 Joseph Marie Jacquard's mechanical loom

Figure 2.5 a recent version of Jacquard's mechanical loom

The analytic study of the stability of Watt's centrifugal governor was initiated by Maxwell in 1868 and completed by Wichnegradsky in 1876. The study of closed-loop controllers owes much to the frequential approach of Nyquist, Bode, Black, Nichols, Hall and Evans, who gave their name to some plots and published most of their results at the end of the second world war.

2.1.3 Third Period: from the Middle of the 20th Century until Now

The third period of automation control began in the 50s. The birth of computers and numeric calculators revolutionized the world of automation control. The calculation power available gave birth to the "modern" automation control methods. The following noticeable facts can be cited:

- the introduction of the *state representation*, which is particularly adapted to the use of numeric calculators for the study and control of complex and multivariable systems, by the Hungarian mathematician Kalman in 1960: the different states of the system at different instants are considered;
- the development of study methods for *nonlinear systems* (by Kockenburger and Cypkin) and *sampled-data systems* (by Jury and Ragazzini): nonlinear systems are systems whose output is not proportional to its input (such systems do not respect the superposition principle), and sampled-data systems are discrete systems, i.e. systems whose information are taken into account only at some specific instants; and
- the consideration of *random phenomena* in recent theories such as Kalman's and Bucy's.

2.2 Modeling

2.2.1 Types of Inputs and Outputs

The command system of continuous systems is able to deal with input/output variables which can be *analog* or *numeric*, i.e. these variables vary:

- in a *continuous* way (Figure 2.6): the variable v_i belongs to $\mathbb{R}$ and is bounded by $[v_{i_{\text{mini}}}, v_{i_{\text{maxi}}}]$ and there is no discontinuities in the signal;
- in a *quasi continuous* way (Figure 2.7): the variable v_i belongs to $\mathbb{N}$.

2.2.2 Positioning of the Studies

Types of Modeling

Two main types of modeling exist, which are related to two types of model:

- a **knowledge model**, which is based on a *deductive approach* and is built from the mathematical manipulation of equations which are obtained from models whose uncertainty is supposed to be limited enough; and

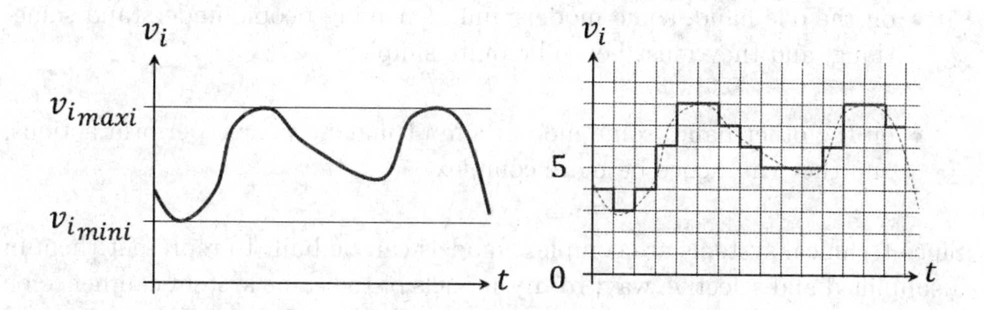

Figure 2.6 an analog variable **Figure 2.7** a numeric variable

- a **behavioral model**, which is based on an *inductive approach* and is built from measures whose experimental protocol is supposed to be trustable enough for the results obtained to be used.

Choice of a Type of Model

These two types of model **cannot** be exchanged and both are used in various scientific fields, amongst which physics, chemistry and industrial science. Indeed:

- the *knowledge model* is priviledged in physics courses in France and in a few other countries, and it is perfectly adapted to the prediction of the performances of simple or highly predictive systems, even though its use is impossible as soon as the system considered is complex; and

- the *behavioral model* is priviledged in physics courses in many countries, and it is perfectly adapted to the analysis of complex systems or to behavioral uncertainties which are too high, even though its use is impossible to estimate the evolution of the performances of a system or to adjust a system.

None of these models is "better" than the other, each one is more or less adapted to the problem studied.

Complexity of the Models

One must keep in mind that the usefulness of a model, whether it is a knowledge model or a behavioral model, depends on the purpose of its use. Indeed:

- on the one hand, some models aim at making people understand something, and they must hence be quite simple;
- on the other hand, some models aim at making people perform actions, and they can hence be more complex.

Since technical systems are complex, models can be built to represent them in a simplified and selective way; **many** models of the same system can hence be built by means of different representation methods and tools. The mathematical complexity of a model does not make it the best model: this complexity needs to be adapted to the needs of the study.

2.2.3 Modeling by Means of Functional Diagrams

Forward Chaining

This model considers that the command system sends orders to the process according to the instructions that it receives on its inputs, only. The command system can hence be modeled by the functional diagram in Figure 2.8, which corresponds to the structure of an automated system that was presented in chapter 1.

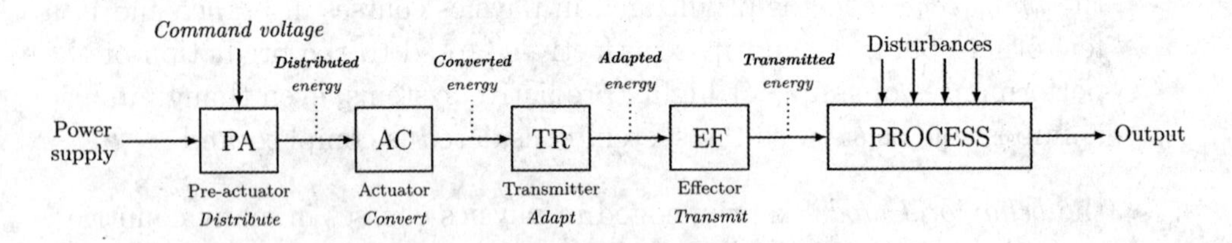

Figure 2.8 functional diagram of a forward chaining structure

It can be noticed that all the elements of this diagram are disturbed, but the resulting disturbances are generally only represented on the process.

Forward chaining systems allow a **natural stability** of the process, but these systems have a dynamics which is limited and hard to optimize. Besides, the use of a knowledge model is possible but often difficult since it is difficult to model losses in a simple way. However, the use of a behavioral model is possible and easy.

Closed-Loop Control

If the process does not reach the expected performances, a sensor and a regulator are implanted in the functional diagram. Two other important elements are implanted:

- an **adapter**, which converts the instruction into a data which can be treated by the regulator (most often, this data is a voltage); and
- a **conditioner**, which is located after the sensor and allows to treat its signal: it generally comprises an electronic device which has a filtering function.

In such a closed-loop structure, the order is elaborated according to the difference ε between the instruction and the output of the system (which is known as the *error*), and the feedback loop of the output allows the command system to generate orders by **continuously** taking into account their effects: this structure can hence be modeled by the functional diagram in Figure 2.9.

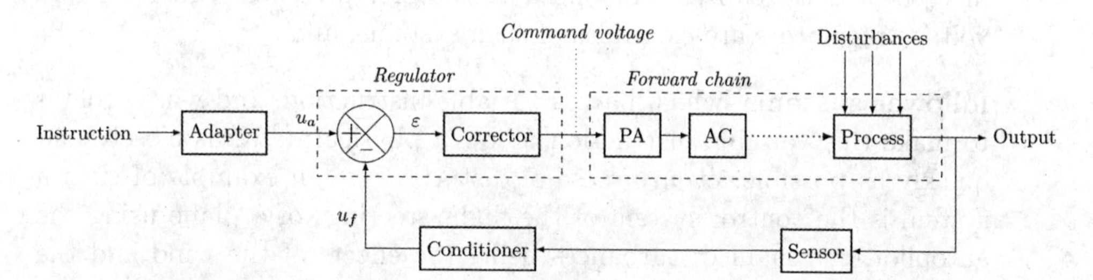

Figure 2.9 functional diagram of a closed-loop structure

Three main functions can hence be identified in such a closed-loop structure:

- an observation of the output by the sensor;
- a reflection of this output by the comparator and the corrector, whose adjustment is the purpose of the automation control course; and
- an action of the forward chain according to the orders generated by the regulator.

The feedback loop between the output and the input of the system allows the command system to generate orders while taking into account their effects. A corrector, located before the forward chain, generates orders according to the "error" ε between the instruction and the output of the system. If this corrector is correctly designed, this structure allows to:

- ensure the stability of the process;
- compensate the uncertainties about the physical characteristics of the process;
- attenuate the effects of the disturbances; and
- obtain quick and accurate responses.

However, the presence of this feedback loop makes the system more complex and deeply modifies its behavior, which becomes difficult to predict without a thorough analysis. Sometimes, this feedback loop can even cause instabilities ... which is counterproductive.

Closed-loop systems are generally divided into two categories:

- **regulators**, which have a constant instruction, and whose role is to maintain a physical data as close as possible to this instruction, even in case of disturbances; car speed regulators, pressure regulators and voltage regulators are examples of such systems; and
- **follower systems**, which have a variable instruction, and whose role is to make this variable instruction follow a physical data, as closely and quickly as possible, even in case of disturbances; an example of such a system is the control system of the flight steerings of a plane using the autopilot, the main disturbances being the effects of the wind and the presence of turbulences.

Conclusion

The compact structure presented in Figure 2.10 will be retained in the remainder of this chapter.

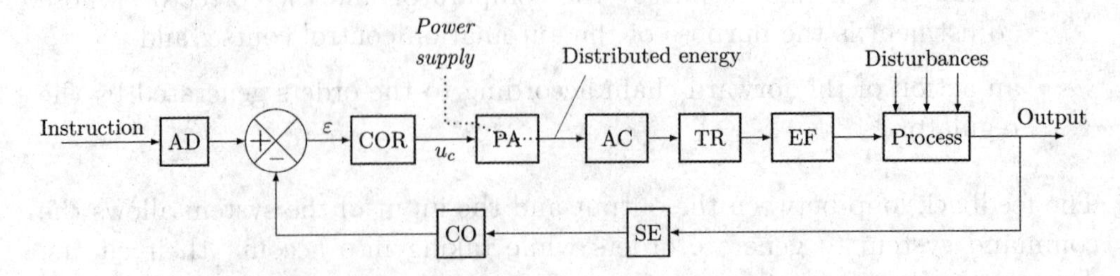

Figure 2.10 functional diagram of a closed-loop structure

In this diagram, the power supply is most often not represented and the other elements are as follows:

- AD = adapter: it converts the physical instruction into a data which can be treated by the calculator (generally, it is a voltage whose magnitude is low);
- COR = corrector: its adjustment aims at making the performances of the corrected system compatible with the specifications of the system; it treats the error ε and generates the command voltage u_c which pilots the pre-actuator;
- PA = pre-actuator: it distributes the power provided by a power supply (which is, most often, not represented) to the actuator;
- AC = actuator: it converts the power distributed by the pre-actuator into a power which has the type needed by the process (mechanical power, thermal power, ...);
- TR = power transmitter: it adapts the power for it to have characteristics which are compatible with the expected performances of the process;
- EF = effector: it transmits the adapted power by means of an element located at the interface with the process (rod, conveyor, ...);
- SE = sensor: it provides an image (which is most often electrical) of the output signal with a resolution (accuracy of the measure) and a dynamics (rapidity of the measure) which are as good as possible;
- CO = conditioner of the sensor: it treats the signal provided by the sensor in order to make it treatable by the calculator.

2.3 Hypotheses Related to the Studies

2.3.1 Continuity

Definition

Definition 10 (Continuous System)

A system is **continuous** if the signals studied are *analog*: a change of state will be performed by means of intermediate values between these two states.

Most systems are continuous, at least from a macroscopic point of view.

However, artificial systems are rarely continuous: for instance, a computer system is not continuous because it can only treat samples from the given continuous signals. Such a system is called a sampled-data system; if the sampling is accurate enough, then the signal will be quasi continuous.

Mathematical Consequence

Such a continuous evolution can be modeled by means of a differential equation which links the output signal $o(t)$ (and n of its successive derivatives) to the input signal $i(t)$ (and m of its successive derivatives). This relation can be expressed under the following form:

$$\boxed{f\left(o(t), \dot{o}(t), \ldots, o^{(n)}(t)\right) = g\left(i(t), \dot{i}(t), \ldots, i^{(m)}(t)\right)}$$

The condition $\boxed{n \geqslant m}$ always holds for physical systems, for them to respect the causality hypothesis.

2.3.2 Linearity

Concept of Operating Point

The characteristic input/output curve in equilibrium is called the **static characteristic curve**: it is obtained by means of successive measurements in steady state, which is only possible with stable systems. To obtain this curve, a constant value I_i is injected at the input of the system, and the output of the system is measured as soon as it has stabilized: the input I_i and the corresponding stabilized output O_i represent a pair (I_i, O_i) which is a specific **operating point** for the system.

The determination of these operating points allows to obtain the curve in Figure 2.11. It can be noticed on this curve that the output becomes non-null only after the input has reached a given value, and that the output no longer varies after the input has reached a given value.

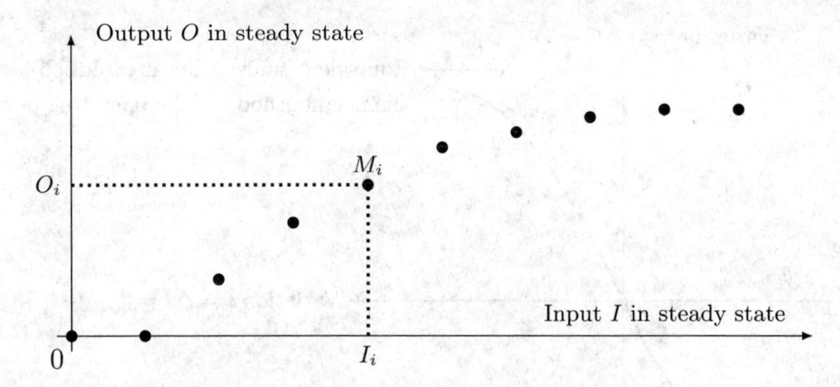

Figure 2.11 static characteristic curve

Local Linearization

To make the study of such a system easier, we are going to determine a linear model for it. However, such a model is impossible to determine globally, so we are going to determine local models.

To obtain this simplified linear model, the most classic method consists in choosing an operating point (e.g. M_1) around which the evolution will not follow the curve, but the local tangent to the curve at this operating point, as illustrated in Figure 2.12. We thus get an equivalent *linearized evolution* by considering the variation of the input data as the input, and the variation of the output data as the output.

By using this linearized model, we will hence get the linear relation $\boxed{o = K_1\, i}$ in steady state at the neighborhood of the operating point M_1, where:

- the input $i = \Delta I = I - I_1$ and the output $o = \Delta O = O - O_1$ correspond to the **variation** of the input and the output with respect to the operating point M_1; and
- the coefficient $K_i = \left[\dfrac{\Delta O}{\Delta I}\right]_{M_i}$ is the **local static gain** whose unit is $[K_i] = [o]/[i]$ and whose value depends on the operating point.

This evolution will have a **range of validity** which depends on the system studied and in which it is possible to merge the real curve and its local model.

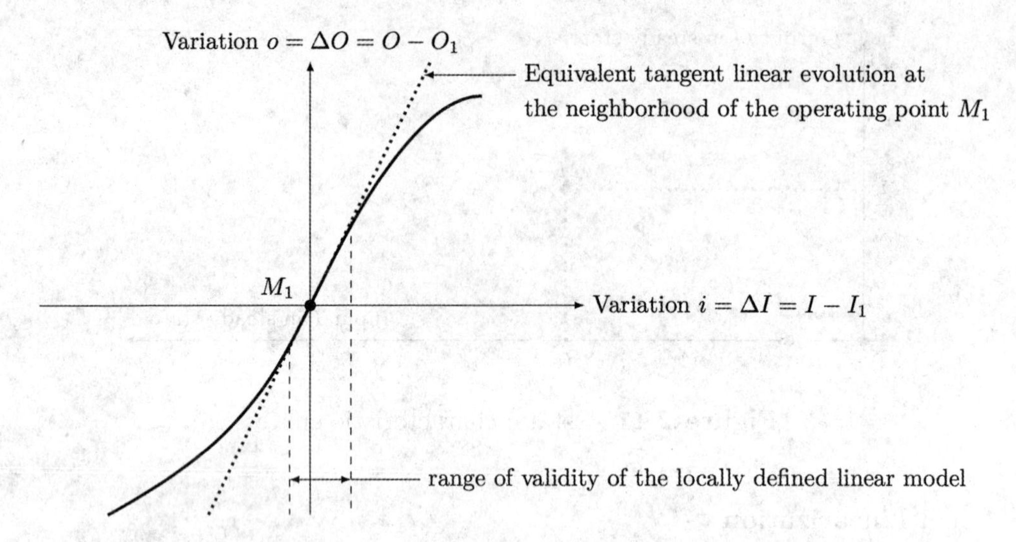

Figure 2.12 local linearization around the operating point M_1

Case of Linear Systems

In the case of linear systems, the static characteristic curve is a line which goes through the origin of the frame, as illustrated in the curve in Figure 2.13, which means that the output is proportional to the input in steady state. The proportionality coefficient between the output and the input is called the **static gain**. It corresponds to the slope of the curve and is defined as $\boxed{K = \frac{o}{i}}$.

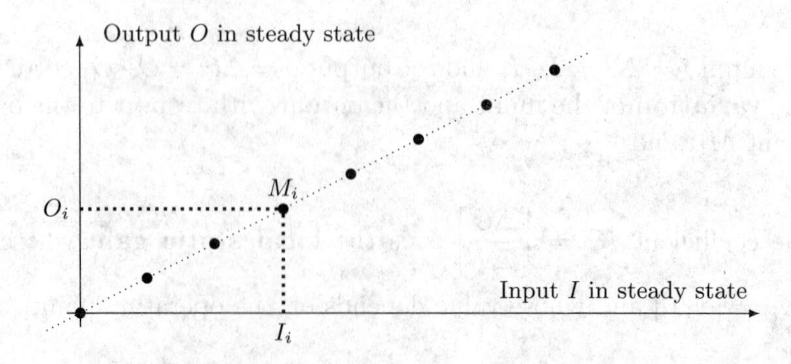

Figure 2.13 static characteristic curve in the case of a linear system

Mathematical Consequences

Proportionality Principle If the system provides an output signal $o(t)$ for an input signal $i(t)$, then it will provide an output signal $\lambda\, o(t)$ for an input signal $\lambda\, i(t)$ $\forall \lambda \in \mathbb{R}$:

$$i(t) \longrightarrow \boxed{\text{System}} \longrightarrow o(t) \qquad \Longrightarrow \qquad \lambda\, i(t) \longrightarrow \boxed{\text{System}} \longrightarrow \lambda\, o(t).$$

Let's consider two input signals i_1 and $i_2 = 2.5\, i_1$ of a system:

- if the system is **linear**, its output signals o_1 and o_2 are perfectly *homothetic*, as illustrated in Figure 2.14;

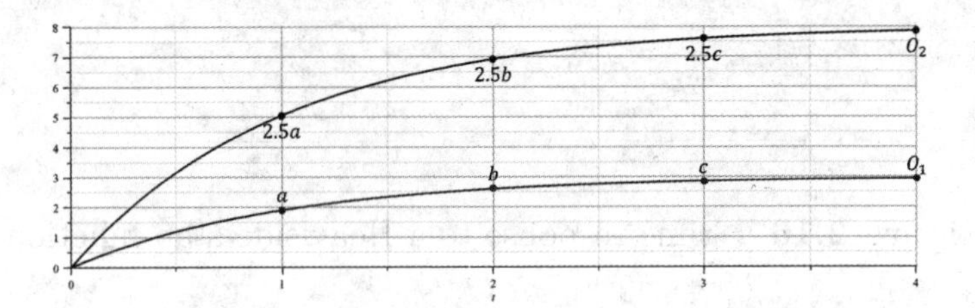

Figure 2.14 homothetic output signals

- if the system is **nonlinear**, its output signals o_1 and o_2 are not homothetic, as illustrated in Figure 2.15.

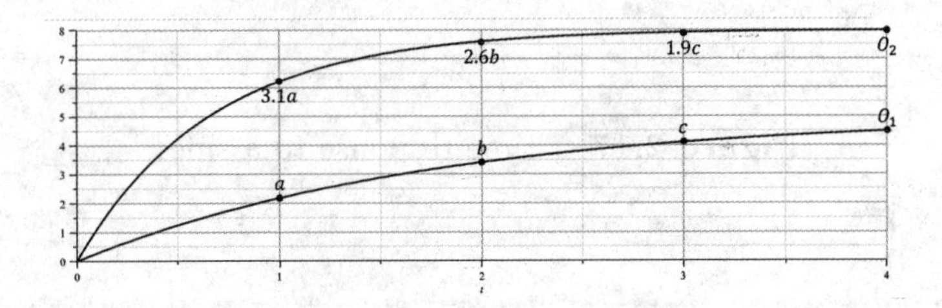

Figure 2.15 non-homothetic output signals

Superposition Principle If the system provides the output signals $o_1(t)$ and $o_2(t)$ for two input signals $i_1(t)$ and $i_2(t)$, it will provide the output signal $o_1(t) + o_2(t)$ for the input signal $i_1(t) + i_2(t)$:

$$i_1(t) \longrightarrow \boxed{\text{System}} \longrightarrow o_1(t) \qquad \text{AND} \qquad i_2(t) \longrightarrow \boxed{\text{System}} \longrightarrow o_2(t)$$
$$\Downarrow$$
$$i_1(t) + i_2(t) \longrightarrow \boxed{\text{System}} \longrightarrow o_1(t) + o_2(t)$$

- response to a first input signal (Heaviside step function, in Figure 2.16) and to a second input signal (sine, in Figure 2.17):

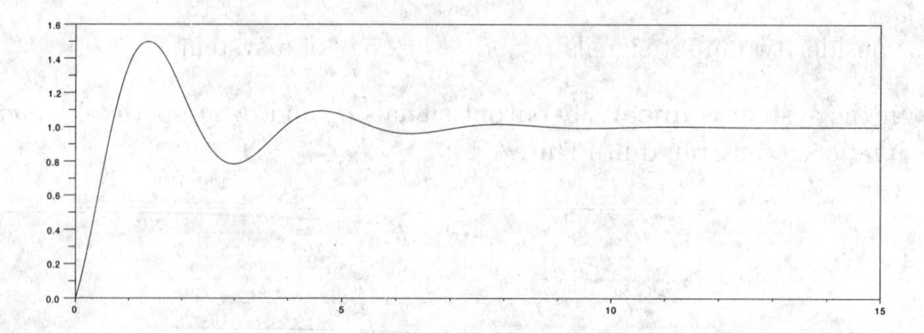

Figure 2.16 time response to a Heaviside step function

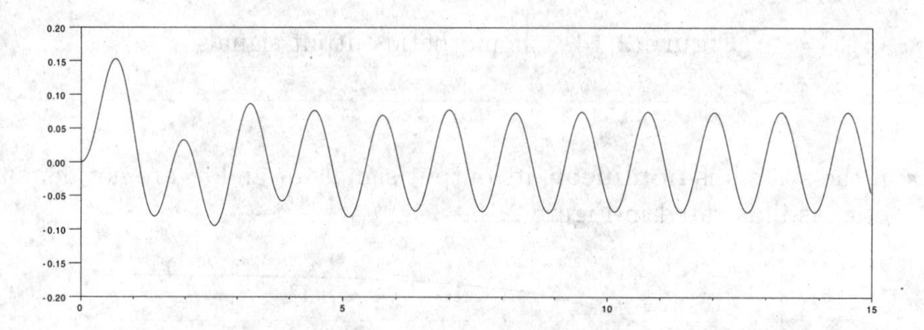

Figure 2.17 time response to a sine

- sum of the two output signals = response to the combined input signal (Figure 2.18):

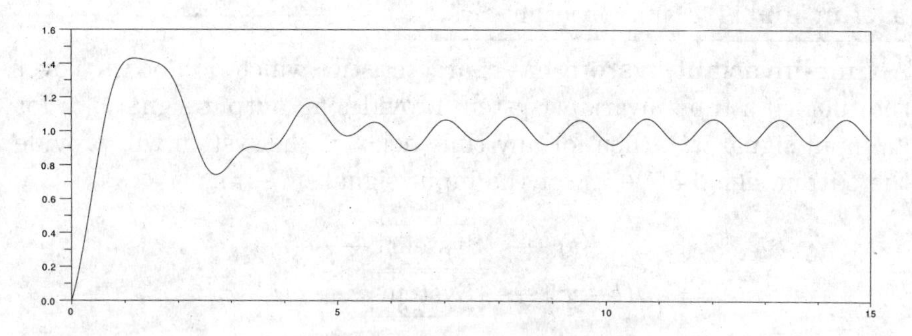

Figure 2.18 time response to the combined input signal

Corollary We will hence have, $\forall \lambda, \mu \in \mathbb{R}$:

$$\lambda\, i_1(t) + \mu\, i_2(t) \longrightarrow \boxed{\text{System}} \longrightarrow \lambda\, o_1(t) + \mu\, o_2(t)$$

Mathematical Consequences The differential equations (i.e. the system is linear) which model the relation between an input signal $i(t)$ and an output signal $o(t)$ are **linear**, which means that an effect is proportional to its cause. In this case, the equations are linear combinations of the successive derivatives of the input signal $i(t)$ and of the output signal $o(t)$ under the following form:

$$\sum_{j=1}^{n} a_j(t)\, o^{(j)}(t) = \sum_{k=1}^{m} b_k(t)\, i^{(k)}(t) \tag{2.1}$$

For instance, the differential equation $\ddot{\theta}(t) + \omega_0^2\, \theta(t) = 0$ is linear whereas the differential equation $\ddot{\theta}(t) + \omega_0^2 \sin \theta(t) = 0$ is nonlinear.

2.3.3 Time Invariance

Definition

Definition 11 (Time-invariant System)

A **time-invariant system** has characteristics which do not vary over the time: if a time-invariant system provides an output signal $o(t)$ for an input signal $i(t)$, then for any time delay τ, the system will provide the output signal $o(t-\tau)$ for the input signal $i(t-\tau)$:

$$i(t) \longrightarrow \boxed{\text{System}} \longrightarrow o(t)$$
$$\Longrightarrow \quad i(t-\tau) \longrightarrow \boxed{\text{System}} \longrightarrow o(t-\tau)$$

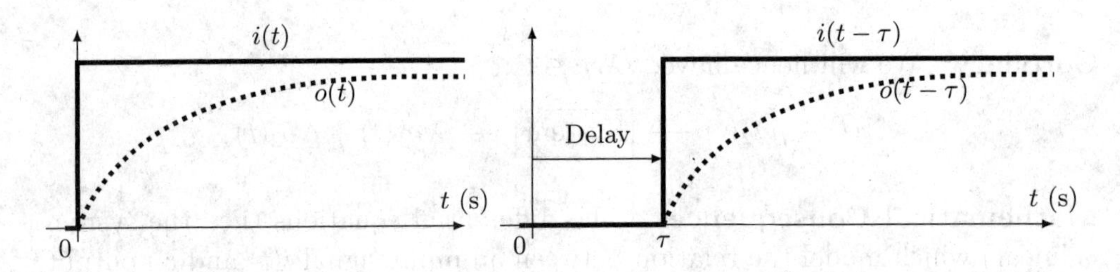

Figure 2.19 time behavior of a time-invariant system

Mathematical Consequence

The linear (i.e. the system is linear) differential equations (i.e. the system is continuous) have **constant coefficients**.

2.4 Performances

2.4.1 Steady State Performances

Stability

Two definitions of the stability of a process exist, even though none of both is fully satisfactory. However, whatever the retained definition, the notion of stability is intrinsic: it does not depend on the chosen input signal. This implies that if a system is instable, this instability can be emphasized with any of both definitions and for any input signal.

These two definitions are as follows:

1. **First definition**: theoretical stability.

   If a stable system is sollicited by an impulse signal, as illustrated in Figure 2.20, then its output signal asymptotically tends to 0, which corresponds to a stop of the system.

   - for a **stable system**, the output signal asymptotically tends to 0, as illustrated in Figure 2.20.

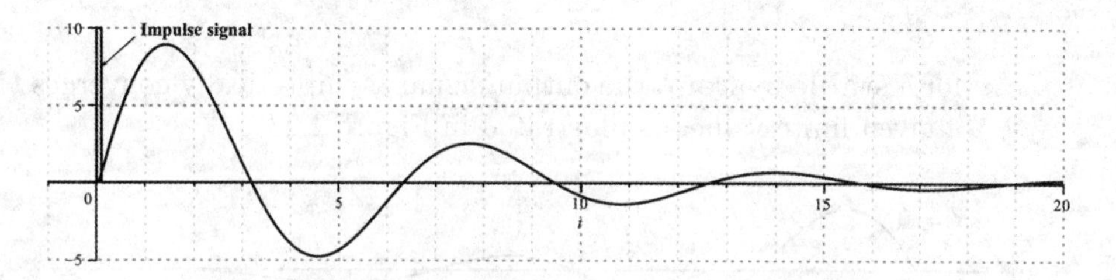

Figure 2.20 output signal of a stable system

   - for a **system at the limit of stability**, the output signal oscillates without converging nor diverging, as illustrated in Figure 2.21.

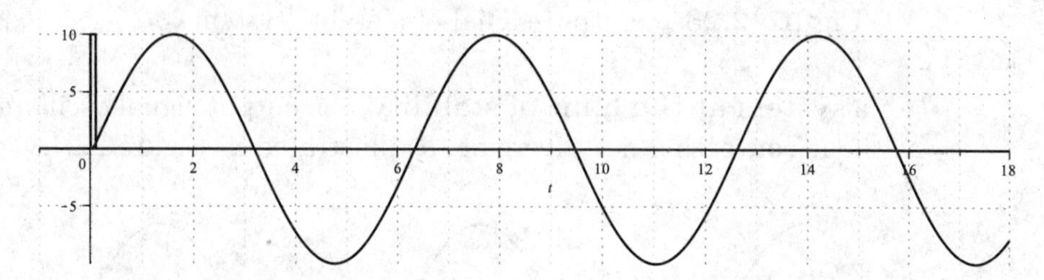

Figure 2.21 output signal of a system at the limit of stability

   - for an **instable system**, the output signal diverges, as illustrated in Figure 2.22.

   Even if this definition is interesting from a theoretical point of view, it is practically very difficult (or even impossible) to use because a system is unlikely not to be sollicited, since there are always disturbances.

2. **Second definition**: BIBO (Bounded Input, Bounded Output) stability.

   The output signal of a stable system remains bounded when it is sollicited by a bounded input signal, which corresponds to a non-divergence of the output signal.

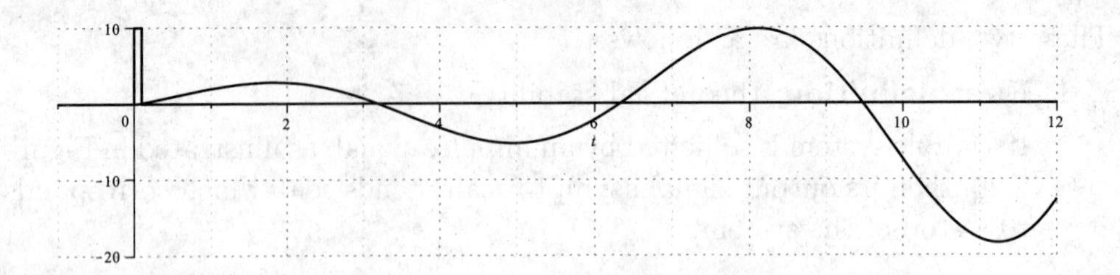

Figure 2.22 output signal of an instable system

- for a **stable system**, the output signal asymptotically converges to a given finite value, as illustrated in Figure 2.23.

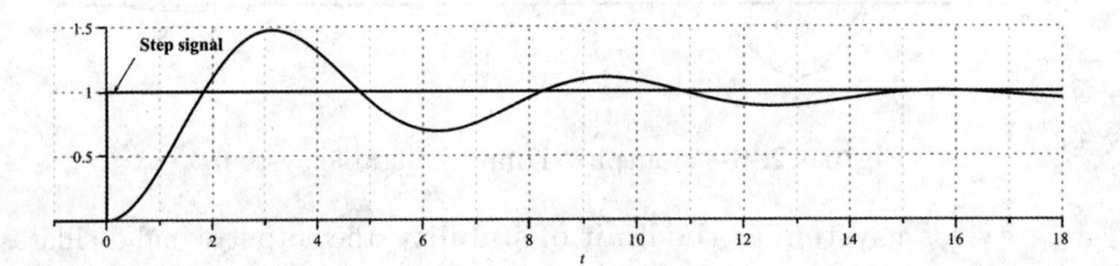

Figure 2.23 output signal of a stable system

- for a **system at the limit of stability**, the output signal oscillates without converging nor diverging, as illustrated in Figure 2.24.

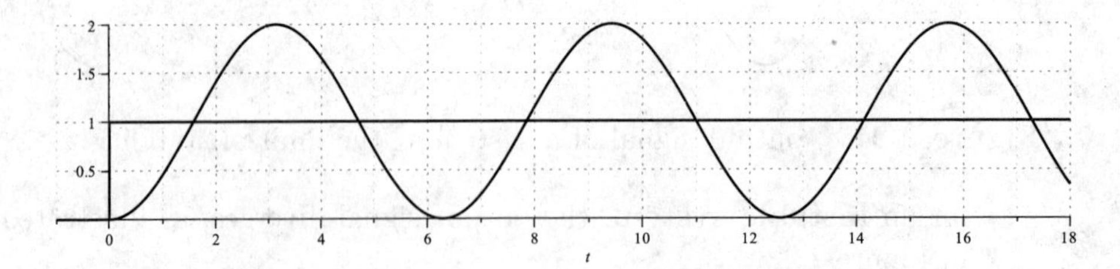

Figure 2.24 output signal of a system at the limit of stability

- for an **instable system**, the output signal diverges, as illustrated in Figure 2.25.

Because of the resolution of the sensors and of the non-linearities in the command, control systems often have an output signal which is not still

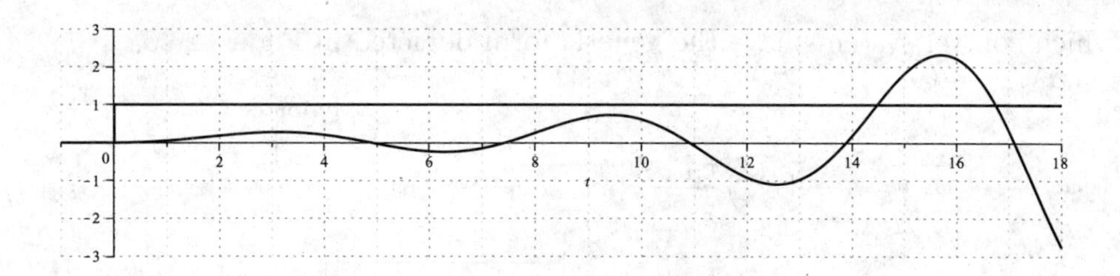

Figure 2.25 output signal of an instable system

and oscillates between two values when reaching the goal value. This type of system is of course stable (as its output signal does not diverge and remains bounded), but it is important to notice that this oscillation is not due to the approach of the limit of stability but to the command.

Precision and Robustness (or unsensitivity)

These two performances can be defined as follows:

- the **precision** of a system can be defined as the ability of this system to reach its input instruction: if the response of the system corresponds to the expected instruction, then the system is precise, otherwise, the system is not precise; the corresponding study is called the **study in pursuit**.
- the **robustness** of a system can be defined as the ability of this system to resist to disturbances: if a disturbance has no influence on the response of the system, then the system is robust, otherwise, the system is not robust; the corresponding study is called the **study in regulation**.
- if the system is supposed to be linear, the global study of a disturbed system corresponds to the application of the superposition theorem to these two successive studies; the corresponding study is called the **complete study**.

The precision and robustness allow to estimate the ability of a system to reach a target value (precision) or to resist to its environment (robustness): as a consequence, these two performances make sense (and even interest) only for control systems and should hence be defined only in this study framework[1] for

[1] Robustness does not make sense for non-looped systems (which are naturally not robust) and precision (if it can be defined) can be respected by simply changing the instruction provided.

which control systems have the generic form depicted in Figure 2.26.

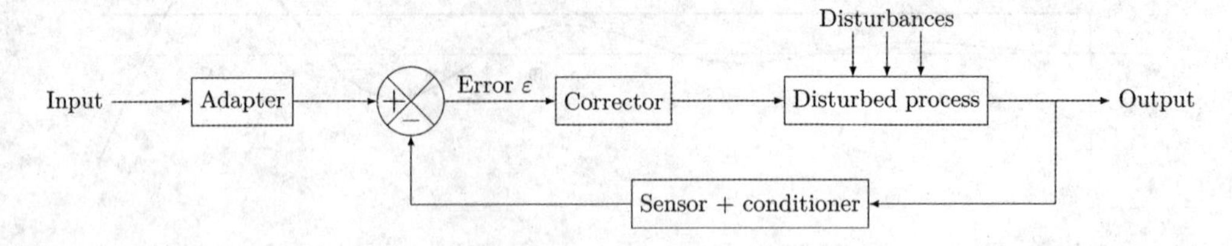

Figure 2.26 generic form of the functional diagram for this study framework

Since the inputs, outputs and disturbances can be of different types, a common data is chosen to define the precision and/or the robustness of a system: the output "error" ε of the comparator. The system hence is precise/robust if this error, which corresponds to the steady-state error (also called **static error**), is null, i.e. $\boxed{\varepsilon_\infty = \lim_{t \to +\infty} \varepsilon(t) = 0}$, in the following cases:

1. only the instruction is taken into account (study in pursuit);
2. only the disturbance is taken into account (study in regulation);
3. both the instruction and the disturbance are taken into account (superposition theorem applied to the two previous studies).

For instance, if we consider a system which is expected to tend to the value 1 with a Heaviside step input function (defined as $\forall t \geqslant 0, i(t) = \alpha_1 \in \mathbb{R}$) and a Heaviside step disturbance function (defined as $\forall t \geqslant 0, p(t) = \alpha_2 \in \mathbb{R}$) at the instant $t = 20$ s, the following cases may happen:

- if the system is **precise and robust**, then the static error is null in pursuit (instruction) and in regulation (disturbance)(Figure 2.27).

- if the system is **precise but not robust**, then the static error is null in pursuit (instruction), but not in regulation (disturbance) (Figure 2.28).

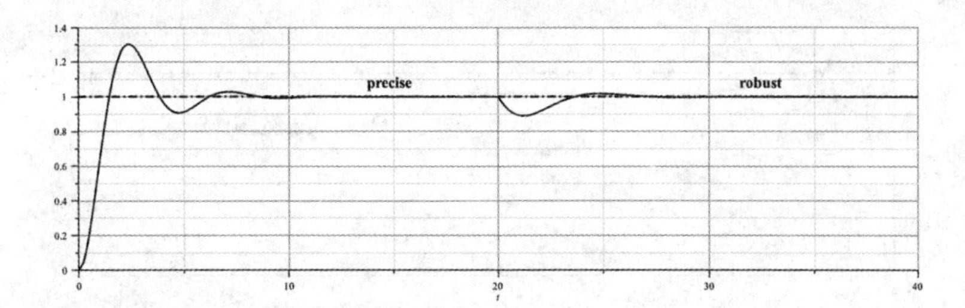

Figure 2.27 example of a system which is precise and robust

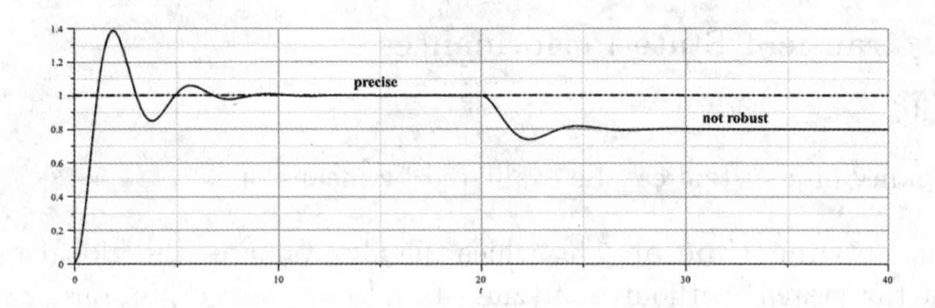

Figure 2.28 example of a system which is precise but not robust

- if the system is **not precise but robust**, then the static error is not null in pursuit (instruction), but it is null in regulation (disturbance)(Figure 2.29).

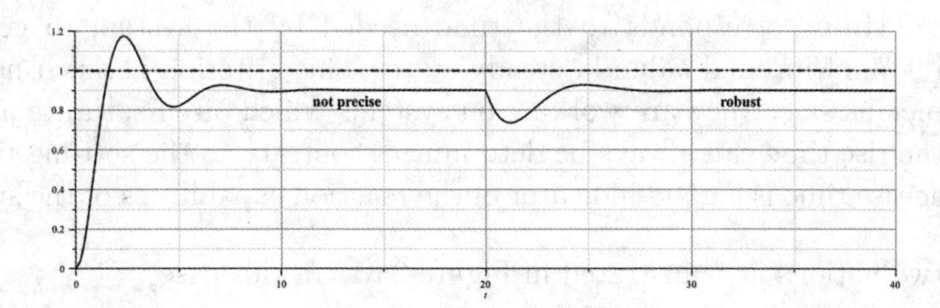

Figure 2.29 example of a system which is not precise but robust

- if the system is **neither precise nor robust**, then the static error is non-null in pursuit (instruction) and in regulation (disturbance)(Figure 2.30).

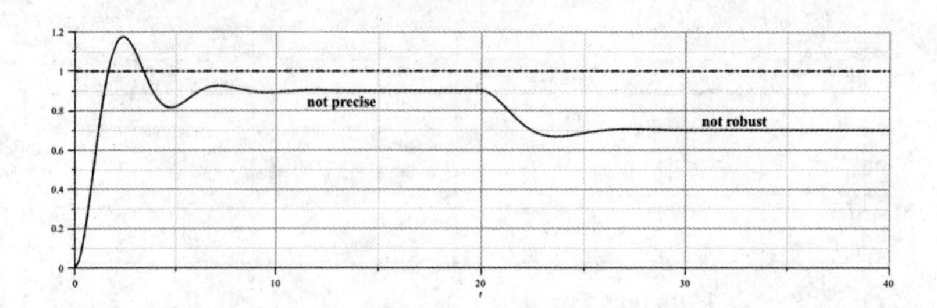

Figure 2.30 example of a system which is neither precise nor robust

2.4.2 Transient State Performances

Rapidity

The rapidity of a system can be evaluated by means of:

- the **settling time at 5%**, which mainly evaluates the global rapidity of the system without analyzing its intermediate evolutions, e.g. the presence of oscillations;
- the **rise time** of the signal, which evaluates the "reactivity" of the system before the possible occurrence of oscillations.

The **settling time at 5%** corresponds to the instant at which the output signal definitively gets within 5% of its final value in steady state.

The **rise time** corresponds to the time needed by the system to go from 10% to 90% of its final value in steady state. This criterion is much used by electronicians since they are working on systems which often saturate and for which the rise time can always be determined, contrary to the settling time at 5%. The rise time is a good indicator of the reaction capabilities of the system.

These two notions are illustrated in Figure 2.31. In this case:

- the settling time at 5% is worth 8.1 s; and
- the rise time is worth 1 s.

Damping

In the case of a temporal study, the damping of a system can be estimated by means of:

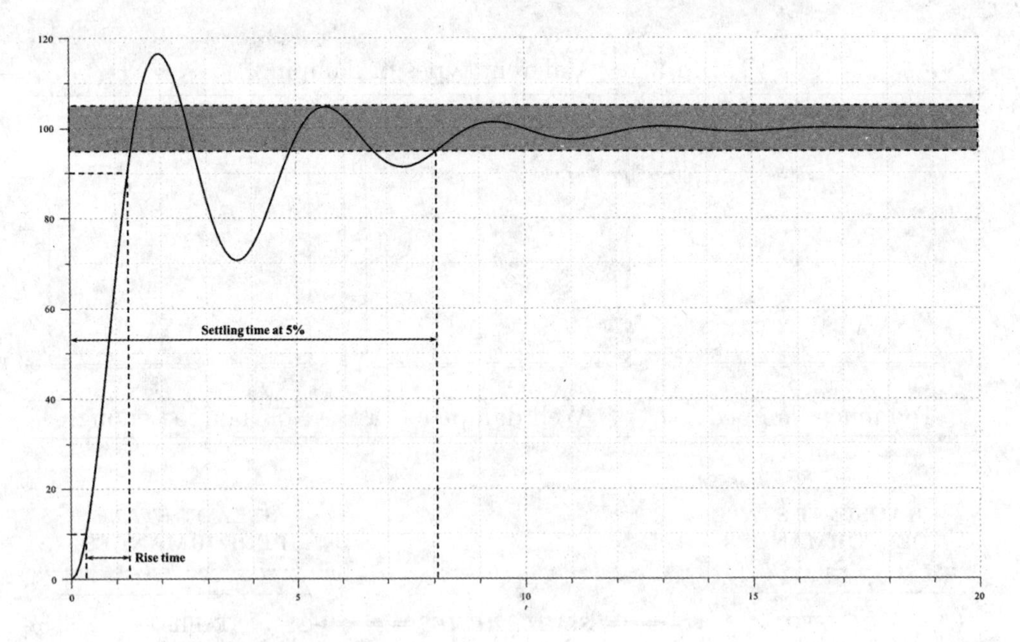

Figure 2.31 settling time and rise time

- the magnitude of the biggest transient oscillation: the biggest transient oscillation often is the first one, and the value of $O_{1\%}$, which corresponds to the ratio between the magnitude of the overshoot and the magnitude of the steady state response, is determined; this aspect will be emphasized in section 2.8.2 in the case of second order systems.

- the number of "visible" oscillations, whose magnitude ratio is practically greater than 5%: in the case of second order systems, this number is called the *quality factor* Q.

For instance, for a system whose target value is the value 1, the curves in Table 2.1 can be observed.

2.4.3 Similarities and Contradictions

The four performance criteria defined in sections 2.4.1 and 2.4.2 are not compatible: the adjustment of a system will hence necessarily result from a compromise which depends on the expected specifications of the system. An analysis of these criteria allows to obtain the diagram in Figure 2.32.

Table 2.1 different types of damping

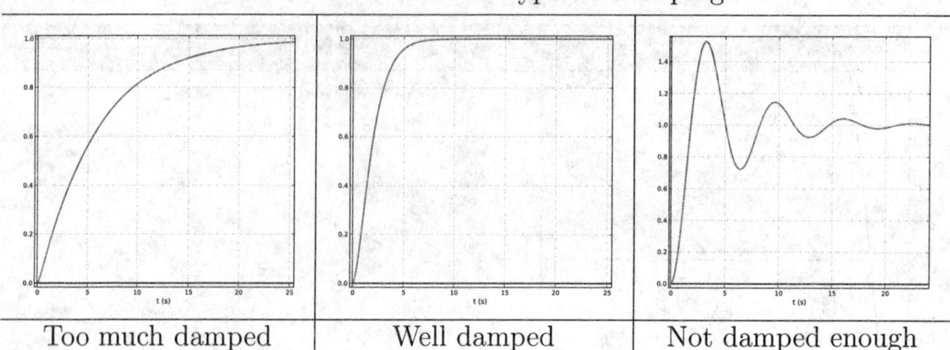

Too much damped	Well damped	Not damped enough

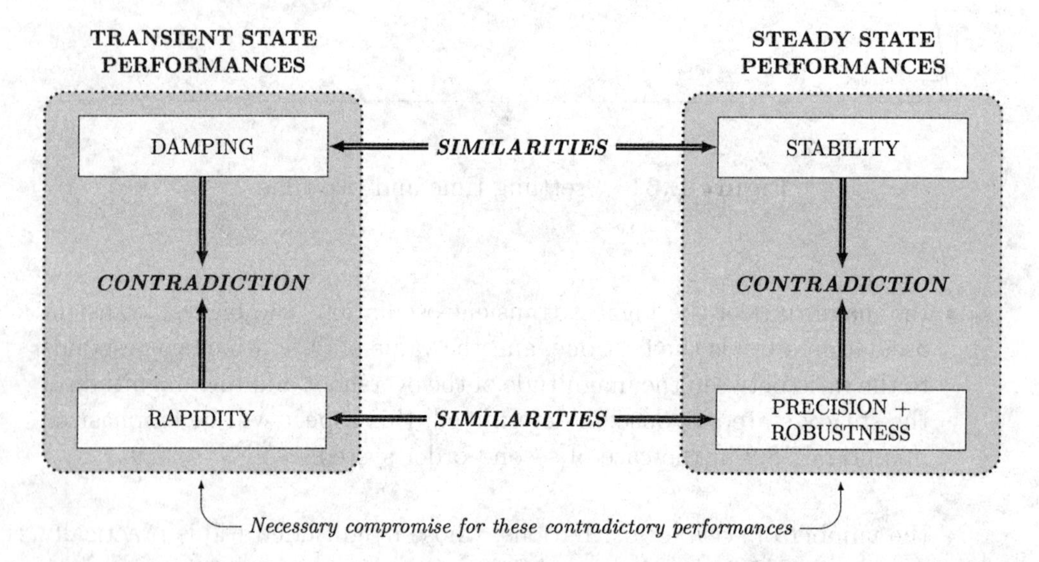

Figure 2.32 similarities and contradictions between the four performance criteria

2.5 Mathematical Tools

2.5.1 Laplace Transform

We are going to consider an approach which allows to solve differential equations by means of simple algebraic operations. This approach is based on an initial transform called the **Laplace transform** (which owes its name to the French mathematician, astronomer and physicist Pierre-Simon de Laplace,

1749-1827) and noted $\mathscr{L}$, and on a final inverse transform noted $\mathscr{L}^{-1}$ to get the temporal solution of the differential equation, as illustrated in the diagram in Figure 2.33.

This approach will allow to transform the resolution of a differential equation into a search for the roots of a polynomial, which is generally simpler. However, it will make us work with functions which are fully symbolic and are hence not real.

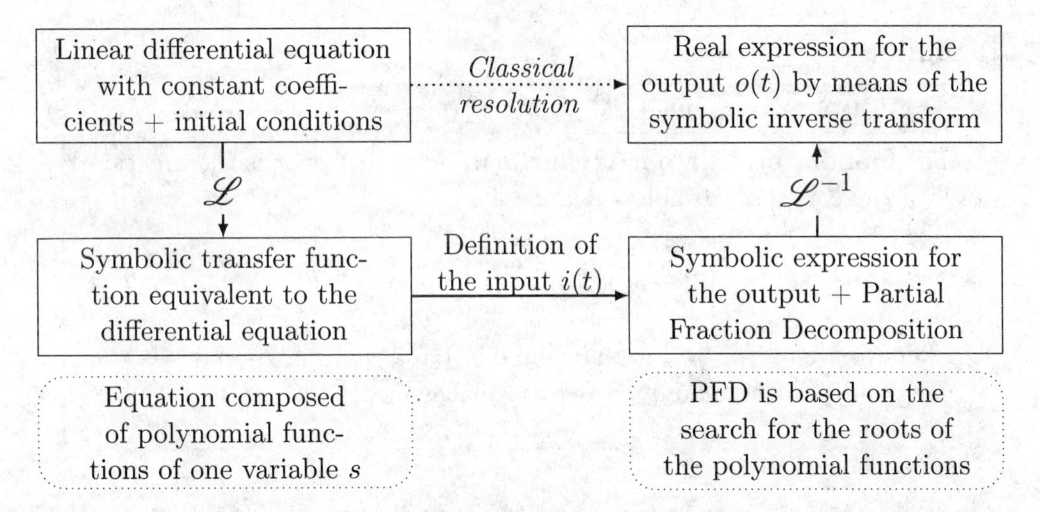

Figure 2.33 solution of differential equations by means of the Laplace transform

2.5.2 Definition and Consequences

The Causality Hypothesis

All the functions studied are supposed to be causal, i.e. they satisfy the condition $\boxed{\forall t < 0, f(t) = 0}$. "Something happens" at the instant $t = 0$. The effect is visible, even though the cause may be unknown. This hypothesis is illustrated in Figure 2.34.

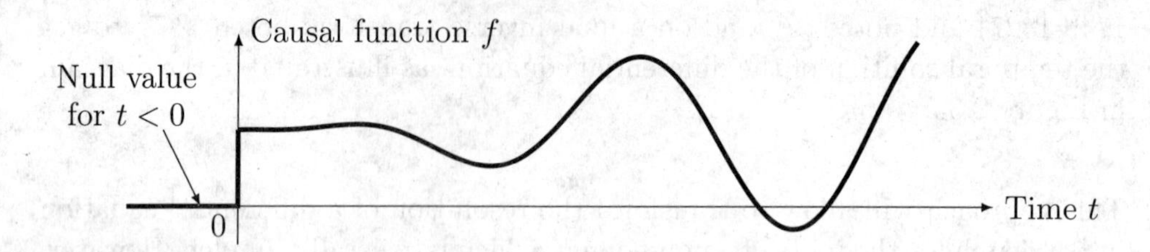

Figure 2.34 a causal function

Definition

Definition 12 (Monolateral Laplace Transform)

The **monolateral Laplace transform** of a function f is the function F of the complex variable s defined as:

$$F(s) = \mathscr{L}[f(t)] = \int_0^{+\infty} f(t)\, e^{-st}\, dt \qquad (2.2)$$

where $s = \sigma + j\omega \in \mathbb{C}$ is the symbolic Laplace variable and j is the imaginary unit commonly noted i in mathematics ($j^2 = -1$).

From a mathematical point of view, this definition implies highly restrictive hypotheses. Indeed, for this function F to be defined, the integral needs to converge. However, the functions f which will be used in our models will naturally respect these hypotheses, so the image function F will always exist.

Of course, a function F of the complex variable s will not necessarily correspond to the Laplace transform of a function f of the temporal variable t. It can be mathematically shown that a function F is the Laplace transform of a function f if and only if $\lim_{s \to +\infty} F(s) = 0$, where s is considered as a variable of the function F, without taking into account its real and imaginary parts.

The following comments can be made:

- the function F is called the image of the function f by the Laplace integral transform.
- in mathematics, the **bilateral Laplace transform** of a function f is

used and defined on $\mathbb{R}$ as:

$$F(s) = \int_{-\infty}^{+\infty} f(t)\, e^{-st}\, dt = \int_{-\infty}^{0} f(t)\, e^{-st}\, dt + \int_{0}^{+\infty} f(t)\, e^{-st}\, dt \quad (2.3)$$

However, since we are considering causal functions only (for which $\forall t < 0, f(t) = 0$), only the integral on $[0, +\infty)$ is considered, and hence the monolateral Laplace transform of f.

- the functions of the temporal variable t are generally noted in lower case letters whereas the functions of the symbolic Laplace variable s are generally noted in upper case letters: we hence have $\Omega(s) = \mathscr{L}[\omega(t)]$.

There exists a function named the **convolution product** of two functions f and g and noted $*$ such that $\boxed{\mathscr{L}[f(t) * g(t)] = \mathscr{L}[f(t)] \cdot \mathscr{L}[g(t)]}$. The convolution product of two functions of time f and g is defined as:

$$(f * g)(t) = \int_{0}^{t} f(\tau) g(t - \tau)\, d\tau$$

Mathematical Properties

The correspondence between a function f of the temporal variable $t \in \mathbb{R}^+$ and the corresponding function F of the complex Laplace variable $s \in \mathbb{C}$ is biunivocal, i.e. the elements of both sets are related with each other, as illustrated in Figure 2.35.

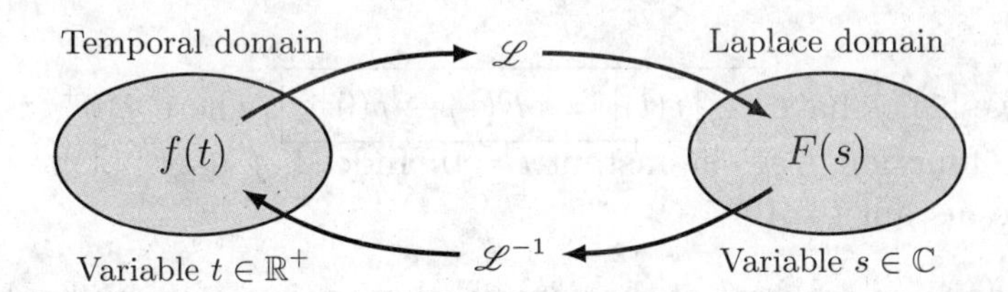

Figure 2.35 correspondence between the temporal domain and the Laplace domain

The Laplace transform hence is **bijective**. Any of these two domains can be chosen, and it will even be possible to alternatively change of domain to improve problem resolution. In this chapter, we will only work in one of these domains, and we will change of domain only when it is necessary.

Besides, for any real coefficients λ and μ and for any functions f and g:

$$\begin{aligned}\mathscr{L}\,[\lambda\, f(t) + \mu\, g(t)] &= \int_0^{+\infty} [\lambda\, f(t) + \mu\, g(t)]\; e^{-st}\, dt \\ &= \underbrace{\lambda \int_0^{+\infty} f(t)\, e^{-st}\, dt}_{\mathscr{L}[f(t)]} + \underbrace{\mu \int_0^{+\infty} g(t)\, e^{-st}\, dt}_{\mathscr{L}[g(t)]} \end{aligned} \tag{2.4}$$

We hence have $\mathscr{L}\,[\lambda\, f(t) + \mu\, g(t)] = \lambda\, \mathscr{L}[f(t)] + \mu\, \mathscr{L}[g(t)]$, which allows to conclude that the Laplace transform is **linear**.

Laplace Transform of a Derivative

The Laplace transform of a derivative can be determined as follows:

- if the Laplace transform F of a function f is known, then the Laplace transform of its **first derivative** can be determined as follows:

$$\mathscr{L}[\dot{f}(t)] = \int_0^{+\infty} \dot{f}(t)\, e^{-st}\, dt$$

  This integral is under the form $\int_0^{+\infty} u\, dv$ with $\begin{cases} u = e^{-st} \Rightarrow du = -s\, e^{-st}\, dt \\ dv = \dot{f}(t)\, dt \Rightarrow v = f(t) \end{cases}$

  Thanks to an integration by parts, we can determine that:

$$\mathscr{L}[\dot{f}(t)] = [u\, v]_0^{+\infty} - \int_0^{+\infty} v\, du = \underbrace{\left[f(t)\, e^{-st}\right]_0^{+\infty}}_{0 - f(0^+)} + s \underbrace{\int_0^{+\infty} f(t)\, e^{-st}\, dt}_{F(s)}$$

  and we hence have $\boxed{\mathscr{L}[\dot{f}(t)] = s\, F(s) - f(0^+)}$, where $f(0^+)$ is the value of the function f at the instant $t = 0$. Indeed, f may not be continuous at the instant $t = 0$.

- the Laplace transform of the **second derivative** of f can be determined as follows:

$$\begin{aligned}\mathscr{L}[\ddot{f}(t)] &= s\mathscr{L}[\dot{f}(t)] - \dot{f}(0^+) \\ &= s\,(s\, F(s) - f(0^+)) - \dot{f}(0^+) \\ &= s^2\, F(s) - s\, f(0^+) - \dot{f}(0^+)\end{aligned}$$

  We hence have $\boxed{\mathscr{L}[\ddot{f}(t)] = s^2\, F(s) - s\, f(0^+) - \dot{f}(0^+)}$.

- by recurrence, it can be shown that the Laplace transform of the **n-th derivative** of f has the following expression:

$$\forall n \in \mathbb{N}^*, \mathscr{L}[f^{(n)}(t)] = s^n\, F(s) - \sum_{i=0}^{n-1} s^i\, f^{(n-1-i)}(0^+) \tag{2.5}$$

The most classical case corresponds to the case of null initial conditions, which are also called **Heaviside initial conditions**: in this case, we have $\boxed{\forall n \in \mathbb{N}, f^{(n)}(0^+) = 0}$. If these conditions are respected, we will hence have $\boxed{\forall n \in \mathbb{N}, \mathscr{L}[f^{(n)}(t)] = s^n\, F(s)}$ and differentiating a function f n times in the temporal domain hence is equivalent to multiplying $F(s) = \mathscr{L}[f(t)]$ by s n times.

Laplace Transform of an Antiderivative

The Laplace transform of an antiderivative can be determined as follows:

- **Integration of order 1**. Let's suppose that the Laplace transform F of a function f is known, and let's determine the Laplace transform $\mathscr{L}[g(t)]$ of the antiderivative g of f defined as $g(t) = \displaystyle\int_0^t f(x)\,dx$. According to the expression for the Laplace transform of a derivative, we have:

$$F(s) = \mathscr{L}[f(t)] = \mathscr{L}[\dot{g}(t)] = s\,\mathscr{L}[g(t)] - g(0^+)$$

Since $g(0^+) = \displaystyle\int_0^{0+} f(x)\,dx = 0$, we can conclude that:

$$\mathscr{L}[\dot{g}(t)] = s\,\mathscr{L}[g(t)]$$

and hence that:

$$F(s) = s\,\mathscr{L}\left[\int_0^t f(x)\,dx\right]$$

As a consequence:

$$\boxed{\mathscr{L}\left[\int_0^t f(x)\,dx\right] = \frac{F(s)}{s}}$$

- **Integration of order 2**. According to the expression above, we have:

$$\mathscr{L}\left[\int_0^t \int_0^v f(u)\,du\,dv\right] = \frac{\mathscr{L}\left[\int_0^t f(u)\,du\right]}{s}$$

with $\mathscr{L}\left[\int_0^t f(u)\,du\right] = \dfrac{F(s)}{s}$. As a consequence:

$$\mathscr{L}\left[\int_0^t \int_0^v f(u)\,du\,dv\right] = \frac{F(s)}{s^2} \tag{2.6}$$

- **Integration of order n**. By recurrence, it can be shown that the Laplace transform of the antiderivative of order n of f has the following expression:

$$\boxed{\mathscr{L}\left[\underbrace{\int\int\int\ldots\int}_{n \text{ integrations}} f\ldots\right] = \frac{F(s)}{s^n}}$$

and integrating a function f n times in the temporal domain hence is equivalent to dividing $F(s) = \mathscr{L}[f(t)]$ by s n times.

In the remainder of this chapter, we will assume that the Heaviside initial conditions are respected: this is always possible as soon as the study is carried out at the neighborhood of an operating point. We hence have $\forall n \in \mathbb{N}, f^{(n)}(0^+) = 0$ and $\forall n \in \mathbb{N}, \mathscr{L}\left[f^{(n)}(t)\right] = s^n\,F(s)$.

2.5.3 Initial and Final Value Theorems

Introduction

The proofs of the two next theorems are quite long and painful and are not of much interest here. They will hence not be detailed in this chapter.

These two theorems allow to know the reaction of the system (and hence of its response) when $t \to 0^+$ and $t \to +\infty$.

Initial Value Theorem

Theorem 1 (Initial Value Theorem)

Let's consider a function f which is not discontinuous at $t = 0$. This function respects:

$$\lim_{t \to 0} f(t) = \lim_{s \to +\infty} s\,F(s) \tag{2.7}$$

It can be noticed that the first derivative of the function f respects:

$$\lim_{t \to 0} \dot{f}(t) = \lim_{s \to +\infty} s\,\left[s\,F(s) - f(0^+)\right]$$

where $f(0^+)$ is the value of the function f at $t = 0$.

Final Value Theorem

This theorem can be used if there is no divergence of the real function in steady state, which implies that the system needs to be stable.

The roots of the denominator of the transfer function are called the **poles** of the transfer function: we will show in the second volume that a system is stable if the real part of ALL the poles of its transfer function is strictly negative. All the poles hence need to be located as illustrated in Figure 2.36.

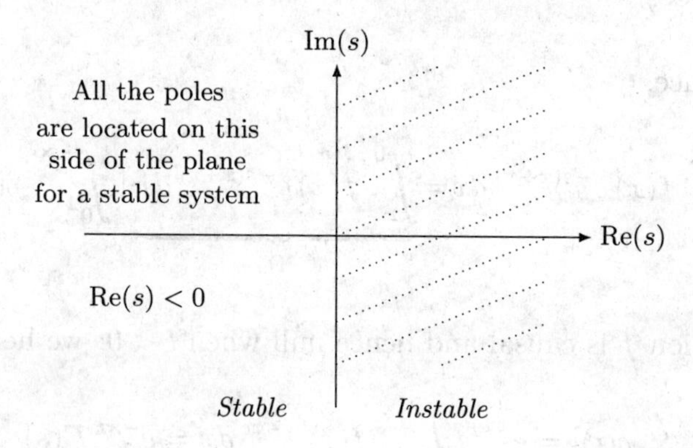

Figure 2.36 condition on the poles for a system to be stable

Theorem 2 (Final Value Theorem)

Let's consider a function f which is not divergent (i.e. the system is stable). This function respects:

$$\lim_{t\to+\infty} f(t) = \lim_{s\to 0} s\,F(s) \tag{2.8}$$

It can be noticed that the first derivative of the function f respects:

$$\lim_{t\to+\infty} \dot{f}(t) = \lim_{s\to 0} s\,\left[s\,F(s) - f(0^+)\right]$$

where $f(0^+)$ is the value of the function f at $t = 0$.

2.5.4 Theorem of the Delay

The Laplace transform F of a function f is supposed to be known. Let's determine the expression for the Laplace transform of the delayed function ρ defined as $\rho(t) = f(t-\tau)$, with $\tau \geqslant 0$, according to τ and F. By definition,

$$\mathscr{L}[\rho(t)] = \int_0^{+\infty} \rho(t)\,e^{-st}\,dt = \int_0^{+\infty} f(t-\tau)\,e^{-st}\,dt$$

Let's pose $x = t - \tau$, which is equivalent to $t = x + \tau$. We hence have $dx = dt$ and $x \in [-\tau, +\infty)$.

As a consequence,

$$\mathscr{L}[\rho(t)] = \int_{-\tau}^{+\infty} f(x)\,e^{-s(x+\tau)}\,dx = \underbrace{\int_{-\tau}^{0} f(x)\,e^{-s(x+\tau)}\,dx}_{0} + \int_0^{+\infty} f(x)\,e^{-s(x+\tau)}\,dx$$

since the function f is causal and hence null when $t < 0$: we hence have

$$\mathscr{L}[\rho(t)] = e^{-s\tau} \underbrace{\int_0^{+\infty} f(x)\,e^{-sx}\,dx}_{F(s)} = e^{-s\tau}\,F(s)$$

Theorem 3 (Theorem of the Delay)

The Laplace transform of a delayed function is defined as (Figure 2.37):

$$\mathscr{L}[f(t-\tau)] = e^{-s\tau}\,\mathscr{L}[f(t)] \text{ with } \tau \geqslant 0 \tag{2.9}$$

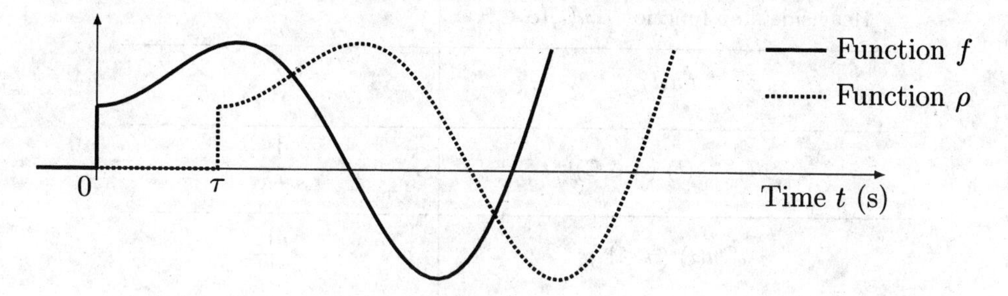

Figure 2.37 illustration of the theorem of the delay

2.5.5 Table of Classical Laplace Transforms

The most common causal functions and their corresponding Laplace transforms are presented in Table 2.2. These Laplace transforms can be determined as follows:

1. $f(t) = \delta(t)$. First, we need to determine

$$\mathscr{L}[\delta_T(t)] = \int_0^{+\infty} \delta_T(t)\, e^{-st}\, dt$$

according to T, where δ_T is the function defined as
$\begin{cases} \delta_T(t) = 0\,\forall t < 0 \text{ and } \forall t > T \\ \delta_T(t) = \dfrac{1}{T}\,\forall t \in [0, T] \end{cases}$:

$$\mathscr{L}[\delta_T(t)] = \int_0^{+\infty} \delta_T(t)\, e^{-st}\, dt = \int_0^{T} \delta_T(t)\, e^{-st}\, dt + \underbrace{\int_T^{+\infty} \delta_T(t)\, e^{-st}\, dt}_{0}$$

$$= \int_0^{T} \frac{1}{T}\, e^{-st}\, dt = \frac{1}{T}\left[-\frac{e^{-st}}{s}\right]_0^T = \frac{1-e^{-sT}}{s\,T}$$

It will be admitted that $\mathscr{L}[\delta(t)] = \mathscr{L}\left[\lim\limits_{T\to 0} \delta_T(t)\right] = \lim\limits_{T\to 0} \mathscr{L}[\delta_T(t)]$. If we

Table 2.2 classical Laplace transforms

$f(t)$	$F(s) = \int_0^{+\infty} f(t)\, e^{-st}\, dt$
Dirac delta function $\delta(t)$	1
Heaviside step function $\alpha\, u(t) (\alpha \in \mathbb{R})$	$\frac{\alpha}{s}$
$t\, u(t)$	$\frac{1}{s^2}$
$e^{-\alpha t}\, u(t)\ (\alpha \in \mathbb{C})$	$\frac{1}{s+\alpha}$
$t^n\, u(t)\ (n \in \mathbb{N})$	$\frac{n!}{s^{n+1}}$
$t\, e^{-\alpha t}\, u(t)\ (\alpha \in \mathbb{C})$	$\frac{1}{(s+\alpha)^2}$
$\cos(\omega t) u(t)\ (\omega \in \mathbb{R}^+)$	$\frac{s}{s^2+\omega^2}$
$\sin(\omega t) u(t)\ (\omega \in \mathbb{R}^+)$	$\frac{\omega}{s^2+\omega^2}$
$e^{-\alpha t} \cos(\omega t) u(t)$	$\frac{s+\alpha}{(s+\alpha)^2+\omega^2}$
$e^{-\alpha t} \sin(\omega t) u(t)$	$\frac{\omega}{(s+\alpha)^2+\omega^2}$

pose $x = s\,T$ and if we suppose that $x \to 0$, we have $e^{-x} \underset{x \to 0}{\sim} 1 - x$ and we hence get:

$$\mathscr{L}[\delta(t)] = \lim_{x \to 0} \frac{1 - e^{-x}}{x} = 1 \tag{2.10}$$

2. $f(t) = \alpha\, u(t)$. On the domain studied, the Heaviside unit step function $u(t)$ is equal to 1. We hence have:

$$\mathscr{L}[\alpha\, u(t)] = \int_0^{+\infty} \alpha\, e^{-st}\, dt = \left[-\frac{\alpha\, e^{-st}}{s} \right]_0^{+\infty} = \frac{\alpha}{s} \tag{2.11}$$

3. $f(t) = t\,u(t)$. By definition:

$$\mathscr{L}[t\,u(t)] = \int_0^{+\infty} t\,e^{-st}\,dt$$

We can pose $\begin{cases} u = t \Rightarrow du = dt \\ dv = e^{-st}\,dt \Rightarrow v = -\dfrac{e^{-st}}{s} \end{cases}$ and determine, by integration by parts, that:

$$\mathscr{L}[t\,u(t)] = \underbrace{\left[-\frac{t\,e^{-st}}{s}\right]_0^{+\infty}}_{0} + \frac{1}{s}\int_0^{+\infty} e^{-st}\,dt = \left[-\frac{e^{-st}}{s^2}\right]_0^{+\infty} = \frac{1}{s^2} \quad (2.12)$$

4. $f(t) = e^{-\alpha t}\,u(t)$. By definition:

$$\begin{aligned} \mathscr{L}[e^{-\alpha t}\,u(t)] &= \int_0^{+\infty} e^{-\alpha t}\,e^{-st}\,dt = \int_0^{+\infty} e^{-(\alpha+s)t}dt \\ &= \left[-\frac{e^{-(\alpha+s)t}}{\alpha+s}\right]_0^{+\infty} = \frac{1}{\alpha+s} \end{aligned} \quad (2.13)$$

5. $f(t) = t^n\,u(t)$. By definition:

$$\mathscr{L}[t^n\,u(t)] = \int_0^{+\infty} t^n\,e^{-st}\,dt$$

We can pose $\begin{cases} u = t^n \Rightarrow du = n\,t^{n-1}dt \\ dv = e^{-st}\,dt \Rightarrow v = -\dfrac{e^{-st}}{s} \end{cases}$ and determine, by integration by parts, that:

$$\mathscr{L}[t^n\,u(t)] = \underbrace{\left[-\frac{t^n\,e^{-st}}{s}\right]_0^{+\infty}}_{0} + \frac{n}{s}\int_0^{+\infty} t^{n-1}\,e^{-st}\,dt = \frac{n}{s}\int_0^{+\infty} t^{n-1}\,e^{-st}\,dt$$

We can pose $\begin{cases} u = t^{n-1} \Rightarrow du = (n-1)\,t^{n-2}dt \\ dv = e^{-st}\,dt \Rightarrow v = -\dfrac{e^{-st}}{s} \end{cases}$ and determine, by integration by parts, that:

$$\mathscr{L}[t^n\, u(t)] = \frac{n}{s}\underbrace{\left[-\frac{t^{n-1}\, e^{-st}}{s}\right]_0^{+\infty}}_{0} + \frac{n(n-1)}{s^2}\int_0^{+\infty} t^{n-2}\, e^{-st}\, dt$$

$$= \frac{n(n-1)}{s^2}\int_0^{+\infty} t^{n-2}\, e^{-st}\, dt$$

By generalizing, we can determine that:

$$\mathscr{L}[t^n\, u(t)] = \frac{n!}{s^n}\int_0^{+\infty} t^0\, e^{-st}\, dt = \frac{n!}{s^n}\int_0^{+\infty} e^{-st}\, dt = \frac{n!}{s^n}\left[-\frac{e^{-st}}{s}\right]_0^{+\infty}$$

$$= \frac{n!}{s^{n+1}} \tag{2.14}$$

6. $f(t) = t\, e^{-\alpha t}\, u(t)$. By definition:

$$\mathscr{L}[t\, e^{-\alpha t}\, u(t)] = \int_0^{+\infty} t\, e^{-\alpha t}\, e^{-st}\, dt = \int_0^{+\infty} t\, e^{-(\alpha+s)t}\, dt$$

We can pose $\begin{cases} u = t \Rightarrow du = dt \\ dv = e^{-(\alpha+s)t}\, dt \Rightarrow v = -\dfrac{e^{-(\alpha+s)t}}{\alpha+s} \end{cases}$ and determine, by integration by parts, that:

$$\mathscr{L}[t\, e^{-\alpha t}\, u(t)] = \underbrace{\left[-\frac{t\, e^{-(\alpha+s)t}}{\alpha+s}\right]_0^{+\infty}}_{0} + \frac{1}{\alpha+s}\int_0^{+\infty} e^{-(\alpha+s)t}\, dt$$

$$= \left[-\frac{e^{-(\alpha+s)t}}{(\alpha+s)^2}\right]_0^{+\infty} = \frac{1}{(\alpha+s)^2} \tag{2.15}$$

7. $f_1(t) = \cos(\omega t)\, u(t)$ and $f_2(t) = \sin(\omega t)\, u(t)$. According to Euler's formula, we have

$$\cos(\omega t) = \frac{e^{j\omega t} + e^{-j\omega t}}{2} \text{ and } \sin(\omega t) = \frac{e^{j\omega t} - e^{-j\omega t}}{2j}$$

As a consequence:

$$\begin{cases} \mathscr{L}[\cos(\omega t)\,u(t)] = \dfrac{1}{2}(\mathscr{L}[e^{j\omega t}\,u(t)] + \mathscr{L}[e^{-j\omega t}\,u(t)]) \\ \qquad = \dfrac{1}{2}\left(\dfrac{1}{s-j\omega} + \dfrac{1}{s+j\omega}\right) = \dfrac{s}{s^2+\omega^2} \\ \mathscr{L}[\sin(\omega t)\,u(t)] = \dfrac{1}{2j}(\mathscr{L}[e^{j\omega t}\,u(t)] - \mathscr{L}[e^{-j\omega t}\,u(t)]) \\ \qquad = \dfrac{1}{2j}\left(\dfrac{1}{s-j\omega} - \dfrac{1}{s+j\omega}\right) = \dfrac{\omega}{s^2+\omega^2} \end{cases}$$

8. $f_1(t) = e^{-\alpha t}\cos(\omega t)\,u(t)$ and $f_2(t) = e^{-\alpha t}\sin(\omega t)\,u(t)$. According to Euler's formula, we have

$$\cos(\omega t) = \frac{e^{j\omega t} + e^{-j\omega t}}{2} \text{ and } \sin(\omega t) = \frac{e^{j\omega t} - e^{-j\omega t}}{2j}$$

As a consequence:

$$\begin{cases} \mathscr{L}[e^{-\alpha t}\cos(\omega t)\,u(t)] = \dfrac{1}{2}(\mathscr{L}[e^{(-\alpha+j\omega)t}\,u(t)] + \mathscr{L}[e^{-(\alpha+j\omega)t}\,u(t)]) \\ \qquad = \dfrac{1}{2}\left(\dfrac{1}{(s+\alpha)-j\omega} + \dfrac{1}{(s+\alpha)+j\omega}\right) \\ \qquad = \dfrac{s+\alpha}{(s+\alpha)^2+\omega^2} \\ \mathscr{L}[e^{-\alpha t}\sin(\omega t)\,u(t)] = \dfrac{1}{2j}(\mathscr{L}[e^{(-\alpha+j\omega)t}\,u(t)] - \mathscr{L}[e^{-(\alpha+j\omega)t}\,u(t)]) \\ \qquad = \dfrac{1}{2j}\left(\dfrac{1}{(s+\alpha)-j\omega} - \dfrac{1}{(s+\alpha)+j\omega}\right) \\ \qquad = \dfrac{\omega}{(s+\alpha)^2+\omega^2} \end{cases}$$

2.6 Transfer Function

The behavior of a linear continuous-time time-invariant system can be described by a linear differential equation with constant coefficients under the following form:

$$\underbrace{a_0\,o(t)+a_1\,\dot{o}(t)+a_2\,\ddot{o}(t)+\cdots+a_n\,o^{(n)}(t)}_{\sum_{k=0}^{n} a_k\,o^{(k)}(t)}$$
$$=\underbrace{b_0\,i(t)+b_1\,\dot{i}(t)+b_2\,\ddot{i}(t)+\cdots+b_m\,i^{(m)}(t)}_{\sum_{l=0}^{m} b_l\,i^{(l)}(t)}$$

with $n \geqslant m$. In the Laplace domain (if the initial conditions are supposed to be null), we hence have:

$$(a_0+a_1\,s+a_2\,s^2+\cdots+a_n\,s^n)O(s)=(b_0+b_1\,s+b_2\,s^2+\cdots+b_m\,s^m)I(s)$$

We can hence determine that: $\boxed{H(s)=\dfrac{O(s)}{I(s)}=\dfrac{b_0+b_1\,s+b_2\,s^2+\cdots+b_m\,s^m}{a_0+a_1\,s+a_2\,s^2+\cdots+a_n\,s^n}}$.

This polynomial fraction is called the **transfer function** of the system and there is an equivalence between transfer functions and differential equations, as illustrated in Figure 2.38.

Differential equation | Transfer function

$$\sum_{k=0}^{n} a_k\,o^{(k)}(t)=\sum_{l=0}^{m} b_l\,i^{(l)}(t) \quad \xrightarrow{\mathscr{L}} \quad \xleftarrow{\mathscr{L}^{-1}} \quad H(s)=\frac{O(s)}{I(s)}=\frac{b_0+b_1\,s+b_2\,s^2+\cdots+b_m\,s^m}{a_0+a_1\,s+a_2\,s^2+\cdots+a_n\,s^n}$$

Figure 2.38 equivalence between transfer functions and differential equations

The Laplace transform hence allows to obtain *polynomial equations* which are much easier to manipulate than differential equations.

2.7 Block Diagram

2.7.1 Structure

From the Functional Diagram to the Block Diagram

The functional diagram described in section 2.2.3 is a diagram which is defined in terms of functions (by means of infinitive verbs such as "convert the energy",

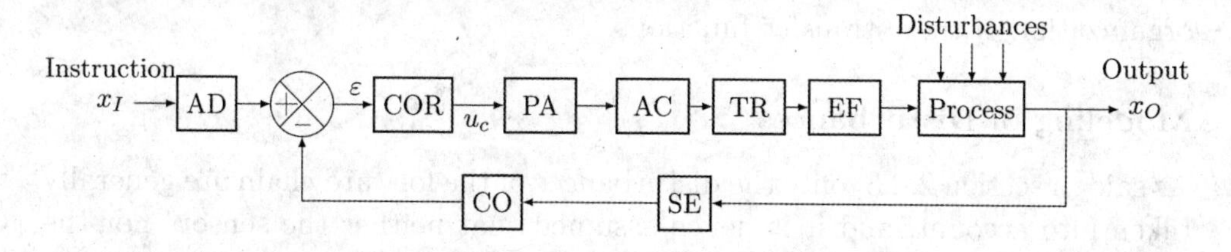

Figure 2.39 generic form of a functional diagram

etc.) or of structure (by indicating the different physical elements which belong to the system) under the form depicted in Figure 2.39.

A functional diagram is built from the functional chain of the system (pre-actuator, actuator, sensor, etc., as presented in chapter 1) after an observation of the structure of the system. The evolution of each of these components can be modeled by one or many laws, depending on the goal. These laws can be obtained either by the manipulation of equations defined at the neighborhood of **a specific operating point** (knowledge model) or by the choice of a model after an observation (behavioral model), and none of these two methods is better than the other. This diversity of laws will allow to have many different transfer functions and thus **many block diagrams**, which represent the evolution of the functional diagram. The generic form of such a block diagram is depicted in Figure 2.40.

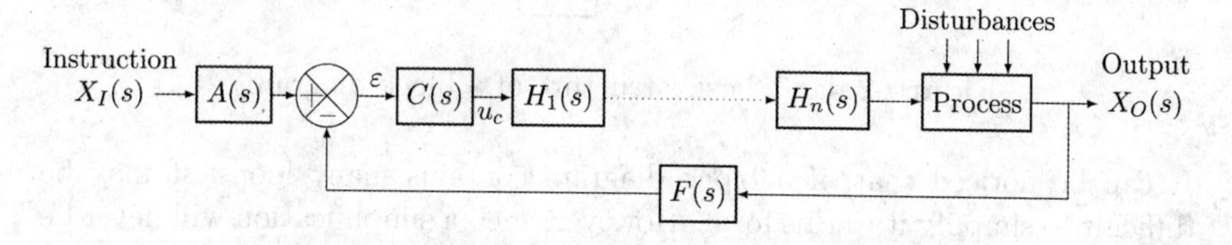

Figure 2.40 generic form of a block diagram

Elements of a Block Diagram

The behavior of the elements of the system can be described by means of differential equations which can be linearized at the neighborhood of an operating point. A transfer function can hence be associated with each differential equation by means of the symbolic Laplace transform (most often, the Heaviside conditions are supposed to be verified). A block diagram describes the organization of these transfer functions.

Modeling of Disturbances

As said in section 2.2.3, only the disturbances of the forward chain are generally taken into account, and it is hence assumed that neither the sensors nor the correctors are significatively disturbed.

Fortunately, practically, this assumption is valid, at least from a macroscopic point of view: if this is not the case, it is still possible to consider an equivalent disturbance to take into account the additional ones. The disturbance will most often affect the conversion of the energy, and hence the actuator (even though this may not be the case).

2.7.2 Basic Structure of a Block Diagram

A control system can generally be represented by the block diagram in Figure 2.41, sometimes after a few manipulations.

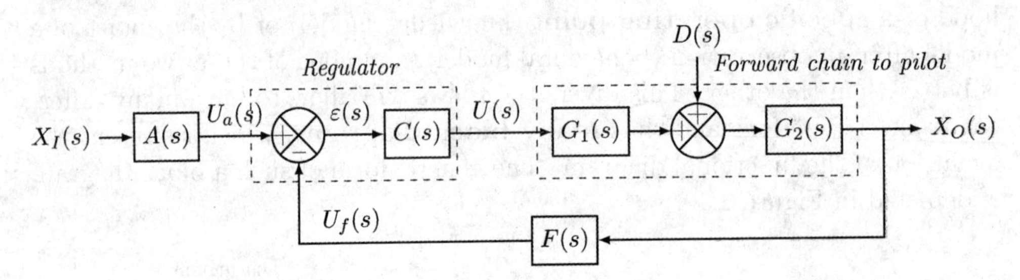

Figure 2.41 basic structure of a block diagram

It can be noticed that, if a block diagram contains many loops, it may be difficult to simplify it to this form. However, such a simplification will never be impossible. This generic form will be used for performances analysis, whatever the initial structure.

The Laplace transforms of the block diagram in Figure 2.41 are as follows:

- $X_I(s) = \mathscr{L}[x_I(t)]$ is the Laplace transform of the instruction signal;
- $X_O(s) = \mathscr{L}[x_O(t)]$ is the Laplace transform of the output signal of the controlled process;
- $D(s) = \mathscr{L}[d(t)]$ is the Laplace transform of the disturbance signal;

- $U_f(s) = \mathscr{L}[u_f(t)]$ is the Laplace transform of the feedback signal;
- $U_a(s) = \mathscr{L}[u_a(t)]$ is the Laplace transform of the (adapted) instruction signal;
- $U(s) = \mathscr{L}[u_c(t)]$ is the Laplace transform of the corrected signal;
- $\varepsilon(s) = \mathscr{L}[\varepsilon(t)] = U_a(s) - U_f(s)$ is the Laplace transform of the error signal.

The transfer functions of the block diagram in Figure 2.41 are as follows:

- $A(s)$ is the adaptation block, which is generally adjusted to get a null error when the instruction signal and the output signal of the system are the same[2];
- $C(s)$ is the correction block, whose adjustment is the purpose of the automation control course;
- $G_1(s)$ is the energy transformation block (which generally corresponds to the pre-actuator and the actuator);
- $G_2(s)$ is the energy adaptation block (which generally corresponds to the power transmitter and the effector);
- $F(s)$ is the block in charge of the measurement and treatment of the controlled data (sensor + conditioner)[3].

Black's Formula

For the classical form including a regulator composed of the serial association of a substractor and a corrector, the basic closed-loop form is the one depicted in Figure 2.42, after grouping the different transfer functions.

The block diagram allows to write the following equations:

$$\begin{cases} X_O(s) = F_1(s)\,\varepsilon(s) \\ \varepsilon(s) = X_I(s) - U_f(s) \\ U_f(s) = F_2(s)\,X_O(s) \end{cases}$$

[2]In this case, for the error $\varepsilon(s) = A(s)\,X_I(s) - F(s)\,X_O(s)$ to be null when $X_O(s) = X_I(s)$, we necessarily have $A(s) = F(s)$.

[3]This block is generally modeled by a pure gain (which is usual), by a pure gain with a pure delay (which is very rare and yet coherent) or by a first order system.

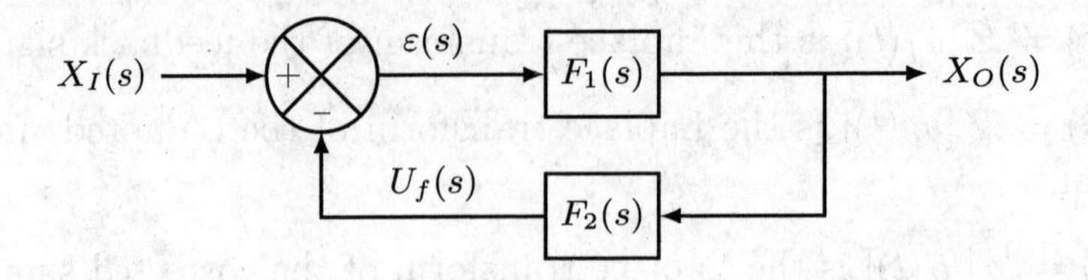

Figure 2.42 basic closed-loop form of a block diagram

We hence have:

$$\begin{aligned} X_O(s) &= F_1(s)\,\varepsilon(s) \\ &= F_1(s)\,X_I(s) - F_1(s)\,U_f(s) \\ &= F_1(s)\,X_I(s) - F_1(s)\,F_2(s)\,X_O(s) \end{aligned}$$

and the transfer function can be determined as:

$$\boxed{H(s) = \frac{X_O(s)}{X_I(s)} = \frac{F_1(s)}{1 + F_1(s)\,F_2(s)}}$$

2.7.3 Transfer Functions

Forward Chain

The **forward chain transfer function** is the function

$$\boxed{H_{FC}(s) = \frac{X_O(s)}{\varepsilon(s)} = C(s)\,G_1(s)\,G_2(s)}$$

It is the transfer function of the part of the block diagram between the output of the comparator and the data measured by the sensor, as illustrated in Figure 2.43.

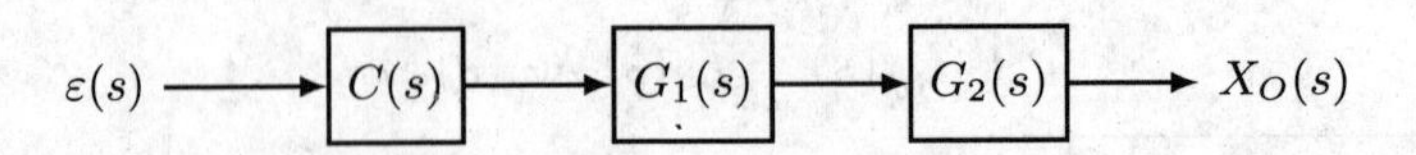

Figure 2.43 forward chain of the block diagram in Figure 2.41

Open Loop

The **open loop transfer function** is the function

$$H_{OL}(s) = \frac{U_f(s)}{\varepsilon(s)} = C(s)\, G_1(s)\, G_2(s)\, F(s)$$

It is the transfer function of the part of the block diagram between the output of the comparator and the output signal of the sensor. It is as if the loop were opened, as illustrated in Figure 2.44, where "OL" stands for "Open Loop".

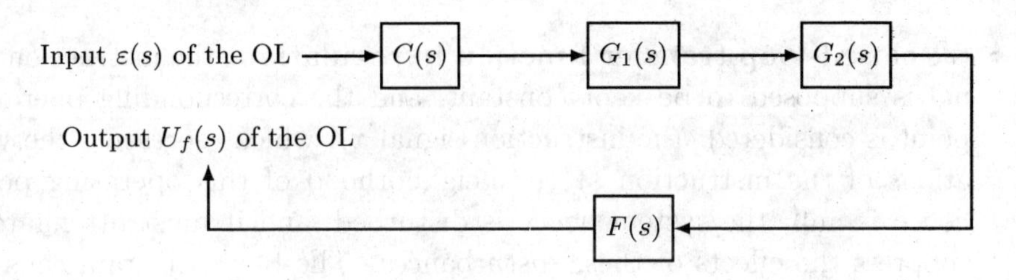

Figure 2.44 open loop of the block diagram in Figure 2.41

Closed-Loop Control and Tracking

Two closed-loop transfer functions exist:

- the **closed-loop transfer function in pursuit**: the disturbances are supposed to be null, and the instruction can vary or not; if the instruction is constant, then the system is *regulated*; if the instruction is variable, then the system is *controlled*; in this latter case, the control system aims at forcing the output signal to follow the input signal while respecting some constraints. The block diagram thus becomes the one depicted in Figure 2.45.

  The **closed-loop transfer function in pursuit** can then be defined as:

$$H_I(s) = \left.\frac{X_O(s)}{X_I(s)}\right|_{D(s)=0} = A(s)\,\frac{C(s)G_1(s)G_2(s)}{1 + C(s)G_1(s)G_2(s)F(s)}$$

  The main constraints which must be respected by the control system are related to precision and rapidity, and they also aim at limiting energy flows or at suppressing vibrations.

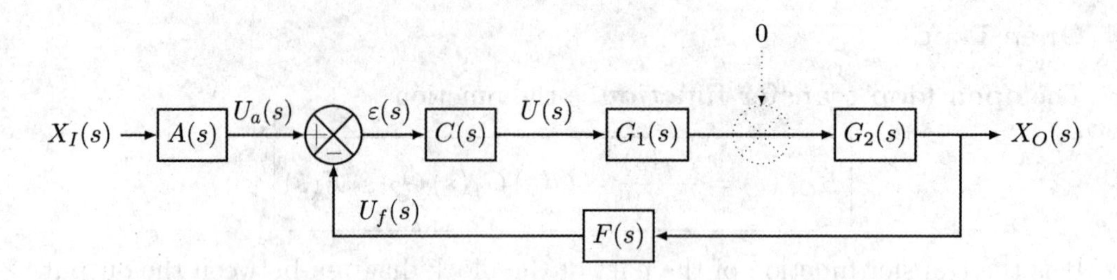

Figure 2.45 simplified block diagram when the disturbances are null

- the **closed-loop transfer function in regulation**: the instruction signal is supposed to be kept constant, and the corresponding operating point is considered: the instruction signal x_I, which represents the variations of the instruction at the neighborhood of this operating point, hence is null; the system hence is disturbed, and it must attenuate or suppress the effects of these disturbances. The block diagram thus becomes the one depicted in Figure 2.46, which is equivalent to the block diagram in Figure 2.47.

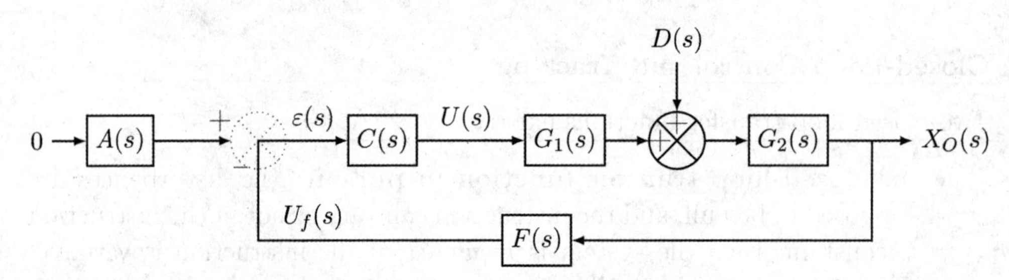

Figure 2.46 simplified block diagram when the instruction signal is null

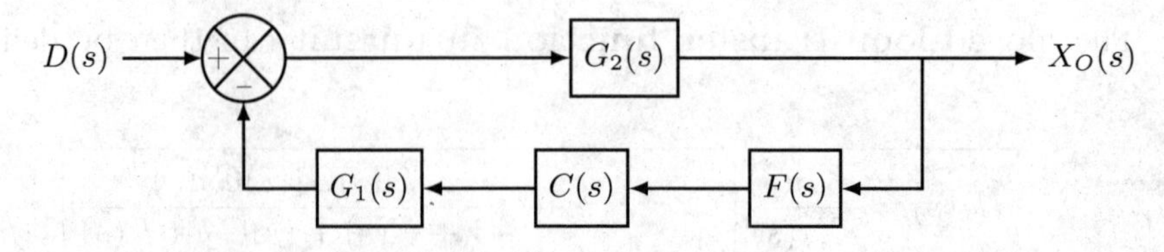

Figure 2.47 block diagram equivalent to the one in Figure 2.46

The **closed-loop transfer function in regulation** can then be defined as:

$$\boxed{H_D(s) = \left.\frac{X_O(s)}{D(s)}\right|_{X_I(s)=0} = \frac{G_2(s)}{1 + C(s)G_1(s)G_2(s)F(s)}}$$

The superposition principle allows to write that

$$\boxed{X_O(s) = H_I(s)\,X_I(s) + H_D(s)\,D(s)}$$

The superposition principle can be used because the studied systems are supposed to be linear. If this was not the case, this principle could not be used. The global performances of the system can hence be deduced from the independent studies in pursuit and in regulation.

2.7.4 Introduction to the Block Diagrams Algebra

Principle

The block diagrams algebra aims at simplifying block diagrams by symbolicly manipulating their blocks in order to get a basic block diagram for which Black's formula can be applied directly.

Motion of a Block

Downstream motion The block A which is *before* the sample point is moved *after* the sample point (Figure 2.48).

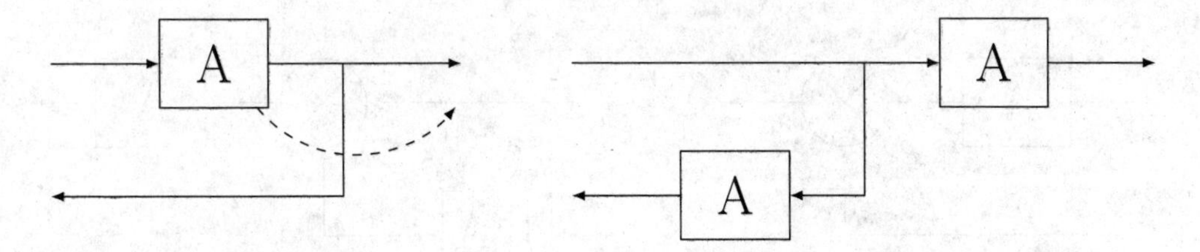

Figure 2.48 downstream motion of the block A

Upstream motion The block A which is *after* the sample point is moved *before* the sample point (Figure 2.49).

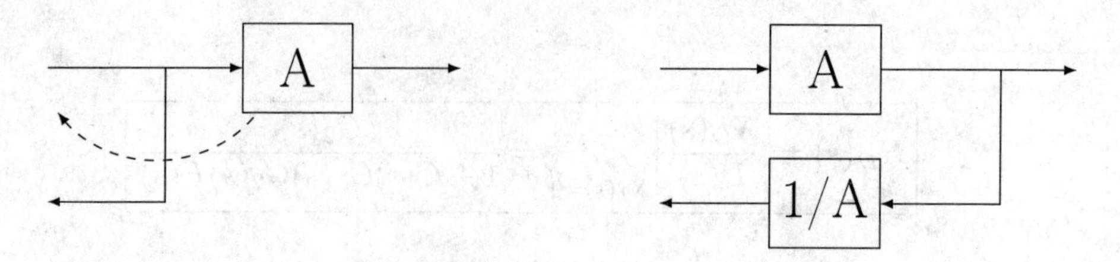

Figure 2.49 upstream motion of the block A

Motion of the Summing Point

Downstream motion The summing point which is *before* the block A is moved *after* this block (Figure 2.50).

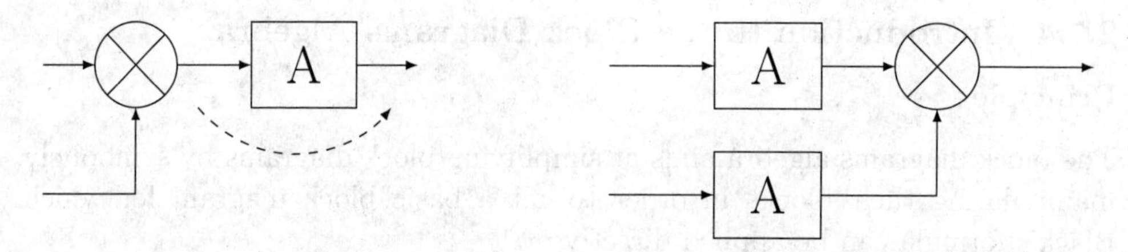

Figure 2.50 downstream motion of the summing point

Upstream motion The summing point which is *after* the block A is moved *before* this block (Figure 2.51).

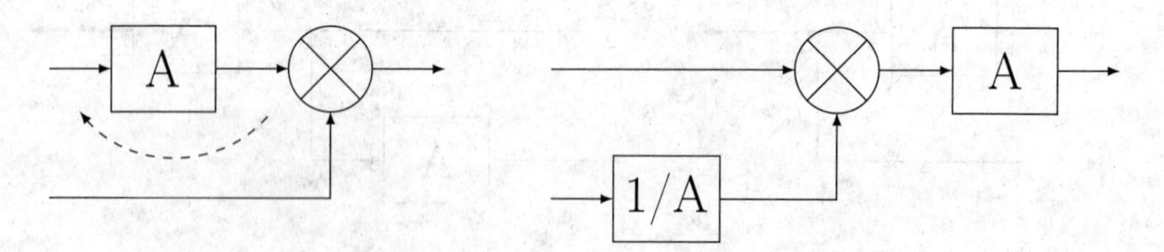

Figure 2.51 upstream motion of the summing point

Consequence: Equivalent Unit Feedback Block Diagram

It is always possible to use this approach to simplify a block diagram to a unit feedback block diagram just by moving the output point. A simple analysis based on the block diagrams algebra allows to determine the equivalent block diagram in Figure 2.52 for the block diagram in Figure 2.41.

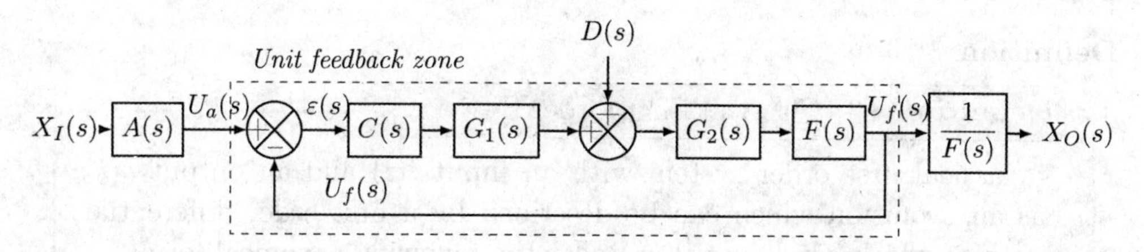

Figure 2.52 equivalent unit feedback block diagram for the block diagram in Figure 2.41

Such a unit feedback block diagram is interesting because it is more realistic for the analysis. Indeed:

- it is generally impossible to observe the output data x_O directly (a temperature, a speed, a force or a pressure are only known thanks to their image u_f provided by the sensors).

- it is equivalent in terms of feedback stability since closed-loop transfer functions have the same denominator, and hence the same poles, which will be useful for the adjustment of damping thanks to the Nichols plot, as we will see in the second volume. Indeed, the closed-loop transfer function can be expressed as $\dfrac{U_f(s)}{U_a(s)} = \dfrac{H_{OL}(s)}{1 + H_{OL}(s)}$, where $H_{OL}(s)$ is the transfer function of the open loop.

However, this theoretical modification of the control system makes us consider it from a very formal point of view which has no relation (but a theoretical one) with the real structure of the control system. One must hence keep in mind that this theoretical modification is mainly a tool to study system performances.

2.8 Time Response

2.8.1 First Order Systems

Definition

Definition 13 (Classical First Order System)

A classical first order system with an input $i(t)$ and an output $o(t)$ has an evolution which can be described by a first order differential equation which can be written under the following canonical form:

$$\tau\,\dot{o}(t) + o(t) = K\,i(t) \tag{2.16}$$

where:

- K is the **static gain** (whose unit is defined as $[K] = [o]/[i]$);
- $\tau > 0$ is the **time constant** of the system (whose unit is s).

In the remainder of this chapter, we will consider that $K > 0$, which corresponds to most of the encountered systems. After transforming this differential equation to the Laplace symbolic domain, the canonical form of the related transfer function can be determined as follows:

$$\boxed{H(s) = \frac{O(s)}{I(s)} = \frac{K}{1+\tau s}}$$

Indeed, if the Heaviside conditions are supposed to be verified, we have $\mathscr{L}[\dot{o}(t)] = s\,O(s) - \underbrace{o(0^+)}_{0}$ if $O(s) = \mathscr{L}[o(t)]$.

Definition 14 (Generalized First Order System)

A generalized first order system with an input $i(t)$ and an output $o(t)$ has an evolution which can be described by a first order differential equation which can be written under the following canonical form:

$$\tau_1\, \dot{o}(t) + o(t) = K\left(\tau_2\, \dot{i}(t) + i(t)\right) \tag{2.17}$$

where:

- K is the **static gain** (whose unit is defined as $[K] = [o]/[i]$);
- $\tau_1 > 0$ and $\tau_2 > 0$ are the **time constants** of the system (whose unit is s).

In the remainder of this chapter, we will consider that $K > 0$, which corresponds to most of the encountered systems. After transforming this differential equation to the Laplace symbolic domain, the canonical form of the related transfer function can be determined as follows:

$$\boxed{H(s) = \frac{O(s)}{I(s)} = K\,\frac{1+\tau_2 s}{1+\tau_1 s}}$$

Some examples of first order systems can be found:

- in physics: RC and RL circuits, active and passive filters (electricity);
- in chemistry: evolution of a concentration (reaction kinetics); and
- in industrial science: behavioral modeling of highly damped systems, proportional integral and proportional derivative controllers (which will be tackled in the second volume).

Step Response

The input signal of the system is a Heaviside step function of magnitude i_0: we hence have $i(t) = i_0\, u(t)$, where $u(t)$ is the Heaviside unit step function (Figure 2.53).

We hence have $I(s) = \mathscr{L}[i(t)] = \dfrac{i_0}{s}$. Consequently, if the system considered is a classical first order system:

$$O(s) = H(s)\, I(s) = \frac{K}{1+\tau s}\,\frac{i_0}{s} = \frac{K\, i_0}{s(1+\tau s)}$$

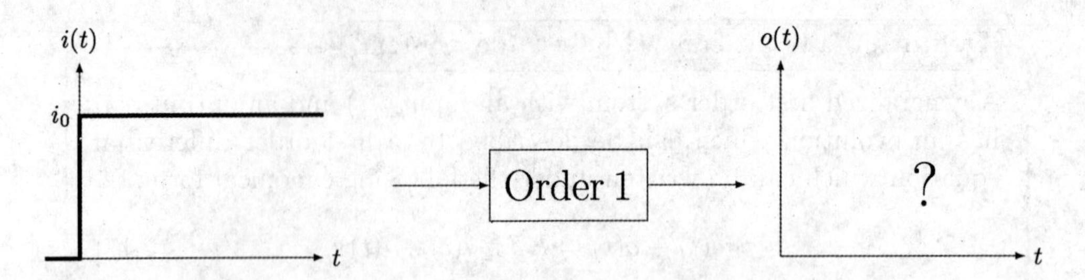

Figure 2.53 step response of a first order system

Thanks to partial fraction decomposition, we hence have:

$$\exists a, b \in \mathbb{R}, \frac{K\, i_0}{s(1+\tau s)} = \frac{a}{s} + \frac{b}{1+\tau s} = \frac{a + (a\tau + b)s}{s(1+\tau s)}$$

By identifying both terms, a and b can be determined as follows:

$$\begin{cases} a = K\, i_0 \\ a\tau + b = 0 \end{cases} \Leftrightarrow \begin{cases} a = K\, i_0 \\ b = -K\, i_0\, \tau \end{cases}$$

As a consequence:

$$O(s) = K\, i_0 \left(\frac{1}{s} - \frac{\tau}{1+\tau s} \right) = K\, i_0 \left(\frac{1}{s} - \frac{1}{s + \frac{1}{\tau}} \right) \tag{2.18}$$

By inverse Laplace transform, we can hence determine that:

$$\boxed{o(t) = K\, i_0 \left(1 - e^{-\frac{t}{\tau}} \right) u(t)}$$

which implies that $o(0) = 0$ and that $\lim_{t \to +\infty} o(t) = K\, i_0$. It can be noticed that $\dot{o}(t) = \frac{K\, i_0}{\tau} e^{-\frac{t}{\tau}}\, u(t)$ and that $\dot{o}(0) = \frac{K\, i_0}{\tau}$. The tangent to the step response at the origin hence is the line which goes through the origin and intersects the asymptot of the output signal at $t = \tau$.

The step response of a classical first order system hence has the shape depicted in Figure 2.54.

Impulse Response

The input signal of the system is a Dirac delta function of area A: we hence have $i(t) = A\, \delta(t)$, and $I(s) = \mathscr{L}[i(t)] = A$. Consequently, if the system considered is a classical first order system:

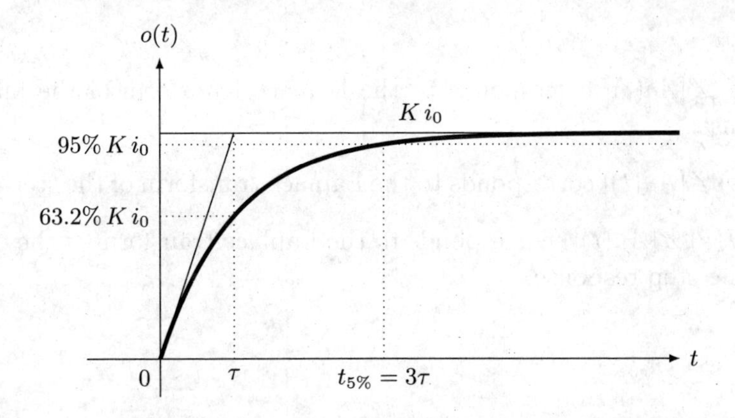

Figure 2.54 step response of a classical first order system

$$O(s) = H(s)\,I(s) = \frac{K\,A}{1+\tau s} = \frac{K\,A}{\tau}\frac{1}{s+\frac{1}{\tau}}$$

By inverse Laplace transform, we can hence determine that:

$$\boxed{o(t) = \frac{K\,A}{\tau}e^{-\frac{t}{\tau}}\,u(t)}$$

which implies that $o(0) = \dfrac{K\,A}{\tau}$ and that $\lim\limits_{t\to+\infty} o(t) = 0$. It can be noticed that $\dot{o}(t) = -\dfrac{K\,A}{\tau^2}e^{-\frac{t}{\tau}}\,u(t)$ and that $\dot{o}(0) = -\dfrac{K\,A}{\tau^2}$. The tangent to the impulse response at $t = 0$ hence is the line which goes through the point $\left(0, \dfrac{K\,A}{\tau}\right)$ and intersects the asymptot of the output signal at $t = \tau$.

The impulse response of a classical first order system hence has the shape depicted in Figure 2.55.
It can be noticed that the impulse response can be obtained directly by differentiating the step response and by replacing i_0 with A. Indeed:

$$\mathscr{L}[A\delta(t)] = A = s\frac{A}{s} = s\mathscr{L}[Au(t)]$$

which implies that:

$$H(s)\mathscr{L}[A\delta(t)] = s\,(H(s)\mathscr{L}[Au(t)])$$

where:

- $H(s)\mathscr{L}[A\delta(t)]$ corresponds to the Laplace transform of the impulse response;
- $H(s)\mathscr{L}[Au(t)]$ corresponds to the Laplace transform of the step response;
- $s\,(H(s)\mathscr{L}[Au(t)])$ corresponds to the Laplace transform of the derivative of the step response.

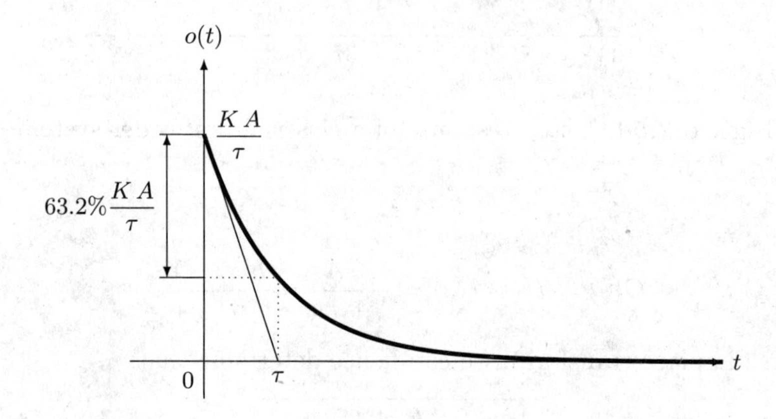

Figure 2.55 impulse response of a classical first order system

Pursuit Response

The input signal of the system is a causal ramp of slope α: we hence have $i(t) = \alpha\, t\, u(t)$, and $I(s) = \mathscr{L}[i(t)] = \dfrac{\alpha}{s^2}$. Consequently, if the system considered is a classical first order system:

$$O(s) = H(s)\,I(s) = \frac{K}{1+\tau s}\,\frac{\alpha}{s^2} = \frac{K\,\alpha}{s^2(1+\tau s)}$$

Thanks to partial fraction decomposition, we hence have:

$$\exists a, b, c \in \mathbb{R},\ \frac{K\,\alpha}{s^2(1+\tau s)} = \frac{a}{s} + \frac{b}{s^2} + \frac{c}{1+\tau s} = \frac{b + (a + b\tau)s + (c + a\tau)s^2}{s^2(1+\tau s)}$$

By identifying both terms, a, b and c can be determined as follows:

$$\begin{cases} b = K\,\alpha \\ a + b\tau = 0 \\ c + a\tau = 0 \end{cases} \Leftrightarrow \begin{cases} b = K\,\alpha \\ a = -K\,\alpha\,\tau \\ c = K\,\alpha\,\tau^2 \end{cases}$$

As a consequence:

$$O(s) = K\,\alpha\left(-\frac{\tau}{s} + \frac{1}{s^2} + \frac{\tau^2}{1+\tau s}\right) = K\,\alpha\left(-\frac{\tau}{s} + \frac{1}{s^2} + \frac{\tau}{s+\frac{1}{\tau}}\right) \tag{2.19}$$

By inverse Laplace transform, we can hence determine that:

$$\boxed{o(t) = K\,\alpha\left(t - \tau + \tau\, e^{-\frac{t}{\tau}}\right) u(t)}$$

which implies that $o(0) = 0$ and that $o(t) \underset{t\to+\infty}{\sim} K\,\alpha(t-\tau)$. It can be noticed that $\dot{o}(t) = K\,\alpha\left(1 - e^{-\frac{t}{\tau}}\right) u(t)$ and that $\dot{o}(0) = 0$. The tangent to the pursuit response at the origin hence is the x axis.
The pursuit response of a classical first order system hence has the shape depicted in Figure 2.56.

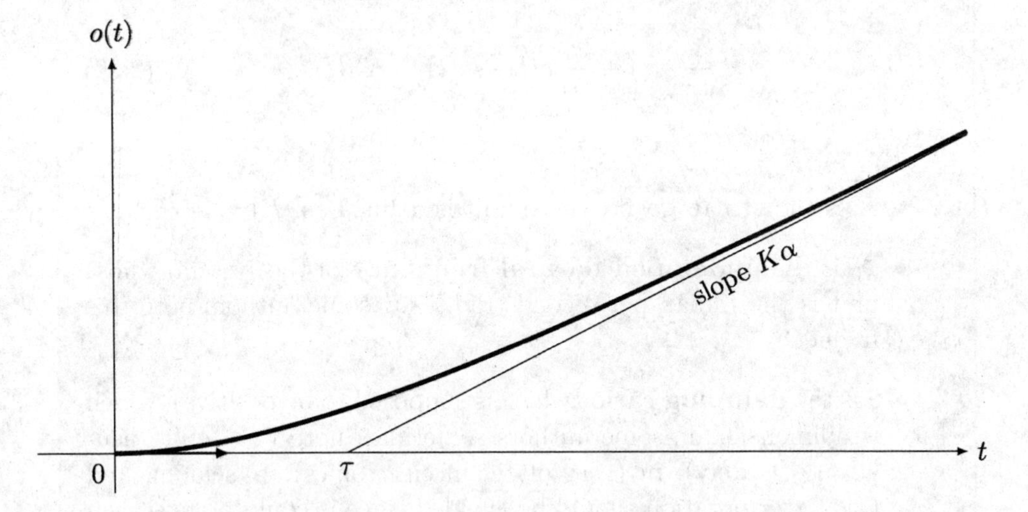

Figure 2.56 pursuit response of a classical first order system

It can be noticed that the pursuit response can be obtained directly by integrating the step response and by replacing i_0 with α. Indeed:

$$\mathscr{L}[\alpha t u(t)] = \frac{\alpha}{s^2} = \frac{1}{s}\frac{\alpha}{s} = \frac{1}{s}\mathscr{L}[\alpha u(t)]$$

which implies that:

$$H(s)\mathscr{L}[\alpha t u(t)] = \frac{1}{s}\left(H(s)\mathscr{L}[\alpha u(t)]\right)$$

where:

- $H(s)\mathscr{L}[\alpha t u(t)]$ corresponds to the Laplace transform of the pursuit response;
- $H(s)\mathscr{L}[\alpha u(t)]$ corresponds to the Laplace transform of the step response;
- $\frac{1}{s}\left(H(s)\mathscr{L}[\alpha u(t)]\right)$ corresponds to the Laplace transform of the antiderivative of the step response.

2.8.2 Second Order Systems

Definition 15 (Second Order System)

A second order system with an input $i(t)$ and an output $o(t)$ has an evolution which can be described by a second order differential equation which can be written under the following canonical form:

$$\frac{1}{\omega_0^2}\,\ddot{o}(t) + \frac{2\xi}{\omega_0}\,\dot{o}(t) + o(t) = K\,i(t) \tag{2.20}$$

where:

- K is the **static gain** (whose unit is defined as $[K] = [o]/[i]$);
- ω_0 is the **undamped natural frequency** of the system, whose unit is the radian per second (rad.s^{-1}): some authors note this frequency ω_n;
- ξ is the **damping ratio** (which is supposed to be positive), which is adimensional: some authors sometimes note it z (quite usually), ζ (rarely), m (very often, in electronics), ε (seldom, and this latter notation should be avoided since it could be mistaken with the error of a control system), λ (mainly used in physics), etc.

In the remainder of this chapter, we will consider that $K > 0$, which corresponds to most of the encountered systems. After transforming this differential equation to the Laplace symbolic domain, the canonical form of the related transfer function can be determined as follows:

$$H(s) = \frac{O(s)}{I(s)} = \frac{K}{1 + 2\xi\dfrac{s}{\omega_0} + \dfrac{s^2}{\omega_0^2}}$$

Indeed, if the Heaviside conditions are supposed to be verified, we have $\forall n \in \mathbb{N}, \mathscr{L}\left[o^{(n)}(t)\right] = s^n\, O(s)$ if $O(s) = \mathscr{L}[o(t)]$.

Some examples of second order systems can be found:

- in physics: RLC circuits, active and passive filters (electricity), dynamics of a material point (mechanics), etc;
- in chemistry: second order kinetic evolutions; and
- in industrial science: dominant mode of oscillating systems (automation control), dynamics of systems of solid bodies (mechanics), etc.

For instance, the position $x(t)$ of the mass of the spring-mass-damper system in Figure 2.57 satisfies the equation $m\,\ddot{x}(t)+c\,\dot{x}(t)+k\,x(t) = 0$, and this system hence is a second order system.

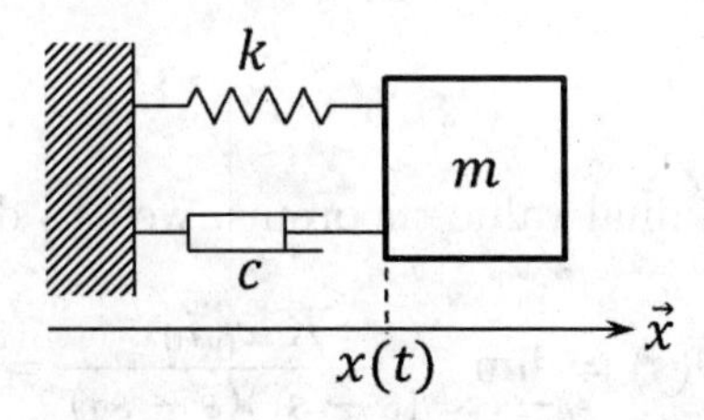

Figure 2.57 a spring-mass-damper system, which can be modeled by a second order system

Step Response

The input signal of the system is a Heaviside step function of magnitude i_0: we hence have $i(t) = i_0\, u(t)$, where $u(t)$ is the Heaviside unit step function. We hence have $I(s) = \mathscr{L}[i(t)] = \dfrac{i_0}{s}$, so:

$$O(s) = H(s)\,I(s) = \frac{K}{1 + 2\xi\dfrac{s}{\omega_0} + \dfrac{s^2}{\omega_0^2}}\,\frac{i_0}{s} = \frac{K\,i_0}{s\left(1 + 2\xi\dfrac{s}{\omega_0} + \dfrac{s^2}{\omega_0^2}\right)}$$

$$= \frac{K\,\omega_0^2\,i_0}{s\left(s^2 + 2\omega_0\xi s + \omega_0^2\right)}$$

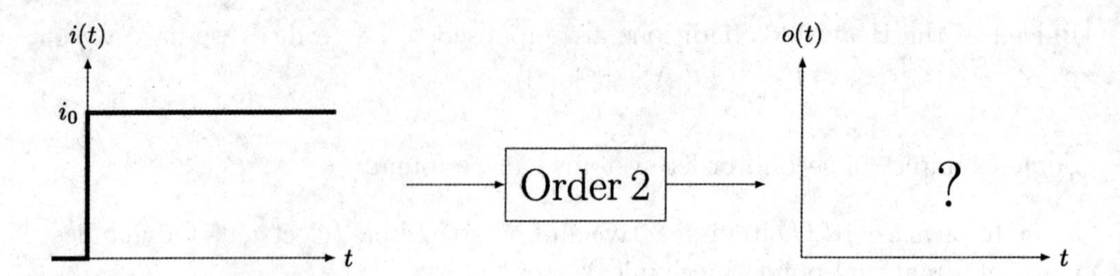

Figure 2.58 step response of a second order system

If we note s_1 and s_2 the two poles of the transfer function, we hence have:

$$O(s) = \frac{K\,\omega_0^2\,i_0}{s(s-s_1)(s-s_2)}$$

with $\begin{cases} s_1 + s_2 = -2\omega_0\xi \\ s_1\,s_2 = \omega_0^2 \end{cases}$.

Thanks to the initial and final value theorems, we can determine that:

- $\lim\limits_{t\to 0} o(t) = \lim\limits_{s\to+\infty} s\,O(s) = \lim\limits_{s\to+\infty} \dfrac{K\,\omega_0^2\,i_0}{(s-s_1)(s-s_2)} = 0^+$
- $\lim\limits_{t\to+\infty} o(t) = \lim\limits_{s\to 0} s\,O(s) = \lim\limits_{s\to 0} \dfrac{K\,\omega_0^2\,i_0}{(s-s_1)(s-s_2)} = K\,i_0$
- $\lim\limits_{t\to 0} \dot{o}(t) = \lim\limits_{s\to+\infty} s\left[s\,O(s) - o(0^+)\right] = \lim\limits_{s\to+\infty} s^2\,O(s)$
  $= \lim\limits_{s\to+\infty} \dfrac{K\,\omega_0^2\,i_0\,s}{(s-s_1)(s-s_2)} = 0^+$
- $\lim\limits_{t\to+\infty} \dot{o}(t) = \lim\limits_{s\to 0} s\left[s\,O(s) - o(0^+)\right] = \lim\limits_{s\to 0} s^2\,O(s)$
  $= \lim\limits_{s\to 0} \dfrac{K\,\omega_0^2\,i_0\,s}{(s-s_1)(s-s_2)} = 0^+$

The discriminant of the polynomial $s^2 + 2\omega_0\xi s + \omega_0^2$ is equal to:

$$\Delta = 4\omega_0^2\xi^2 - 4\omega_0^2 = 4\omega_0^2(\xi^2 - 1)$$

The time response will hence depend on the value of the damping ratio ξ:

- **1st case:** $\xi = 0$. We hence have $\Delta = -4\omega_0^2 < 0$. The two poles of the transfer function hence are conjugate imaginary numbers with $s_1 = j\,\omega_0$ and $s_2 = -j\,\omega_0$. Consequently:

$$O(s) = \frac{K\,\omega_0^2\,i_0}{s(s - j\omega_0)(s + j\omega_0)}$$

Thanks to partial fraction decomposition, we hence have:

$$\exists a, b, c \in \mathbb{R}, \frac{K\,\omega_0^2\,i_0}{s(s - j\omega_0)(s + j\omega_0)} = \frac{a}{s} + \frac{b}{s - j\omega_0} + \frac{c}{s + j\omega_0}$$
$$= \frac{a\omega_0^2 + (b - c)j\omega_0 s + (a + b + c)s^2}{s(s - j\omega_0)(s + j\omega_0)}$$

By identifying both terms, a, b and c can be determined as follows:

$$\begin{cases} a = K\,i_0 \\ b - c = 0 \\ a + b + c = 0 \end{cases} \Leftrightarrow \begin{cases} a = K\,i_0 \\ b = c = -\dfrac{K\,i_0}{2} \end{cases}$$

As a consequence:

$$O(s) = K\,i_0 \left(\frac{1}{s} - \frac{1}{2}\frac{1}{s - j\omega_0} - \frac{1}{2}\frac{1}{s + j\omega_0}\right)$$
$$= K\,i_0 \left(\frac{1}{s} - \frac{1}{2}\frac{s - j\omega_0 + s + j\omega_0}{(s - j\omega_0)(s + j\omega_0)}\right)$$
$$= K\,i_0 \left(\frac{1}{s} - \frac{s}{s^2 + \omega_0^2}\right) \tag{2.21}$$

By inverse Laplace transform, we can hence determine that:

$$\boxed{o(t) = K\,i_0\,(1 - \cos(\omega_0 t))\,u(t)}$$

The system is **at the limit of stability** (as illustrated in Figure 2.24) because the real part of the poles of the transfer function is null (as it was showed in Figure 2.36, and as we will see in the second volume). The time response hence has **undamped oscillations** whose period is $T_p = \dfrac{2\pi}{\omega_0}$, as illustrated in Figure 2.59.

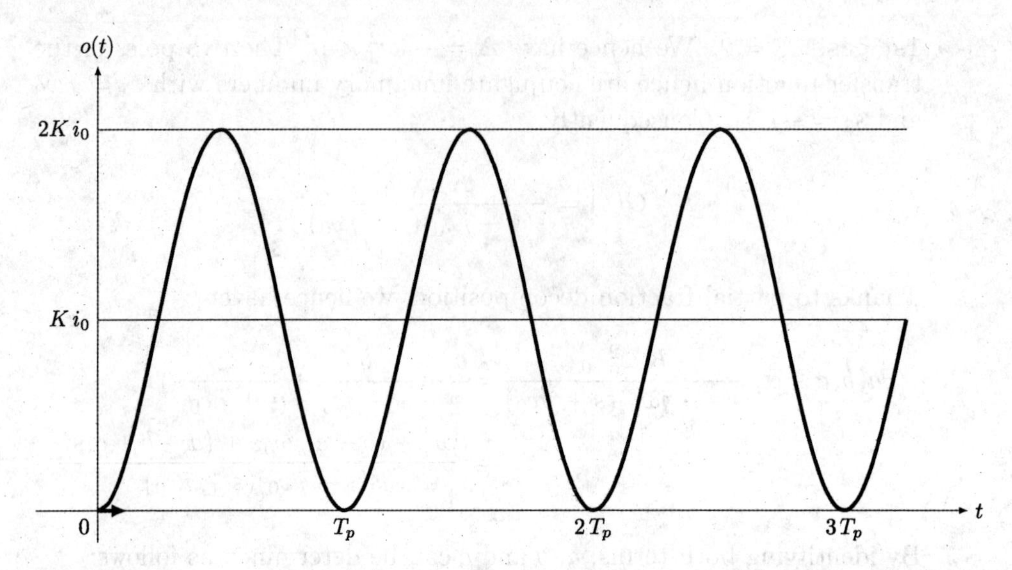

Figure 2.59 time response of a second order system at the limit of stability ($\xi = 0$)

- **2nd case:** $0 < \xi < 1$. We hence have $\Delta = 4\omega_0^2(\xi^2 - 1) < 0$. The two poles of the transfer function hence are conjugate complex numbers with $s_1 = \omega_0\left(-\xi + j\sqrt{1-\xi^2}\right)$ and $s_2 = \omega_0\left(-\xi - j\sqrt{1-\xi^2}\right)$. They can also be expressed under the simpler form $s_1 = -\sigma + j\omega_p$ and $s_2 = -\sigma - j\omega_p$ with $\begin{cases} \sigma = \omega_0\xi \\ \omega_p = \omega_0\sqrt{1-\xi^2} \end{cases}$. Consequently:

$$O(s) = \frac{K\,\omega_0^2\,i_0}{s(s-(-\sigma+j\omega_p))(s-(-\sigma-j\omega_p))} = \frac{K\,\omega_0^2\,i_0}{s((s+\sigma)-j\omega_p)((s+\sigma)+j\omega_p)}$$

Thanks to partial fraction decomposition, we hence have:

$$\exists a, b, c \in \mathbb{R},$$
$$O(s) = \frac{a}{s} + \frac{b}{(s+\sigma)-j\omega_p} + \frac{c}{(s+\sigma)+j\omega_p} = \frac{a(\omega_p^2+\sigma^2) + ((2a+b+c)\sigma + (b-c)j\omega_p)\,s + (a+b+c)s^2}{s((s+\sigma)-j\omega_p)((s+\sigma)+j\omega_p)}$$

By identifying both terms, a, b and c can be determined as follows:

$$\begin{cases} a(\omega_p^2+\sigma^2) = a\omega_0^2 = K\,\omega_0^2\, i_0 \\ (2a+b+c)\sigma + (b-c)j\omega_p = 0 \\ a+b+c=0 \end{cases}$$

$$\Leftrightarrow \begin{cases} a = K\,i_0 \\ b-c = -\dfrac{a\sigma}{j\omega_p} = -\dfrac{Ki_0\omega_0\xi}{j\omega_0\sqrt{1-\xi^2}} = -\dfrac{Ki_0\xi}{j\sqrt{1-\xi^2}} \\ b+c = -K\,i_0 \end{cases}$$

We can hence determine that:

$$\begin{cases} a = Ki_0 \\ b = -\dfrac{K\,i_0}{2}\left(1-j\dfrac{\xi}{\sqrt{1-\xi^2}}\right) \\ c = -\dfrac{K\,i_0}{2}\left(1+j\dfrac{\xi}{\sqrt{1-\xi^2}}\right) \end{cases}$$

As a consequence:

$$\begin{aligned} O(s) &= Ki_0\left(\frac{1}{s}-\frac{1}{2}\left(1-j\frac{\xi}{\sqrt{1-\xi^2}}\right)\frac{1}{(s+\sigma)-j\omega_p}\right. \\ &\quad \left. -\frac{1}{2}\left(1+j\frac{\xi}{\sqrt{1-\xi^2}}\right)\frac{1}{(s+\sigma)+j\omega_p}\right) \\ &= Ki_0\left(\frac{1}{s}-\frac{s+\sigma}{(s+\sigma)^2+\omega_p^2}+j\frac{\xi}{\sqrt{1-\xi^2}}\frac{j\omega_p}{(s+\sigma)^2+\omega_p^2}\right) \\ &= Ki_0\left(\frac{1}{s}-\frac{s+\sigma}{(s+\sigma)^2+\omega_p^2}-\frac{\xi}{\sqrt{1-\xi^2}}\frac{\omega_p}{(s+\sigma)^2+\omega_p^2}\right) \end{aligned} \tag{2.22}$$

By inverse Laplace transform, we can hence determine that:

$$\begin{aligned} o(t) &= Ki_0\left(1-e^{-\sigma t}\cos(\omega_p t)-\frac{\xi e^{-\sigma t}}{\sqrt{1-\xi^2}}\sin(\omega_p t)\right)u(t) \\ &= Ki_0\left(1-\frac{e^{-\xi\omega_0 t}}{\sqrt{1-\xi^2}}\left(\xi\sin(\omega_p t)+\sqrt{1-\xi^2}\cos(\omega_p t)\right)\right)u(t) \end{aligned} \tag{2.23}$$

If we pose $\begin{cases} \sin\beta = -\xi \\ \cos\beta = \sqrt{1-\xi^2} \end{cases} \Rightarrow \beta = -\arctan\dfrac{\xi}{\sqrt{1-\xi^2}}$, then we can write:

$$
\begin{aligned}
o(t) &= Ki_0\left(1-\frac{e^{-\xi\omega_0 t}}{\sqrt{1-\xi^2}}\left(-\sin\beta\,\sin(\omega_p t)+\cos\beta\,\cos(\omega_p t)\right)\right)u(t)\\
&= Ki_0\left(1-\frac{e^{-\xi\omega_0 t}}{\sqrt{1-\xi^2}}\cos(\omega_p t+\beta)\right)u(t) \qquad (2.24)
\end{aligned}
$$

$$
\boxed{o(t) = Ki_0\left(1-\frac{e^{-\xi\omega_0 t}}{\sqrt{1-\xi^2}}\cos(\omega_p t+\beta)\right)u(t)}
$$

The time response has **regular damped oscillations** whose period is $\boxed{T_p = \dfrac{2\pi}{\omega_p} = \dfrac{2\pi}{\omega_0\sqrt{1-\xi^2}}}$, as illustrated in Figure 2.60.

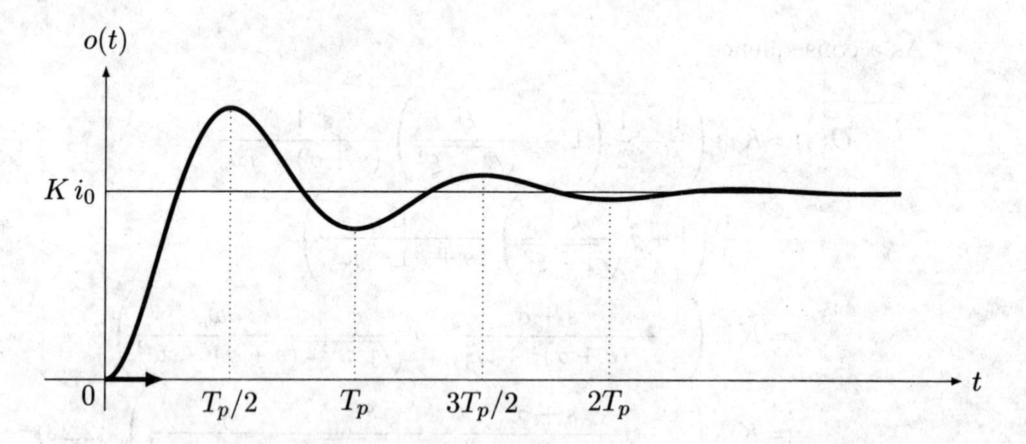

Figure 2.60 time response of a damped second order system ($0 < \xi < 1$)

It can be noticed that:

$$
\dot{o}(t) = K\,i_0\,\omega_0\,\frac{e^{-\xi\omega_0 t}}{\sqrt{1-\xi^2}}\sin(\omega_0\sqrt{1-\xi^2}t)\,u(t)
$$

as it will be detailed for the impulse response when $0 < \xi < 1$. As a consequence:

$$\dot{o}(t) = 0 \Leftrightarrow \sin(\omega_0\sqrt{1-\xi^2}t) = 0 \Leftrightarrow \omega_0\sqrt{1-\xi^2}t = k\pi$$
$$\Leftrightarrow t = \frac{k\pi}{\omega_0\sqrt{1-\xi^2}}, k \in \mathbb{N}$$

The first and second overshoots respectively occur "upwards" and "downwards" at the instants:

$$\boxed{t_{O_1} = \frac{T_p}{2} = \frac{\pi}{\omega_0\sqrt{1-\xi^2}}} \quad \text{and} \quad \boxed{t_{O_2} = T_p = \frac{2\pi}{\omega_0\sqrt{1-\xi^2}}}$$

and the n-th overshoot occurs at the instant $\boxed{t_{O_n} = \frac{n\,T_p}{2} = \frac{n\pi}{\omega_0\sqrt{1-\xi^2}}}$

with:

- an "upwards" overshoot (above the asymptot) if n is odd;
- a "downwards" overshoot (below the asymptot) if n is even.

The expression for:

- the first relative overshoot is

$$\boxed{O_{1\%} = 100\left|\frac{o(t_{O_1}) - Ki_0}{Ki_0}\right| = 100\,e^{\left(\frac{-\xi\pi}{\sqrt{1-\xi^2}}\right)}}.$$

- the second relative overshoot is

$$\boxed{O_{2\%} = 100\left|\frac{o(t_{O_2}) - Ki_0}{Ki_0}\right| = 100\,e^{\left(\frac{-2\xi\pi}{\sqrt{1-\xi^2}}\right)} = 100\left(\frac{O_{1\%}}{100}\right)^2}.$$

- ...
- the n-th relative overshoot is

$$\boxed{O_{n\%} = 100\left|\frac{o(t_{O_n}) - Ki_0}{Ki_0}\right| = 100\left(\frac{O_{1\%}}{100}\right)^n = 100\,e^{\left(\frac{-n\xi\pi}{\sqrt{1-\xi^2}}\right)}}.$$

As the determination of the value of these overshoots is quite error-prone, an *abacus* was created to determine directly the value of the successive overshoots from the knowledge of the damping ratio, or to determine the value of the damping ratio from the knowledge of the overshoot. This abacus provides:

– the damping ratio ξ, which is comprised between 0.01 and 1, in abscissa; and

– the relative overshoot in ordinate;

both values being represented on a logarithmic scale. It is depicted in Figure 2.61.

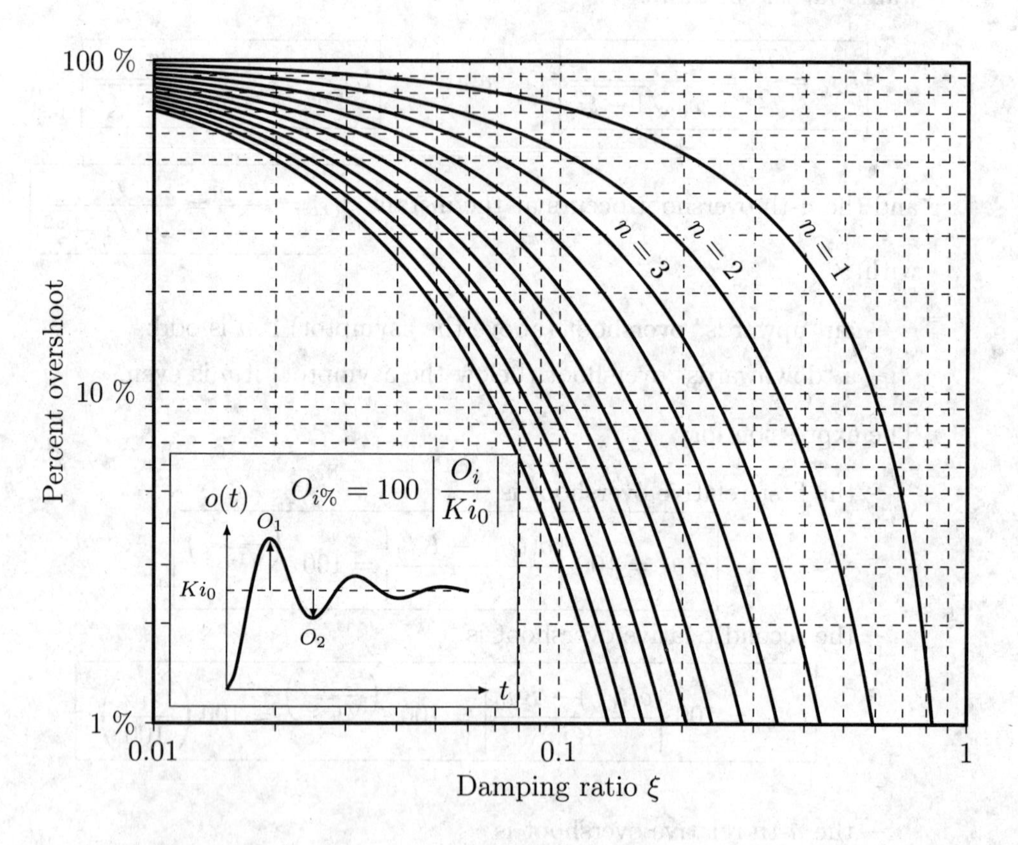

Figure 2.61 abacus for the determination of the overshoots

- **3rd case:** $\xi = 1$. We hence have $\Delta = 0$. The two poles of the transfer function hence are real numbers and are equal with $s_1 = s_2 = -\omega_0$. Consequently:

$$O(s) = \frac{K\,\omega_0^2\, i_0}{s(s+\omega_0)^2}$$

Thanks to partial fraction decomposition, we hence have:

$$\exists a, b, c \in \mathbb{R}, \frac{K\,\omega_0^2\, i_0}{s(s+\omega_0)^2} = \frac{a}{s} + \frac{b}{s+\omega_0} + \frac{c}{(s+\omega_0)^2}$$
$$= \frac{a\omega_0^2 + ((2a+b)\omega_0 + c)s + (a+b)s^2}{s(s+\omega_0)^2}$$

By identifying both terms, a, b and c can be determined as follows:

$$\begin{cases} a = K\, i_0 \\ (2a+b)\omega_0 + c = 0 \\ a+b = 0 \end{cases} \Leftrightarrow \begin{cases} a = Ki_0 \\ b = -Ki_0 \\ c = -K\omega_0 i_0 \end{cases}$$

As a consequence:

$$O(s) = Ki_0 \left(\frac{1}{s} - \frac{1}{s+\omega_0} - \frac{\omega_0}{(s+\omega_0)^2} \right) \tag{2.25}$$

By inverse Laplace transform, we can hence determine that:

$$o(t) = Ki_0 \left(1 - e^{-\omega_0 t} - \omega_0 t e^{-\omega_0 t}\right) u(t)$$

$$\boxed{o(t) = Ki_0 \left(1 - (1+\omega_0 t)e^{-\omega_0 t}\right) u(t)}$$

The system is **critically damped**: it converges to Ki_0 as fast as possible without oscillating, as illustrated in Figure 2.62.

- **4th case:** $\xi > 1$. We hence have $\Delta = 4\omega_0^2(\xi^2 - 1) > 0$. The two poles of the transfer function hence are strictly negative real numbers with $s_1 = \omega_0 \left(-\xi + \sqrt{\xi^2-1}\right)$ and $s_2 = \omega_0 \left(-\xi - \sqrt{\xi^2-1}\right)$. Consequently:

$$O(s) = \frac{K\,\omega_0^2\, i_0}{s\left(s - \omega_0\left(-\xi + \sqrt{\xi^2-1}\right)\right)\left(s - \omega_0\left(-\xi - \sqrt{\xi^2-1}\right)\right)}$$

$$= \frac{K\,\omega_0^2\, i_0}{s\left(s + \omega_0\left(\xi - \sqrt{\xi^2-1}\right)\right)\left(s + \omega_0\left(\xi + \sqrt{\xi^2-1}\right)\right)}$$

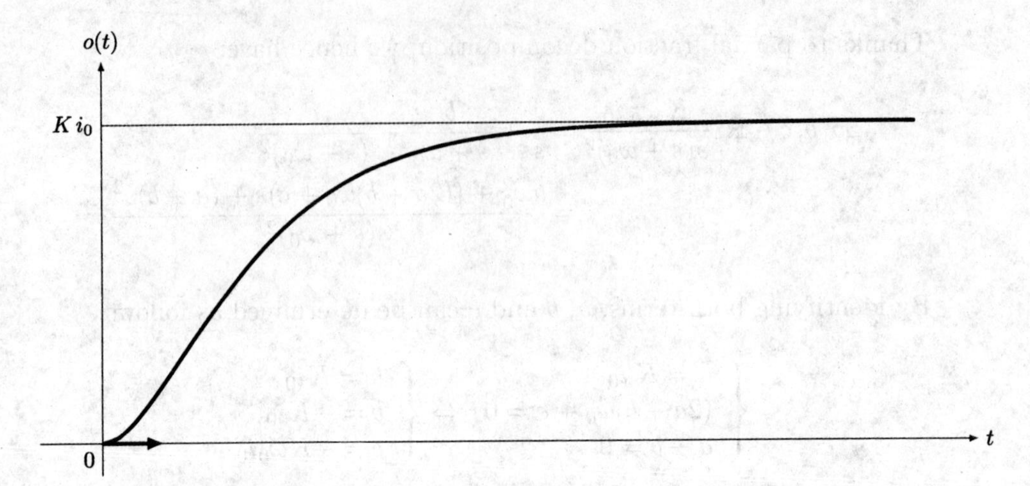

Figure 2.62 time response of a critically damped second order system ($\xi = 1$)

Thanks to partial fraction decomposition, we hence have:

$$\exists a, b, c \in \mathbb{R},$$

$$O(s) = \frac{a}{s} + \frac{b}{s + \omega_0\left(\xi - \sqrt{\xi^2 - 1}\right)} + \frac{c}{s + \omega_0\left(\xi + \sqrt{\xi^2 - 1}\right)}$$

$$= \frac{a\omega_0^2 + \left((2a + b + c)\xi\omega_0 + (b - c)\omega_0\sqrt{\xi^2 - 1}\right)s + (a + b + c)s^2}{s\left(s + \omega_0\left(\xi - \sqrt{\xi^2 - 1}\right)\right)\left(s + \omega_0\left(\xi + \sqrt{\xi^2 - 1}\right)\right)}$$

By identifying both terms, a, b and c can be determined as follows:

$$\begin{cases} a\omega_0^2 = K\,\omega_0^2\,i_0 \\ (2a + b + c)\xi\omega_0 + (b - c)\omega_0\sqrt{\xi^2 - 1} = 0 \\ a + b + c = 0 \end{cases}$$

$$\Leftrightarrow \begin{cases} a = K\,i_0 \\ b - c = -\dfrac{a\xi}{\sqrt{\xi^2 - 1}} = -\dfrac{K i_0 \xi}{\sqrt{\xi^2 - 1}} \\ b + c = -K\,i_0 \end{cases}$$

We can hence determine that:

$$\begin{cases} a = Ki_0 \\ b = -\dfrac{K\,i_0}{2}\left(1+\dfrac{\xi}{\sqrt{\xi^2-1}}\right) \\ c = -\dfrac{K\,i_0}{2}\left(1-\dfrac{\xi}{\sqrt{\xi^2-1}}\right) \end{cases}$$

As a consequence:

$$\begin{aligned} O(s) = Ki_0 & \left(\frac{1}{s} - \frac{1}{2}\left(1+\frac{\xi}{\sqrt{\xi^2-1}}\right)\frac{1}{s+\omega_0\left(\xi-\sqrt{\xi^2-1}\right)}\right. \\ & \left. -\frac{1}{2}\left(1-\frac{\xi}{\sqrt{\xi^2-1}}\right)\frac{1}{s+\omega_0\left(\xi+\sqrt{\xi^2-1}\right)}\right) \\ = Ki_0 & \left(\frac{1}{s} - \left(\frac{\xi+\sqrt{\xi^2-1}}{2\sqrt{\xi^2-1}}\right)\frac{1}{s+\omega_0\left(\xi-\sqrt{\xi^2-1}\right)}\right. \\ & \left. -\left(\frac{\sqrt{\xi^2-1}-\xi}{2\sqrt{\xi^2-1}}\right)\frac{1}{s+\omega_0\left(\xi+\sqrt{\xi^2-1}\right)}\right) \end{aligned} \tag{2.26}$$

By inverse Laplace transform, we can hence determine that:

$$\begin{aligned} o(t) = Ki_0 & \left(1 - \left(\frac{\xi+\sqrt{\xi^2-1}}{2\sqrt{\xi^2-1}}\right)e^{-\omega_0\left(\xi-\sqrt{\xi^2-1}\right)t}\right. \\ & \left. -\left(\frac{\sqrt{\xi^2-1}-\xi}{2\sqrt{\xi^2-1}}\right)e^{-\omega_0\left(\xi+\sqrt{\xi^2-1}\right)t}\right)u(t) \\ = Ki_0 & \left(1 - \frac{\left(\xi+\sqrt{\xi^2-1}\right)e^{-\omega_0\left(\xi-\sqrt{\xi^2-1}\right)t}}{2\sqrt{\xi^2-1}}\right. \\ & \left. -\frac{\left(-\xi+\sqrt{\xi^2-1}\right)e^{-\omega_0\left(\xi+\sqrt{\xi^2-1}\right)t}}{2\sqrt{\xi^2-1}}\right)u(t) \end{aligned} \tag{2.27}$$

This expression is sometimes written by means of hyperbolic sines and cosines, which allows to write it under the form:

$$o(t)=Ki_0\left(1-\frac{e^{-\xi\omega_0 t}}{\sqrt{\xi^2-1}}\left(\xi\sinh(\omega_0\sqrt{\xi^2-1}t)\right.\right.$$
$$\left.\left.+\sqrt{\xi^2-1}\cosh(\omega_0\sqrt{\xi^2-1}t)\right)\right)u(t) \quad (2.28)$$

The system is **non-oscillatory and damped**, as illustrated in Figure 2.63, and it is also called an **overdamped system**.

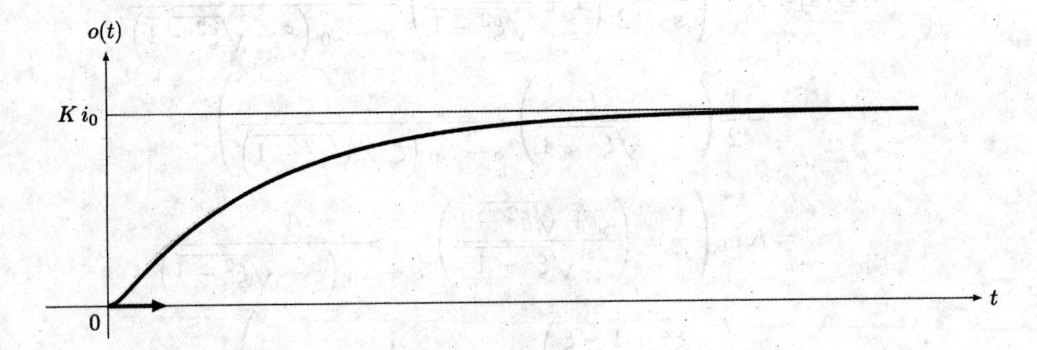

Figure 2.63 time response of an overdamped second order system ($\xi > 1$)

Since $s_1, s_2 \in \mathbb{R}_-^*$, it is possible to pose two time constants $\tau_1 = -\frac{1}{s_1}$ and $\tau_2 = -\frac{1}{s_2}$ with $\tau_1\tau_2 = \frac{1}{s_1 s_2} = \frac{1}{\omega_0^2}$. We hence have $\boxed{H(s) = \frac{K}{(1+\tau_1 s)(1+\tau_2 s)}}$, which implies that:

$$I(s) \longrightarrow \boxed{\frac{K\omega_0^2}{s^2+2\xi\omega_0 s+\omega_0^2}} \longrightarrow O(s)$$
$$\Leftrightarrow I(s) \longrightarrow \boxed{K} \longrightarrow \boxed{\frac{1}{1+\tau_1 s}} \longrightarrow \boxed{\frac{1}{1+\tau_2 s}} \longrightarrow O(s)$$

A damped second order system hence is equivalent to the serial association of two first order systems, and the ratio of their two time constants

$\frac{\tau_1}{\tau_2}$ (with $\tau_1 > \tau_2$) highly increases when ξ increases. Indeed:

$$\frac{\tau_1}{\tau_2} = \frac{s_2}{s_1} = \frac{-\xi - \sqrt{\xi^2 - 1}}{-\xi + \sqrt{\xi^2 - 1}} = \frac{\xi + \sqrt{\xi^2 - 1}}{\xi - \sqrt{\xi^2 - 1}} = \frac{\xi + \xi\sqrt{1 - \frac{1}{\xi^2}}}{\xi - \xi\sqrt{1 - \frac{1}{\xi^2}}}$$

Besides, $(1+X)^a \underset{X\to 0}{\sim} 1 + aX$ so $\left(1 + \frac{1}{X}\right)^a \underset{X\to +\infty}{\sim} 1 + \frac{a}{X}$ and hence $\sqrt{1 - \frac{1}{\xi^2}} \underset{\xi\to +\infty}{\sim} 1 - \frac{1}{2\xi^2}$. As a consequence:

$$\frac{\tau_1}{\tau_2} \underset{\xi\to +\infty}{\sim} \frac{\xi + \xi\left(1 - \frac{1}{2\xi^2}\right)}{\xi - \xi\left(1 - \frac{1}{2\xi^2}\right)} \underset{\xi\to +\infty}{\sim} \frac{2\xi - \frac{1}{2\xi}}{\frac{1}{2\xi}} \underset{\xi\to +\infty}{\sim} 4\xi^2 - 1 \underset{\xi\to +\infty}{\sim} 4\xi^2$$

$\frac{\tau_1}{\tau_2} \underset{\xi\to +\infty}{\sim} 4\xi^2 >> 1$ when the damping ratio becomes high. The general dynamics will hence be imposed by the greatest time constant, and a second order system has a degraded behavior which can be approximated by simpler models[4]. Two strategies can then be adopted:

- **1st strategy**: the influence of τ_2 is considered neglectable, and we hence have $H(s) \sim \frac{K}{1 + \tau_1 s}$ $\rightarrow$ the second order system has a behavior which is equivalent to the behavior of a first order system, as illustrated in Figure 2.64.

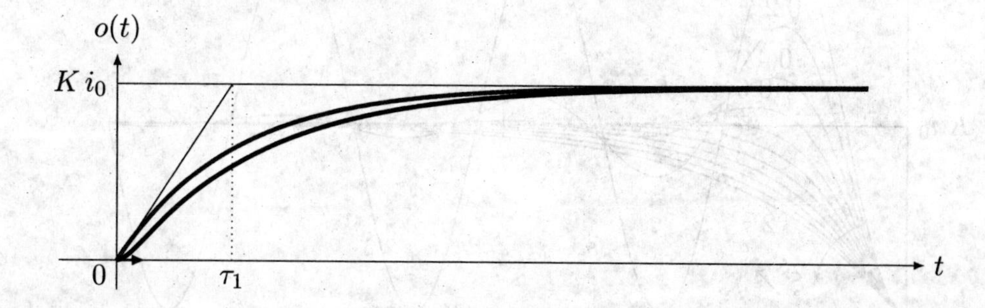

Figure 2.64 first strategy to approximate a damped second order system

[4]This corresponds to the notion of *dominant pole* which will be tackled in the second volume.

- **2nd strategy**: it is assumed that $\dfrac{1}{1+\tau_2 s} \sim e^{-\tau_2 s}$, in which case we have $H(s) \sim \dfrac{Ke^{-\tau_2 s}}{1+\tau_1 s} \rightarrow$ the second order system has a behavior which is equivalent to the behavior of a first order system delayed by τ_2 s (according to the theorem of the delay), as illustrated in Figure 2.65.

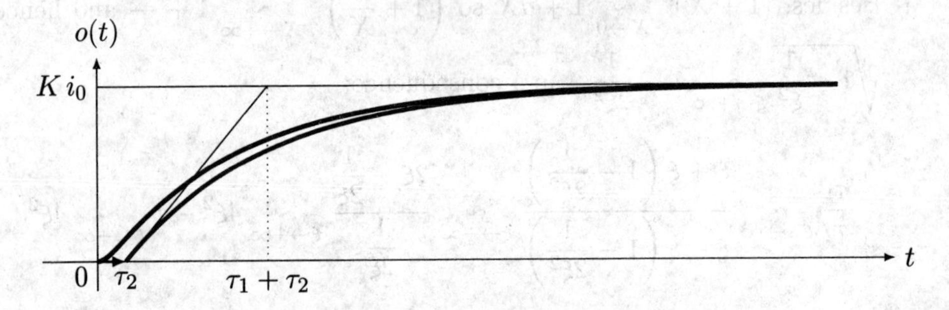

Figure 2.65 second strategy to approximate a damped second order system

- **Conclusion regarding the relation between the value of the damping ratio and the shape of the step response**

  If ξ varies between 0 and 2, we get the step responses in Figure 2.66.

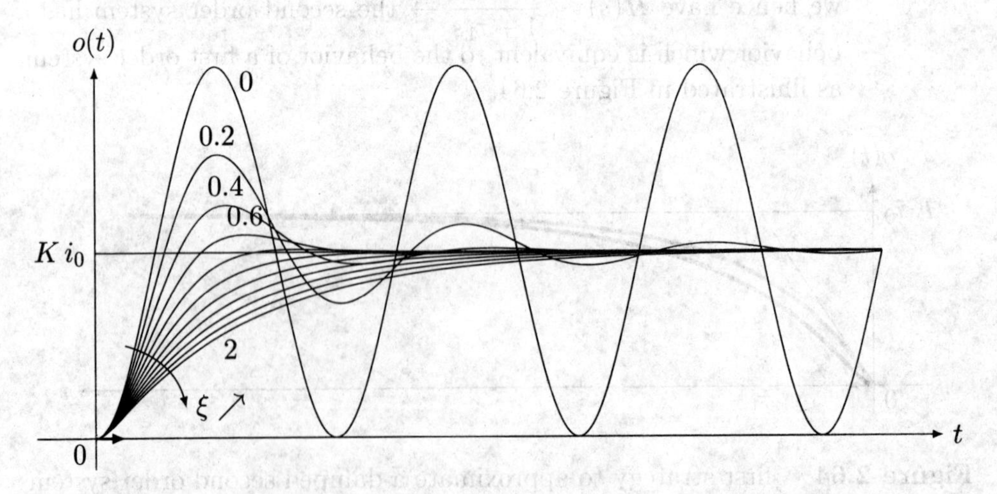

Figure 2.66 relation between the damping ratio ξ and the shape of the step response

- **settling time** at 5%. The calculation of the settling time at 5% is a painful process which justifies the existence of a determination abacus. Indeed, even if this settling time can be determined quite easily when the system is not oscillating by solving the equation $o(t) = 0.95Ki_0$, its determination can become quite difficult as soon as the curve has many oscillations since the equations $o(t) = 0.95Ki_0$ and $o(t) = 1.05Ki_0$ need to be solved separately, the greatest of both solutions being retained. The curve in Figure 2.67 results from a point plotting for different values of ξ:
  - the value of ξ appears in abscissa; and
  - the value of the product $t_{5\%}\omega_0$ appears in ordinate;

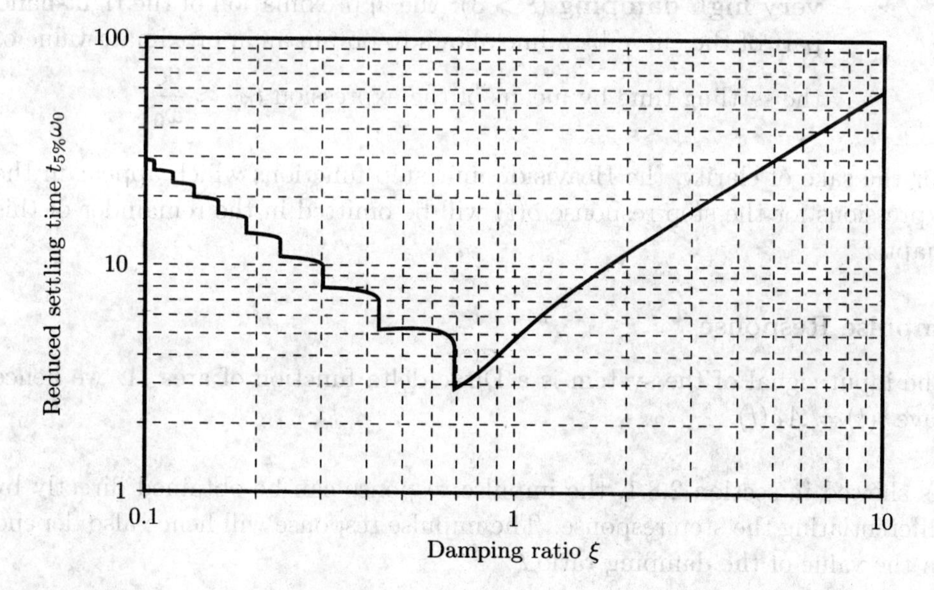

Figure 2.67 abacus for the determination of the settling time at 5% for a second order system

both values being represented on a logarithmic scale. The shape of the curve for $\xi \in [0.1, 0.69]$ is due to the fact that the settling time is calculated successively when the curve intersects the line of equation $o(t) = 0.95Ki_0$ or the line of equation $o(t) = 1.05Ki_0$. The following values can be retained for the settling time:

- **critical damping**: $\xi = 1 \Longrightarrow t_{5\%} \approx \dfrac{5}{\omega_0}$.

This is an approximate value. The real value is $t_{5\%} \approx \dfrac{4.744}{\omega_0}$.

- **optimal damping** (the quickest – almost 40% quicker than critical damping – but some small oscillations may be present, even though they will always be lower than 5%): $\xi = 0.69 \approx 0.7 \Longrightarrow t_{5\%} \approx \dfrac{3}{\omega_0}$.

  This is an approximate value. The real value is $t_{5\%} \approx \dfrac{2.859}{\omega_0}$.
- **very low damping** ($\xi < 10^{-1}$): the approximation of the left-hand part of the curve by a line allows to obtain an approximate value of the settling time by means of the expression $t_{5\%} \approx \dfrac{3}{\xi\omega_0}$.
- **very high damping** ($\xi > 3$): the approximation of the right-hand part of the curve by a line allows to obtain an approximate value of the settling time by means of the expression $t_{5\%} \approx \dfrac{6\xi}{\omega_0}$.

For the sake of clarity, the Heaviside unit step functions which appear in the expressions for the step response $o(t)$ will be omitted in the remainder of this chapter.

Impulse Response

The input signal of the system is a Dirac delta function of area A: we hence have $i(t) = A\,\delta(t)$.

As showed in section 2.8.1, the impulse response can be obtained directly by differentiating the step response. The impulse response will hence also depend on the value of the damping ratio ξ:

- **1st case:** $\xi = 0$.

$$\begin{aligned} o(t) &= \frac{d}{dt}\left[K\,A\,(1 - \cos(\omega_0 t))\right] \\ &= K\,A\,\omega_0 \sin(\omega_0 t) \end{aligned} \tag{2.29}$$

- **2nd case:** $0 < \xi < 1$.

$$o(t) = \frac{d}{dt}\left[K\,A\left(1 - \frac{e^{-\xi\omega_0 t}}{\sqrt{1-\xi^2}}\left(\xi \sin(\omega_0\sqrt{1-\xi^2}t) + \sqrt{1-\xi^2}\cos(\omega_0\sqrt{1-\xi^2}t)\right)\right)\right]$$

$$= K\,A\,\omega_0 \frac{e^{-\xi\omega_0 t}}{\sqrt{1-\xi^2}} \sin(\omega_0\sqrt{1-\xi^2}t) \tag{2.30}$$

- **3rd case:** $\xi = 1$.

$$o(t) = \frac{d}{dt}\left[K\,A\left(1-(1+\omega_0 t)e^{-\omega_0 t}\right)\right]$$

$$= K\,A\,\omega_0^2\, t\, e^{-\omega_0 t} \tag{2.31}$$

- **4th case:** $\xi > 1$.

$$o(t) = \frac{d}{dt}\left[K\,A\left(1 - \frac{\left(\xi+\sqrt{\xi^2-1}\right)e^{-\omega_0\left(\xi-\sqrt{\xi^2-1}\right)t}}{2\sqrt{\xi^2-1}} - \frac{\left(-\xi+\sqrt{\xi^2-1}\right)e^{-\omega_0\left(\xi+\sqrt{\xi^2-1}\right)t}}{2\sqrt{\xi^2-1}}\right)\right]$$

$$= K\,A\,\omega_0 \frac{e^{-\omega_0\left(\xi-\sqrt{\xi^2-1}\right)t} - e^{-\omega_0\left(\xi+\sqrt{\xi^2-1}\right)t}}{2\sqrt{\xi^2-1}} \tag{2.32}$$

- **Conclusion regarding the relation between the value of the damping ratio and the shape of the impulse response**

  If ξ varies between 0 and 2, we get the impulse responses in Figure 2.68.

The impulse response has the same behavior as the step response:

- there are oscillations if $\xi < 1$ (which have the same period as the oscillations of the step response);

- the response does not converge if $\xi = 0$ (at the limit of stability); and
- the initial and final behavior of the impulse response are the same, whatever the damping ratio (except if $\xi = 0$).

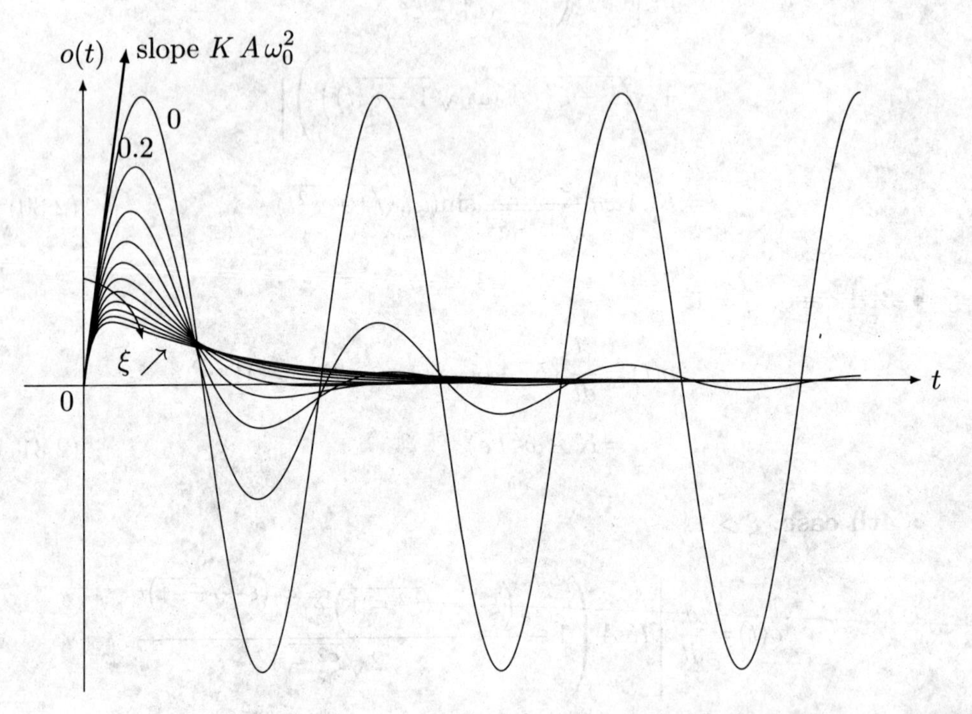

Figure 2.68 relation between the damping ratio ξ and the shape of the impulse response

Pursuit Response

The input signal of the system is a causal ramp of slope α: we hence have $i(t) = \alpha\, t\, u(t)$.

As showed in section 2.8.1, the pursuit response can be obtained directly by integrating the step response. The pursuit response will hence also depend on the value of the damping ratio ξ:

- **1st case:** $\xi = 0$.

$$o(t) = \int_0^t K\,\alpha\,(1-\cos(\omega_0 x))\,dx = K\,\alpha\left(t - \frac{1}{\omega_0}\sin(\omega_0 t)\right) \quad (2.33)$$

- **2nd case:** $0 < \xi < 1$.

$$\begin{aligned} o(t) &= \int_0^t K\alpha\left(1 - \frac{e^{-\xi\omega_0 x}}{\sqrt{1-\xi^2}}\Big(\xi\sin(\omega_0\sqrt{1-\xi^2}x) + \sqrt{1-\xi^2}\cos(\omega_0\sqrt{1-\xi^2}x)\Big)\right)dx \\ &= K\alpha\,t + \frac{K\alpha}{\omega_0}\frac{\xi}{\sqrt{1-\xi^2}}e^{-\xi\omega_0 t} \Big(\xi\sin(\omega_0\sqrt{1-\xi^2}t) + \sqrt{1-\xi^2}\cos(\omega_0\sqrt{1-\xi^2}t)\Big) \\ &\quad + \frac{K\alpha}{\omega_0}e^{-\xi\omega_0 t}\Big(\xi\cos(\omega_0\sqrt{1-\xi^2}t) - \sqrt{1-\xi^2}\sin(\omega_0\sqrt{1-\xi^2}t)\Big) \\ &\quad - 2K\alpha\frac{\xi}{\omega_0} \\ &= K\alpha\left(t + \frac{e^{-\xi\omega_0 t}}{\omega_0}\left(\left(\frac{\xi^2}{\sqrt{1-\xi^2}} - \sqrt{1-\xi^2}\right)\sin(\omega_0\sqrt{1-\xi^2}t) + 2\xi\cos(\omega_0\sqrt{1-\xi^2}t)\right) - \frac{2\xi}{\omega_0}\right) \\ &= K\alpha\left(t + \frac{e^{-\xi\omega_0 t}}{\omega_0}\left(\frac{2\xi^2-1}{\sqrt{1-\xi^2}}\sin(\omega_0\sqrt{1-\xi^2}t) + 2\xi\cos(\omega_0\sqrt{1-\xi^2}t)\right) - \frac{2\xi}{\omega_0}\right) \end{aligned} \quad (2.34)$$

- **3rd case:** $\xi = 1$.

$$\begin{aligned} o(t) &= \int_0^t K\alpha\left(1 - (1+\omega_0 x)e^{-\omega_0 x}\right)dx \\ &= K\alpha\,t - K\alpha\int_0^t (1+\omega_0 x)e^{-\omega_0 x}dx \end{aligned}$$

This integral is under the form $\int_0^t u\,dv$ with $\begin{cases} u=1+\omega_0 x \Rightarrow du=\omega_0\,dx \\ dv=e^{-\omega_0 x}dx \Rightarrow v=-\dfrac{e^{-\omega_0 x}}{\omega_0} \end{cases}$

Thanks to an integration by parts, we can determine that:

$$\begin{aligned} o(t) &= K\alpha\, t - K\alpha\left[-(1+\omega_0 x)\frac{e^{-\omega_0 x}}{\omega_0}\right]_0^t - K\alpha\int_0^t e^{-\omega_0 x}dx \\ &= K\alpha\left(t-\frac{1}{\omega_0}\left(2-(2+\omega_0 t)e^{-\omega_0 t}\right)\right) \end{aligned} \tag{2.35}$$

- **4th case:** $\xi > 1$.

$$\begin{aligned} o(t) &= \int_0^t K\alpha\left(1-\frac{\left(\xi+\sqrt{\xi^2-1}\right)e^{-\omega_0\left(\xi-\sqrt{\xi^2-1}\right)x}}{2\sqrt{\xi^2-1}}\right. \\ &\qquad \left.-\frac{\left(-\xi+\sqrt{\xi^2-1}\right)e^{-\omega_0\left(\xi+\sqrt{\xi^2-1}\right)x}}{2\sqrt{\xi^2-1}}\right)dx \\ &= K\alpha\left(t+\frac{(\xi+\sqrt{\xi^2-1})^2 e^{-\omega_0\left(\xi-\sqrt{\xi^2-1}\right)t}}{2\omega_0\sqrt{\xi^2-1}}\right. \\ &\qquad \left.-\frac{(\xi-\sqrt{\xi^2-1})^2 e^{-\omega_0\left(\xi+\sqrt{\xi^2-1}\right)t}}{2\omega_0\sqrt{\xi^2-1}}-\frac{2\xi}{\omega_0}\right) \end{aligned} \tag{2.36}$$

- **Conclusion regarding the relation between the value of the damping ratio and the shape of the pursuit response**

  If ξ varies between 0 and 2, we get the pursuit responses in Figure 2.69.

The pursuit response has the same behavior as the step response:

- there are oscillations if $\xi < 1$ (which have the same period as the oscillations of the step response); and
- the initial and final behavior of the pursuit response are the same, whatever the damping ratio (except if $\xi = 0$).

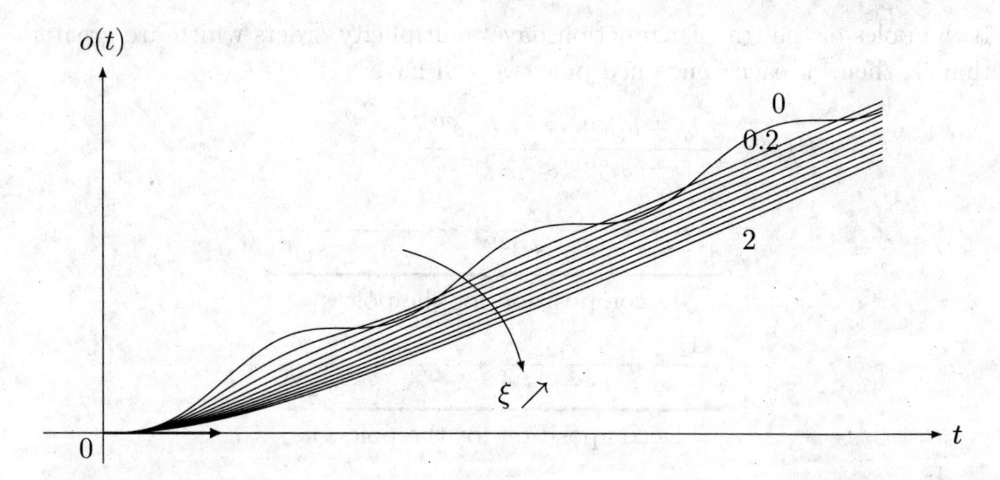

Figure 2.69 relation between the damping ratio ξ and the shape of the pursuit response

2.8.3 Other Systems

Partial Fraction Decomposition

According to the fundamental theorem of algebra, a polynomial $D(s)$ of degree n has n complex roots s_i, which implies that it is possible to perform the following factorization in $\mathbb{C}$:

$$D(s) = a_0 + a_1 s + a_2 s^2 + \cdots + a_n s^n = a_n(s - s_1)(s - s_2)\ldots(s - s_n)$$

If the coefficients a_i of the polynomial are real numbers, then it can be proved that its roots s_i necessarily are real numbers or conjugate complex numbers.

As a consequence, if we consider a transfer function whose denominator is a polynomial of degree n, we will hence have:

$$H(s) = \frac{b_0 + b_1 s + \cdots + b_m s^m}{a_0 + a_1 s + \cdots + a_n s^n} = \frac{b_0 + b_1 s + \cdots + b_m s^m}{a_n(s - s_1)(s - s_2)\ldots(s - s_n)}$$

If all the poles of the transfer function are simple, it can be shown that we have the following decomposition in $\mathbb{C}$:

$$H(s) = \frac{b_0 + b_1 s + \cdots + b_m s^m}{a_n(s - s_1)(s - s_2)\ldots(s - s_n)} = \frac{A_1}{s - s_1} + \frac{A_2}{s - s_2} + \cdots + \frac{A_n}{s - s_n}$$

If the poles of the transfer function have multiplicity orders which are greater than 1, then for each concerned pole, we will have:

$$\begin{aligned}H(s)=&\frac{b_0+b_1s+\cdots+b_ms^m}{a_n(s-s_1)^{\alpha_1}(s-s_2)^{\alpha_2}\ldots}\\=&\underbrace{\frac{A_{11}}{s-s_1}+\frac{A_{12}}{(s-s_1)^2}+\cdots+\frac{A_{1\alpha_1}}{(s-s_1)^{\alpha_1}}}_{\text{Decomposition for the pole } s_1}\\&+\underbrace{\frac{A_{21}}{s-s_2}+\frac{A_{22}}{(s-s_2)^2}+\cdots+\frac{A_{2\alpha_2}}{(s-s_2)^{\alpha_2}}}_{\text{Decomposition for the pole } s_2}+\ldots\end{aligned}$$

where α_1 is the multiplicity order of the pole s_1, etc.

Many efficient methods exist to determine the values of these coefficients. One of them is the method that we have used so far and which consists in expressing all the fractions on the same denominator and identifying the different terms.

Table 2.3 presents some common partial fraction decompositions.

General Methodology and Implications

In all calculations, the Heaviside conditions (null initial conditions for any function and its successive derivatives) will be supposed to hold.

Partial fraction decomposition will hence allow to decompose the transfer function of any system into a sum of partial fractions whose inverse Laplace transform can be determined by means of Table 2.2. The time response of any system can hence be determined.

Another equivalent notation can be determined by considering the transfer function as a product of polynomials: a system of order n can thus be decomposed into a product of first order systems and second order systems, as it will be detailed in section 2.9.4. The study which was carried out in this section regarding first and second order systems hence is pertinent and necessary.

Table 2.3 common partial fraction decompositions

Function $F(s)$	Partial fraction decomposition	$\alpha \in \mathbb{R}$	$\beta \in \mathbb{R}$	$\gamma \in \mathbb{R}$
$\frac{1}{s} \times \frac{K}{1+\tau s}$	$F(s) = K\left(\frac{\alpha}{s} + \frac{\beta}{1+\tau s}\right)$	1	$-\tau$	/
$\frac{1}{s^2} \times \frac{K}{1+\tau s}$	$F(s) = K\left(\frac{\alpha}{s} + \frac{\beta}{s^2} + \frac{\gamma}{1+\tau s}\right)$	$-\tau$	1	τ^2
$\frac{\omega}{s^2+\omega^2} \times \frac{K}{1+\tau s}$	$F(s) = \frac{K\omega}{1+\tau^2\omega^2}\left(\frac{\alpha s+\beta}{s^2+\omega^2} + \frac{\gamma}{1+\tau s}\right)$	$-\tau$	1	τ^2
$\frac{1}{s} \times \frac{K}{(1+\tau_1 s)(1+\tau_2 s)}$	$F(s) = K\left(\frac{\alpha}{s} + \frac{\beta}{1+\tau_1 s} + \frac{\gamma}{1+\tau_2 s}\right)$	1	$\frac{-\tau_1{}^2}{\tau_1-\tau_2}$	$\frac{\tau_2{}^2}{\tau_1-\tau_2}$
$\frac{1}{s} \times \frac{K}{(1+\tau s)^2}$	$F(s) = K\left(\frac{\alpha}{s} + \frac{\beta}{1+\tau s} + \frac{\gamma}{(1+\tau s)^2}\right)$	1	$-\tau$	$-\tau$
$\frac{1}{s} \times \frac{K}{(s+\sigma)^2+\omega^2}$	$F(s) = \frac{K}{\sigma^2+\omega^2}\left(\frac{\alpha}{s} + \frac{\beta s+\gamma}{(s+\sigma)^2+\omega^2}\right)$	1	-1	-2σ

2.9 Frequency Response

2.9.1 Definition and Methods

Definition

Definition 16 (Frequency Response)

The **frequency response** of a linear system is its steady-state response to a sinusoidal input signal of magnitude i_0 and of frequency ω:

$$i(t) = i_0 \sin(\omega t)$$

If the system is linear, then the same type of response will **always** be observed: in **steady state**, the output signal can be assimilated to a sinusoidal signal which is **delayed** by t_D seconds with respect to the input signal and whose magnitude depends on the excitation frequency and on the magnitude of the input signal. As a consequence, if the system is linear, we will have:

$$i_0 \sin(\omega t) \longrightarrow \boxed{H(s)} \longrightarrow \underbrace{o_\tau(t)}_{\to 0} + o_0(\omega)\sin(\omega t + \varphi(\omega))$$

where:

- $o_0(\omega) = \text{fct}(i_0, \omega)$ is the magnitude of the steady-state output signal: practically, this magnitude depends on the magnitude i_0 and on the frequency $f = \dfrac{\omega}{2\pi}$ of the input signal;

- $\varphi(\omega) = \text{fct}(\omega) = -\omega t_D(\omega) \leqslant 0$ corresponds to the phase shift (expressed in rad) of the steady-state output signal with respect to the input signal: practically, this phase shift only depends on the frequency of the input signal; and

- $o_\tau(t)$ is a function which corresponds to the transient state and hence respects $\lim\limits_{t\to+\infty} o_\tau(t) = 0$.

This expression for the output signal can be determined as follows. Any transfer function can be written under the following general form:

$$H(s) = \frac{O(s)}{I(s)} = \frac{K}{s^\alpha} \frac{1 + b_1 s + b_2 s^2 + \cdots + b_m s^m}{1 + a_1 s + a_2 s^2 + \cdots + a_{n-\alpha} s^{n-\alpha}}$$

The different terms of this expression are as follows:

- α is the **class** of the system: practically, $\alpha \in \{0, 1, 2\}$ (consequently, $\alpha \in \mathbb{N}$).

- K is the **gain** of the system; this gain can be:

  - a **static gain** if $\alpha = 0$, in which case its unit is $[K] = [o]/[i]$;

  - a **velocity-feedback gain** if $\alpha = 1$, in which case its unit is $[K] = [\dot{o}]/[i]$;

  - an **acceleration-feedback gain** if $\alpha = 2$, in which case its unit is $[K] = [\ddot{o}]/[i]$.

- $n - \alpha$ is the **order** of the system (some authors consider that this order is n); real systems satisfy the condition $\boxed{m \leqslant n}$.

The transfer function can be expressed under the following form:

$$H(s) = \frac{O(s)}{I(s)} = \frac{K'}{s^\alpha} \times \frac{\prod_{i=1}^{m}(s - z_i)}{\prod_{j=1}^{n-\alpha}(s - p_j)}$$

where the terms $z_i \in \mathbb{C}$ (the roots of the polynomial of the numerator) correspond to the **zeros** of the transfer function and the terms $p_j \in \mathbb{C}$ (the roots of the polynomial of the denominator) correspond to the **poles** of the transfer function. We will consider that these poles are simple, and that the class of the system is null (i.e. $\alpha = 0$).
Let's consider a causal sinusoidal input signal:

$$i(t) = i_0 \sin(\omega t) \longrightarrow \boxed{\mathscr{L}} \longrightarrow I(s) = \frac{i_0 \omega}{s^2 + \omega^2}$$

The rational fraction $O(s) = H(s)I(s)$ can then be decomposed into partial fractions on $\mathbb{C}$ as follows:

$$\begin{aligned} O(s) &= H(s)\,\frac{i_0\omega}{s^2+\omega^2} = H(s)\,\frac{i_0\omega}{(s-j\omega)(s+j\omega)} \\ &= i_0 \left(\frac{A}{s-j\omega} + \frac{B}{s+j\omega} + \sum_{j=1}^{n} \frac{C_j}{s-p_j} \right) \end{aligned}$$

If we note $f(s) = O(s)\,(s - j\omega)$, then we can notice that $f(j\omega) = \frac{H(j\omega)}{2j} i_0 = A\, i_0$, so $A = \frac{H(j\omega)}{2j}$.
If we note $g(s) = O(s)\,(s + j\omega)$, then we can notice that $g(-j\omega) = -\frac{H(-j\omega)}{2j} i_0 = B\, i_0$, so $B = -\frac{H(-j\omega)}{2j}$.
After reducing all the fractions to the same denominator, we get:

$$O(s) = i_0 \left(\frac{H(j\omega) - H(-j\omega)}{2j}\,\frac{s}{s^2+\omega^2} + \frac{H(j\omega)+H(-j\omega)}{2}\,\frac{\omega}{s^2+\omega^2} + \sum_{j=1}^{n} \frac{C_j}{s-p_j} \right)$$

As the coefficients are real numbers, we hence have $H(\overline{z}) = \overline{H(z)}$. As a consequence:

$$\begin{cases} \mathrm{Re}[H(j\omega)] = \dfrac{H(j\omega) + \overline{H(j\omega)}}{2} = \dfrac{H(j\omega) + H(-j\omega)}{2} \\ \mathrm{Im}[H(j\omega)] = \dfrac{H(j\omega) - \overline{H(j\omega)}}{2j} = \dfrac{H(j\omega) - H(-j\omega)}{2j} \end{cases}$$

We hence get:

$$O(s) = i_0 \left(\mathrm{Im}[H(j\omega)] \frac{s}{s^2 + \omega^2} + \mathrm{Re}[H(j\omega)] \frac{\omega}{s^2 + \omega^2} + \sum_{j=1}^{n} \frac{C_j}{s - p_j} \right)$$

If we pose $A(\omega) = |H(j\omega)|$ and $\varphi(\omega) = \arg[H(j\omega)]$, we can immediately determine that:

$$O(s) = i_0 \left(A(\omega) \, \sin\varphi(\omega) \, \frac{s}{s^2 + \omega^2} + A(\omega) \, \cos\varphi(\omega) \, \frac{\omega}{s^2 + \omega^2} + \sum_{j=1}^{n} \frac{C_j}{s - p_j} \right)$$

Since $\mathscr{L}[\sin(\omega t)] = \dfrac{\omega}{s^2 + \omega^2}$ and $\mathscr{L}[\cos(\omega t)] = \dfrac{s}{s^2 + \omega^2}$:

$$\begin{aligned} o(t) &= \mathscr{L}^{-1}[O(s)] \\ &= i_0 \left(A(\omega) \, \sin\varphi(\omega) \, \cos(\omega t) + A(\omega) \, \cos\varphi(\omega) \, \sin(\omega t) + \sum_{j=1}^{n} C_j e^{p_j t} \right) \end{aligned}$$

We hence get $\boxed{o(t) \underset{t \to +\infty}{\sim} A(\omega) \, i_0 \, \sin(\omega t + \varphi(\omega))}$ since $\lim\limits_{t \to +\infty} e^{p_j t} = 0$ if the system is stable.

The steady-state output signal hence is a sinusoidal signal whose frequency is the same as the frequency of the input signal, whose magnitude is $o_0(\omega) = |H(j\omega)| \, i_0$, and which is delayed so that the signal has a phase shift $\varphi(\omega) = \arg[H(j\omega)]$.

Determination of the Magnitude and of the Phase Shift

Let's consider a sinusoidal input signal. In steady state, we have:

$$i(t) = i_0 \, \sin(\omega t) \to \boxed{H(s)} \to o(t) \underset{t \to +\infty}{\sim} o_0(\omega) \, \sin(\omega t + \varphi(\omega))$$

This frequential approach aims at determining, **without any calculation** and for **any frequency** ω, the two following values:

- the gain $A(\omega) = \dfrac{o_0(\omega)}{i_0} > 0$, which is the ratio between the magnitude of the sinusoidal **steady-state** output signal and the magnitude of the exciting signal, from a first plot named **magnitude plot**; and
- the phase $\varphi(\omega) = -\omega t_D(\omega) \leqslant 0$ (where $t_D(\omega)$ is the delay), which is the phase shift between the **steady-state** output signal and the exciting signal, from a second plot named **phase plot**.

The knowledge of $A(\omega)$ and $\varphi(\omega)$ for any value of ω will allow us to determine the steady-state response to an input signal $i(t) = i_0 \sin(\omega_1 t)$, which is $o(t) \underset{t\to+\infty}{\sim} \underbrace{A_1\, i_0}_{o_0(\omega_1)} \sin(\omega_1 t + \varphi_1)$, where $A_1 = A(\omega_1)$ and $\varphi_1 = \varphi(\omega_1)$.

Magnitude and Phase Plots

The magnitude and phase plots can be obtained analytically for any system of transfer function $H(s)$ by posing $s = j\omega$ to determine the isochronous function $H(j\omega)$ of the transfer function $H(s)$, and by determining:

$$\boxed{A(\omega) = \frac{o_0(\omega)}{i_0} = |H(j\omega)|} \qquad \text{and} \qquad \boxed{\varphi(\omega) = \arg[H(j\omega)]}$$

Practically, the isochronous function $H(j\omega)$ is a complex function which can be expressed under the form $H(j\omega) = a(\omega) + jb(\omega)$, and the expressions for the two functions A and φ can be obtained directly as follows:

$$\begin{cases} A(\omega) = \sqrt{a^2(\omega) + b^2(\omega)} \\ \varphi(\omega) = \arctan\left(\dfrac{b(\omega)}{a(\omega)}\right) \end{cases}$$

Comment Regarding Frequencies

Practically, a **non-linear scale** is used to represent the magnitude and phase plots. Such a scale allows to take into account a wider range of frequencies.

The scale which was chosen is the logarithmic scale, which is illustrated in Figure 2.70.

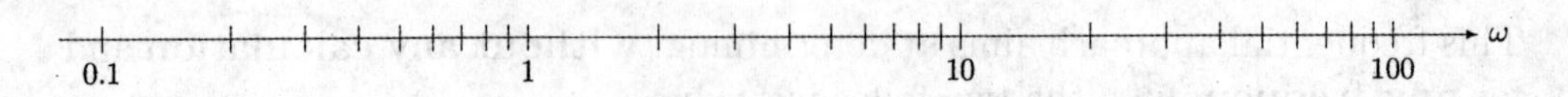

Figure 2.70 the logarithmic scale

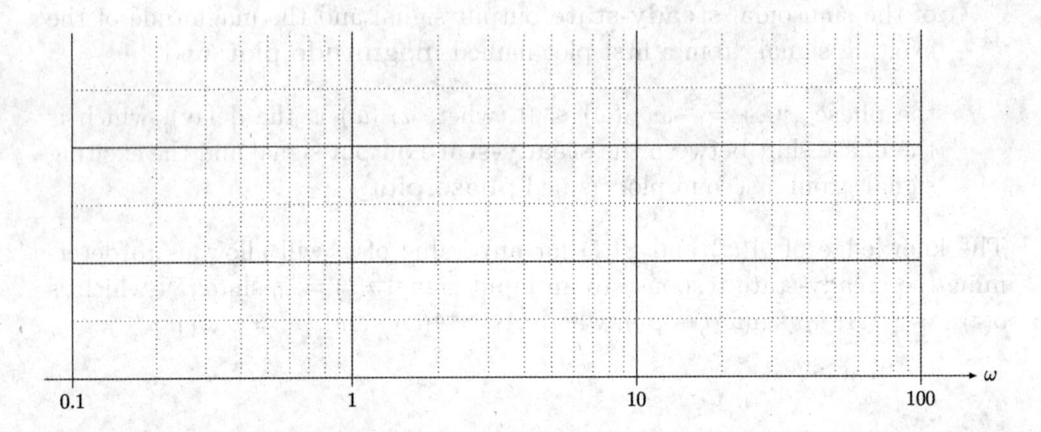

Figure 2.71 an example of semi-logarithmic paper

Frequential plots can hence be represented on special semi-logarithmic papers, as illustrated in Figure 2.71.

As a consequence, the magnitude and phase plots will respectively represent:

$$\begin{cases} A_{\mathrm{dB}}(\omega) = 20 \log A(\omega) = 20 \log |H(j\omega)| \\ \varphi(\omega) = \arg[H(j\omega)] \end{cases}$$

where A_{dB} is called the **magnitude in decibels** (dB).

2.9.2 Frequency Plots

Bode Plots

The **Bode plots** represent, on two plots, the magnitude $A_{\mathrm{dB}}(\omega) = 20 \log \left(\frac{o_0(\omega)}{i_0} \right)$ and the phase shift $\varphi(\omega)$: the abscissa is represented in logarithmic scale, which allows to use a wide range of frequencies on the same plot.

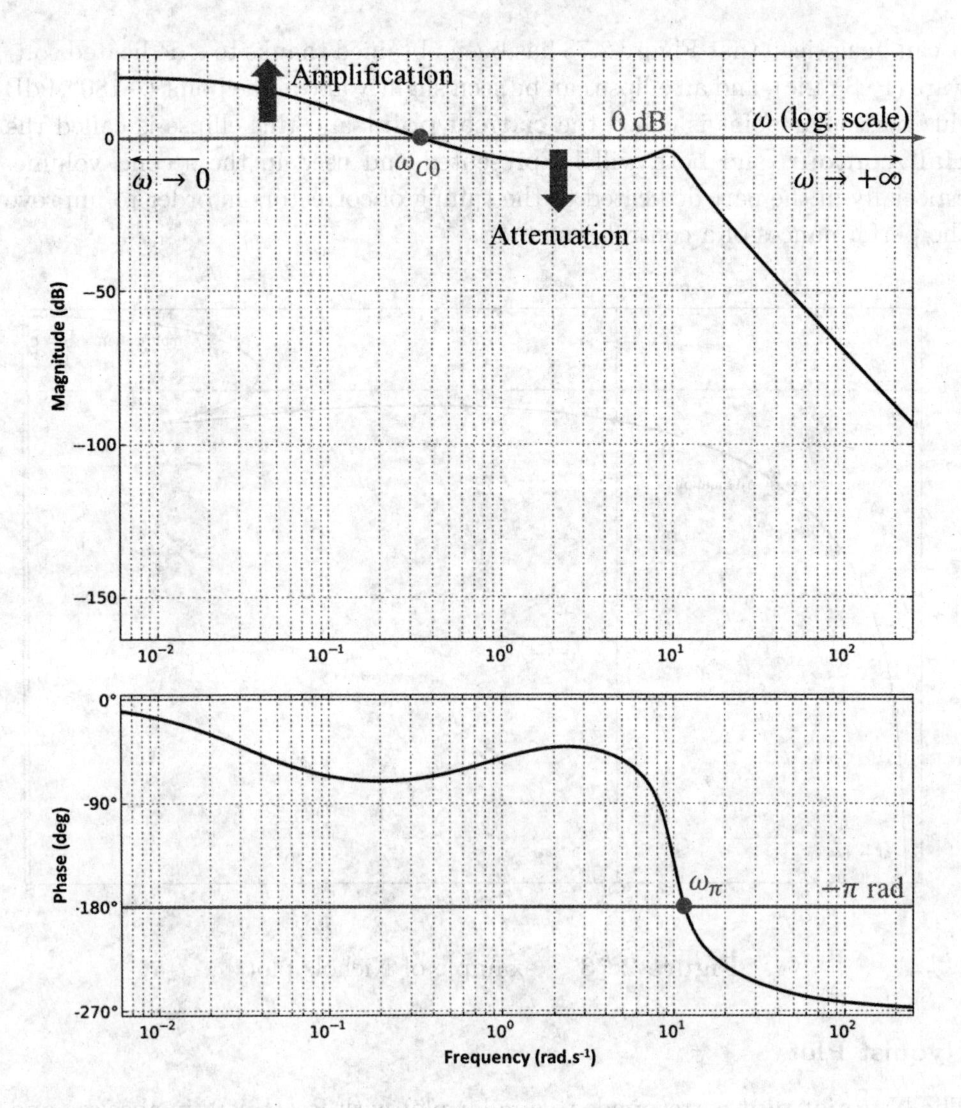

Figure 2.72 example of Bode plots

Nichols Plot

The **Nichols plot**[5] corresponds to a parameterized plot with $\varphi(\omega)$ in abscissa and $A_{\rm dB}(\omega)$ in ordinate.

[5]The Nichols plot is also known as the Black plot in many countries (including France, Spain, Italy, etc.); in these countries, the Nichols plot refers to a special evolution of this plot which will be studied in the second volume.

It can be noticed that Figure 2.73 has been obtained thanks to a dedicated software (PySYLic), and an ellipse can be seen slightly above the point $(-180°, 0\text{dB})$: this particular point is called the **critical point** and this ellipse is called the **Hall contour**, and both will be presented and used in the second volume, especially in the part dedicated to the tuning of correctors in order to improve the performances of a control system.

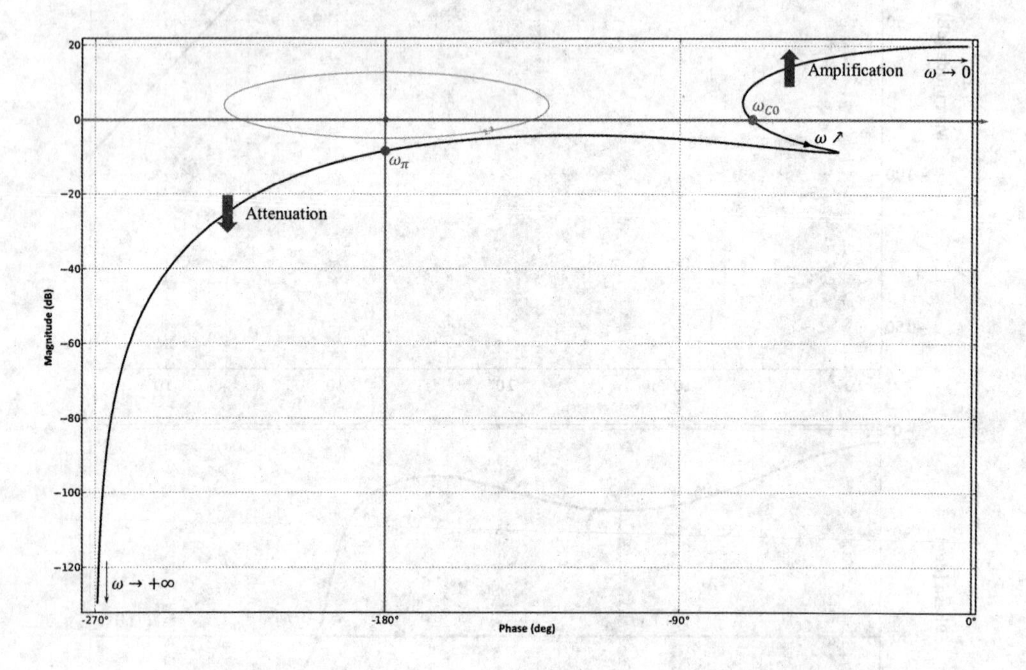

Figure 2.73 example of Nichols plot

Nyquist Plot

The **Nyquist plot** corresponds to a polar plot with $\text{Re}[H(j\omega)]$ in abscissa and $\text{Im}[H(j\omega)]$ in ordinate, which corresponds to the plot of $\dfrac{o_0(\omega)}{i_0} = \text{fct}(\varphi(\omega))$ (Figure 2.74).

2.9.3 Plots of the Frequency Responses of Some Conventional Systems

Integrator System

$$H(s) = \frac{K}{s} \Longrightarrow H(j\omega) = \frac{K}{j\omega}.$$

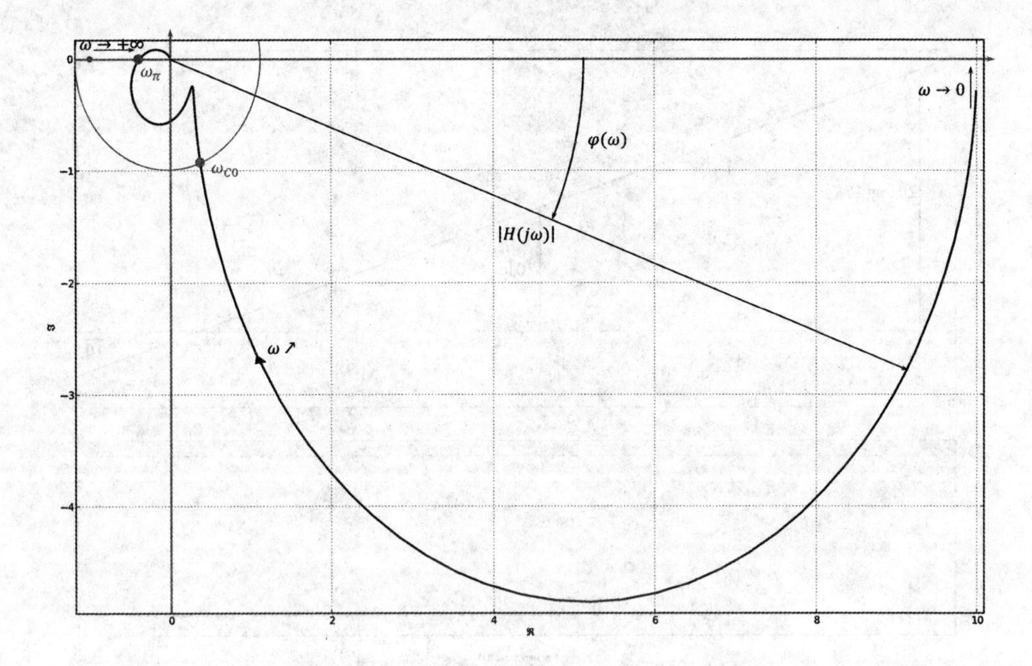

Figure 2.74 example of Nyquist plot

We can hence determine that:

$$\begin{cases} A(\omega) = |H(j\omega)| = \dfrac{K}{\omega} \\ \varphi(\omega) = \arg[H(j\omega)] = -90° \end{cases}$$

$$\Rightarrow \begin{cases} A_{\text{dB}}(\omega) = 20 \log A(\omega) = 20 \log K - 20 \log \omega \\ \varphi(\omega) = -90° \end{cases}$$

Bode plots

The slope of -20 dB/decade means that, when the frequency goes from ω_1 to $10\,\omega_1$ (which corresponds to a decade), the magnitude of the signal is divided by 10.

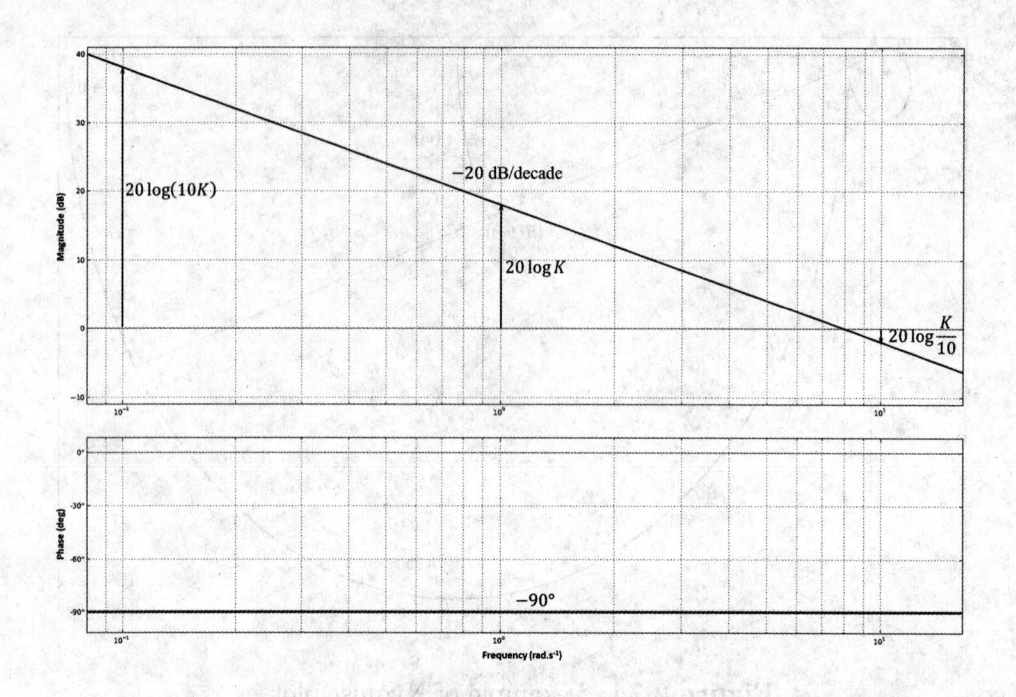

Figure 2.75 Bode plots for an integrator system

Nichols plot and Nyquist plot

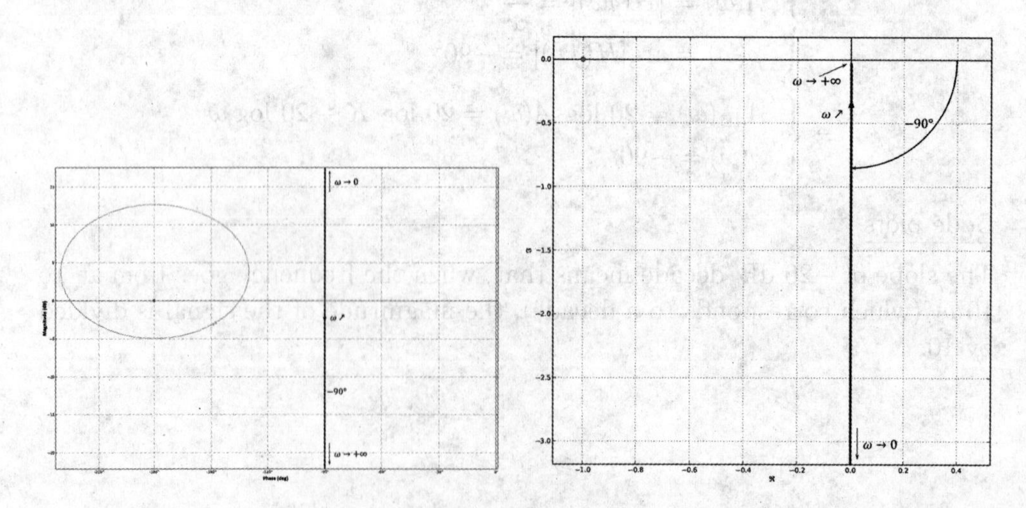

Figure 2.76 Nichols and Nyquist plot for an integrator system

First Order System

$$H(s) = \frac{K}{1+\tau s} \Longrightarrow H(j\omega) = \frac{K}{1+j\tau\omega}.$$

We can hence determine that:

$$\begin{cases} A(\omega) = |H(j\omega)| = \dfrac{K}{\sqrt{1+\tau^2\omega^2}} \\ \varphi(\omega) = \arg[H(j\omega)] = -\arctan(\tau\omega) \end{cases}$$

$$\Rightarrow \begin{cases} A_{\mathrm{dB}}(\omega) = 20\ \log\ A(\omega) = 20\ \log\ K - 10\ \log(1+\tau^2\omega^2) \\ \varphi(\omega) = -\arctan(\tau\omega) \end{cases}$$

Bode plots

When $\omega = \omega_C = \dfrac{1}{\tau}$, we have:

$$\begin{cases} A_{\mathrm{dB}}(\omega_C) = 20\log K - \underbrace{10\log 2}_{\approx 3} \\ \varphi(\omega_C) = -\arctan 1 = -45^\circ \end{cases}$$

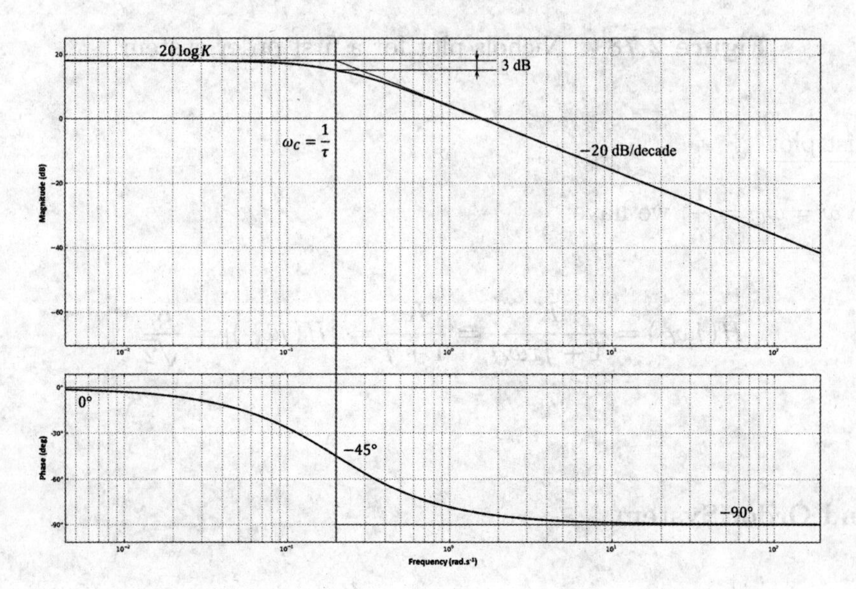

Figure 2.77 Bode plots for a first order system

Nichols plot

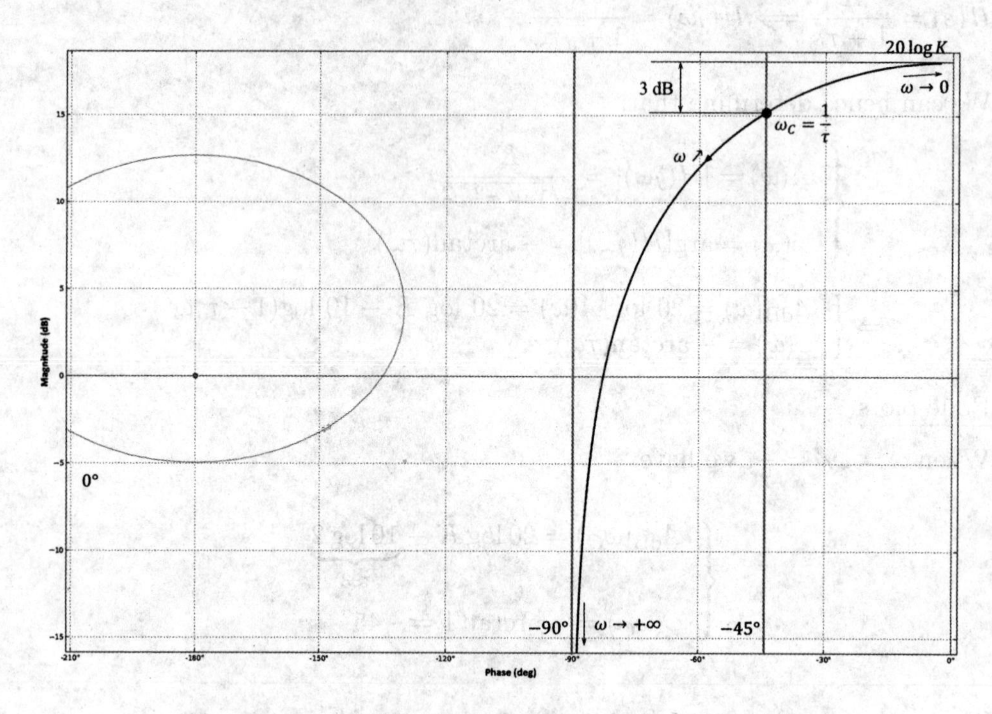

Figure 2.78 Nichols plot for a first order system

Nyquist plot

When $\omega = \omega_C = \dfrac{1}{\tau}$, we have:

$$H(j\omega_C) = \frac{K}{1 + j\tau\omega_C} = \frac{K}{1 + j} \Rightarrow |H(j\omega_C)| = \frac{K}{\sqrt{2}}$$

Second Order System

$$H(s) = \frac{K}{1 + 2\xi\dfrac{s}{\omega_0} + \dfrac{s^2}{\omega_0^2}} \Longrightarrow H(j\omega) = \frac{K}{1 + 2j\xi\dfrac{\omega}{\omega_0} - \dfrac{\omega^2}{\omega_0^2}} = \frac{K}{\left(1 - \frac{\omega^2}{\omega_0^2}\right) + 2j\xi\dfrac{\omega}{\omega_0}}.$$

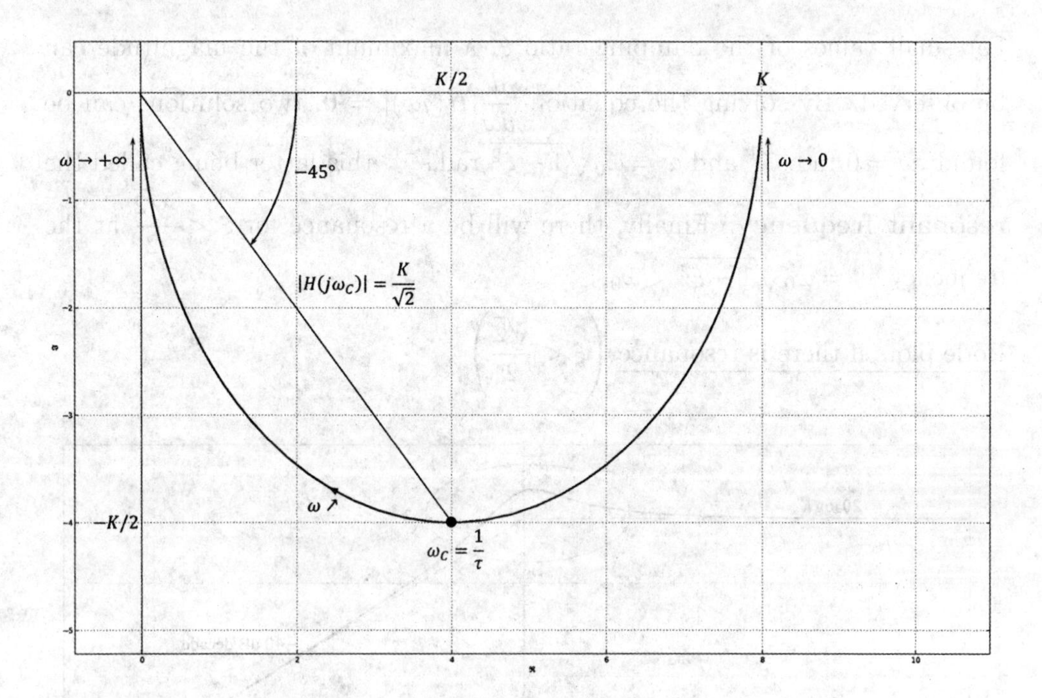

Figure 2.79 Nyquist plot for a first order system

We can hence determine that:

$$\begin{cases} A(\omega) = |H(j\omega)| = \dfrac{K}{\sqrt{\left(1-\frac{\omega^2}{\omega_0^2}\right)^2 + 4\xi^2\dfrac{\omega^2}{\omega_0^2}}} \\ \varphi(\omega) = \arg[H(j\omega)] = -\dfrac{\pi}{2} + \arctan\left(\dfrac{1-\dfrac{\omega^2}{\omega_0^2}}{2\xi\dfrac{\omega}{\omega_0}}\right) \end{cases}$$

$$\Rightarrow \begin{cases} A_{\mathrm{dB}}(\omega) = 20\log A(\omega) = 20\log K - 10\log\left(\left(1-\dfrac{\omega^2}{\omega_0^2}\right)^2 + 4\xi^2\dfrac{\omega^2}{\omega_0^2}\right) \\ \varphi(\omega) = -\dfrac{\pi}{2} + \arctan\left(\dfrac{1-\dfrac{\omega^2}{\omega_0^2}}{2\xi\dfrac{\omega}{\omega_0}}\right) \end{cases}$$

For small values of the damping ratio ξ, a maximum of the magnitude can be observed. By solving the equation $\frac{d}{d\omega}|H(j\omega)| = 0$, two solutions can be found: $\omega = 0$ rad.s^{-1} and $\omega = \omega_0\sqrt{1-\xi^2}$ rad.s^{-1}, this latter being called the **resonant frequency**. Finally, there will be a resonance for $\xi < \frac{\sqrt{2}}{2}$ at the frequency $\omega_r = \omega_0\sqrt{1-\xi^2} < \omega_0$.

Bode plots if there is resonance $\left(\xi < \frac{\sqrt{2}}{2}\right)$

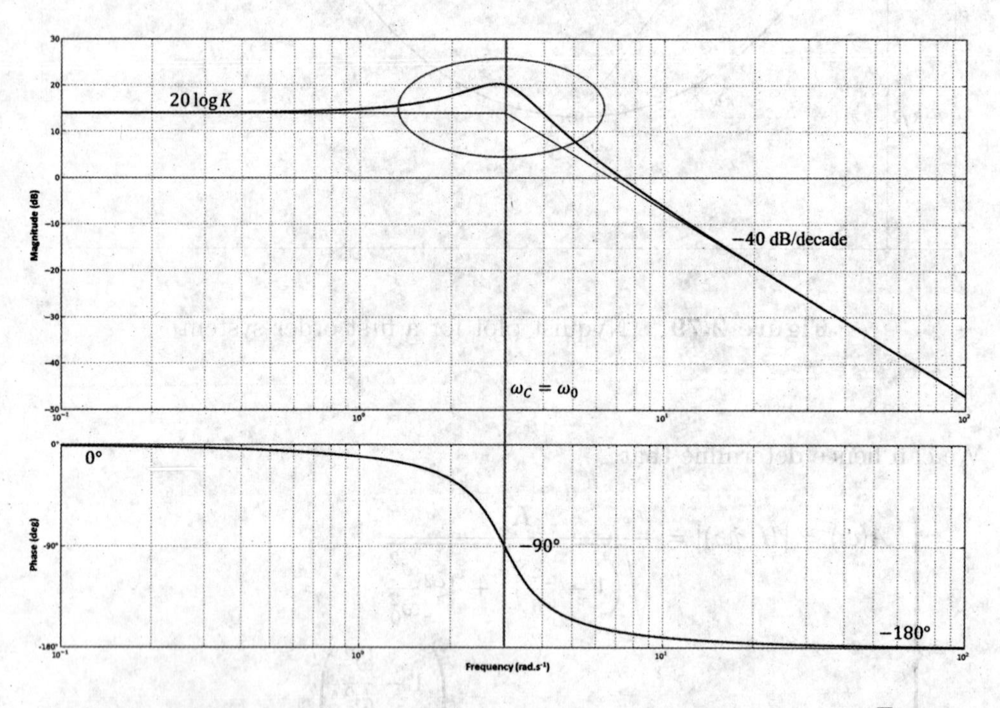

Figure 2.80 Bode plots for a second order system if $\xi < \frac{\sqrt{2}}{2}$

The overamplification is defined as $Q_S = \frac{|H(j\omega_r)|}{|H(0)|}$: a simple calculation allows to determine that $Q_S = \frac{1}{2\xi\sqrt{1-\xi^2}}$. In the same way, the quality factor Q is defined as $Q = \frac{|H(j\omega_0)|}{|H(0)|}$: a simple calculation allows to determine that

$Q = \dfrac{1}{2\xi}$. It can be noticed that the Q factor can **always** be defined (whatever the value of ξ) whereas the Q_S factor can only be defined when $\xi < \dfrac{\sqrt{2}}{2}$, i.e. when there is resonance.

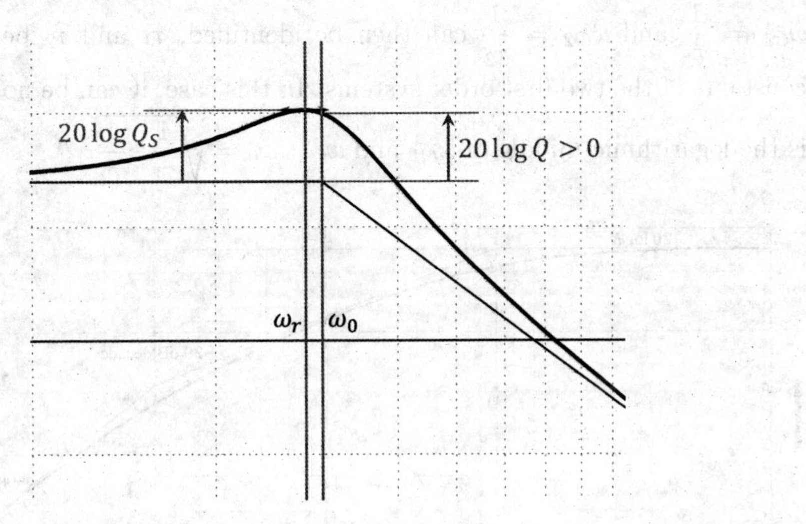

Figure 2.81 Zoom on the circled part in Figure 2.80

Bode plots if there is no resonance $\left(\xi \geqslant \frac{\sqrt{2}}{2}\right)$

When $\xi > 1$, the second order system is equivalent to the serial association of two first order systems, as seen in section 2.8.2. Two particular frequencies $\omega_{C1} = \frac{1}{\tau_1}$ and $\omega_{C2} = \frac{1}{\tau_2}$ can then be identified, τ_1 and τ_2 being the time constants of the two first order systems. In this case, it can be noticed that ω_0 is the logarithmic middle of ω_{C1} and ω_{C2}: $\omega_0 = \sqrt{\frac{1}{\tau_1} \cdot \frac{1}{\tau_2}}$.

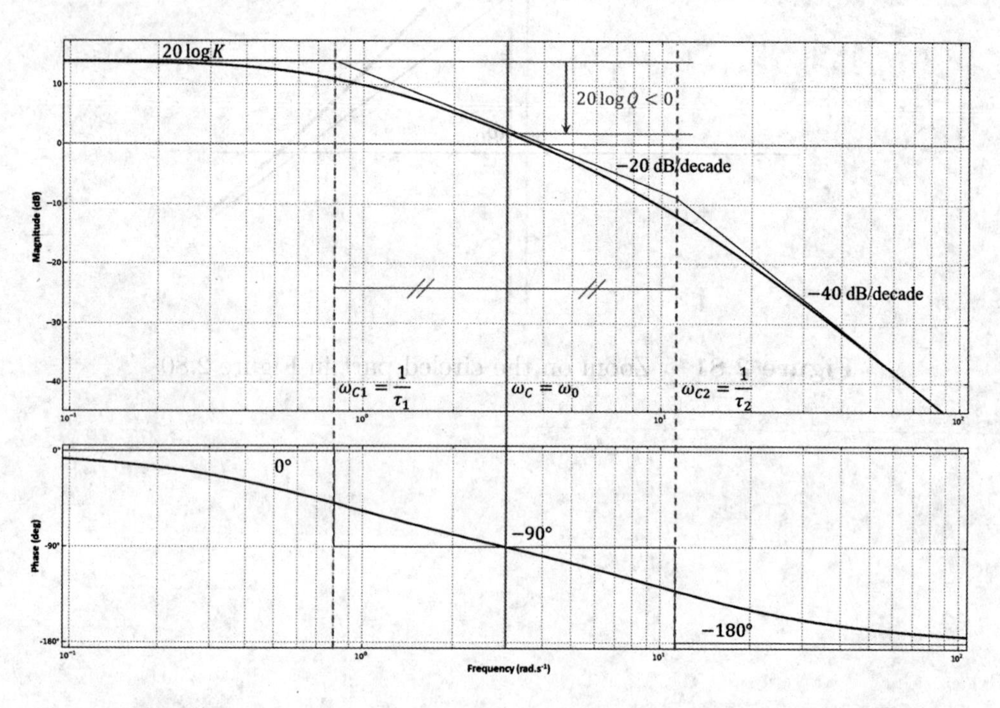

Figure 2.82 Bode plots for a second order system if $\xi \geqslant \frac{\sqrt{2}}{2}$

<u>Nichols plot if there is resonance</u> $\left(\xi < \dfrac{\sqrt{2}}{2}\right)$

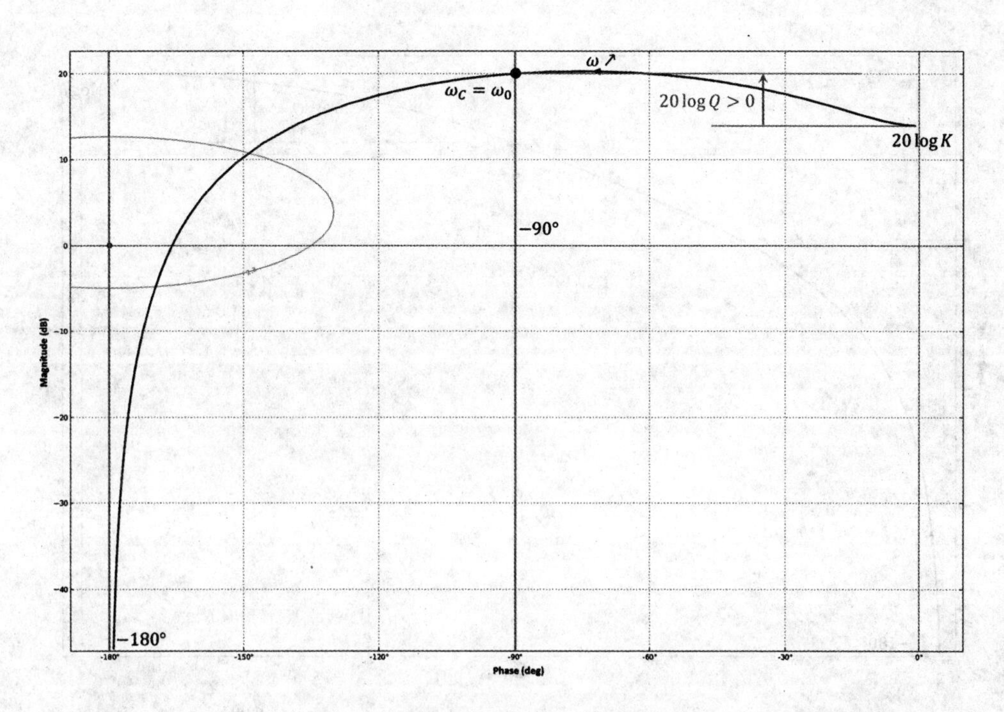

Figure 2.83 Nichols plot for a second order system if $\xi < \dfrac{\sqrt{2}}{2}$

Nichols plot if there is no resonance $\left(\xi \geqslant \dfrac{\sqrt{2}}{2}\right)$

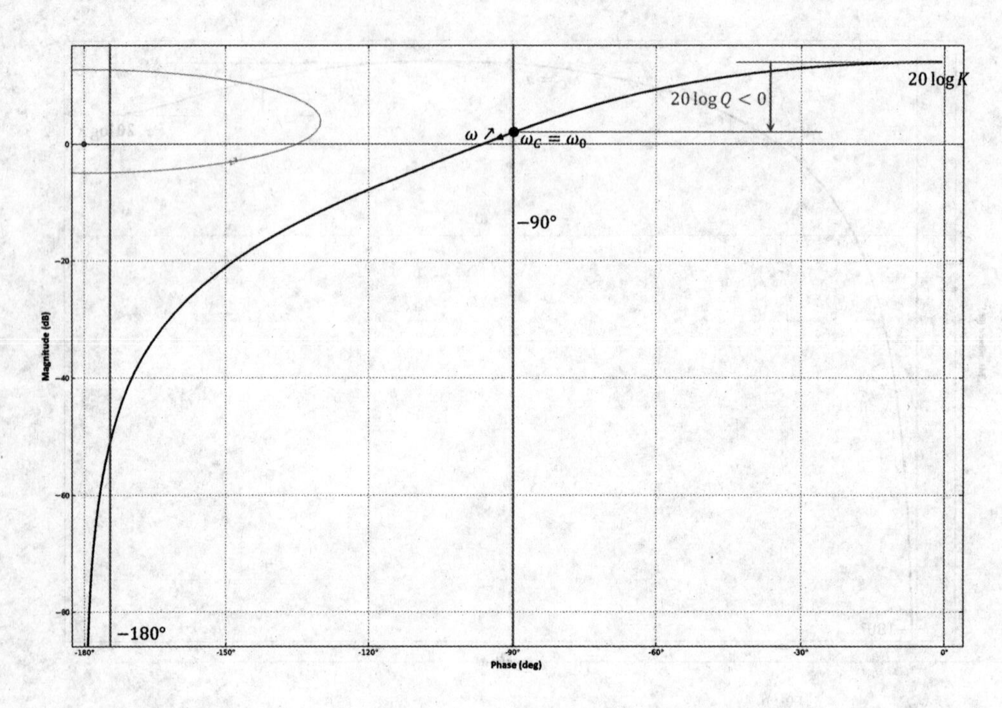

Figure 2.84 Nichols plot for a second order system if $\xi \geqslant \dfrac{\sqrt{2}}{2}$

Nyquist plot if there is resonance $\left(\xi < \frac{\sqrt{2}}{2}\right)$

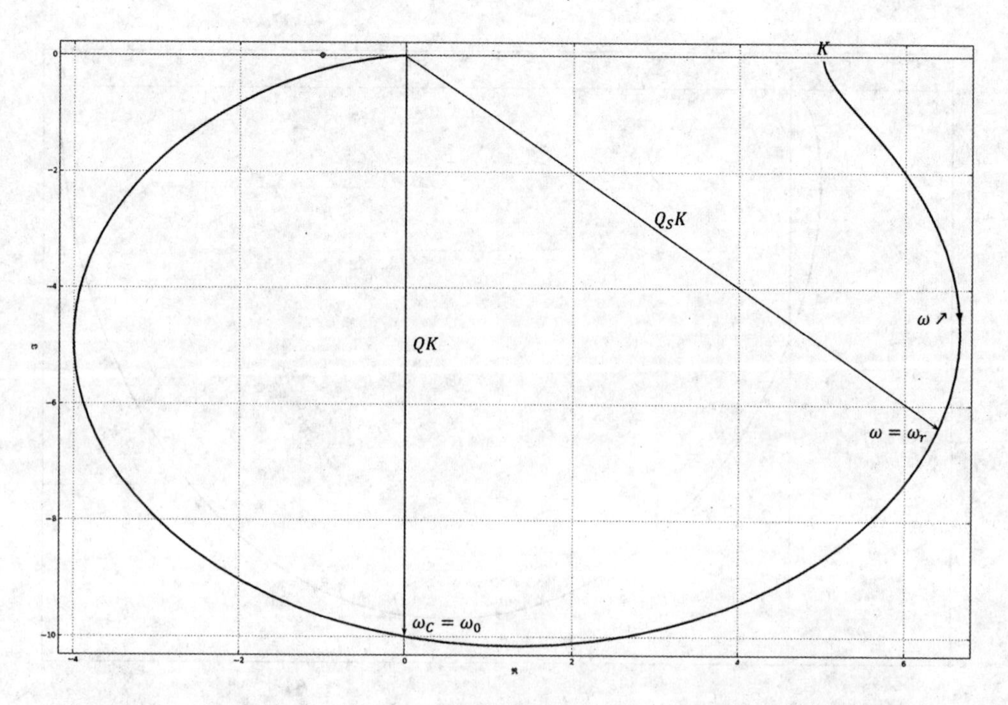

Figure 2.85 Nyquist plot for a second order system if $\xi < \frac{\sqrt{2}}{2}$

Nyquist plot if there is no resonance $\left(\xi \geqslant \frac{\sqrt{2}}{2}\right)$

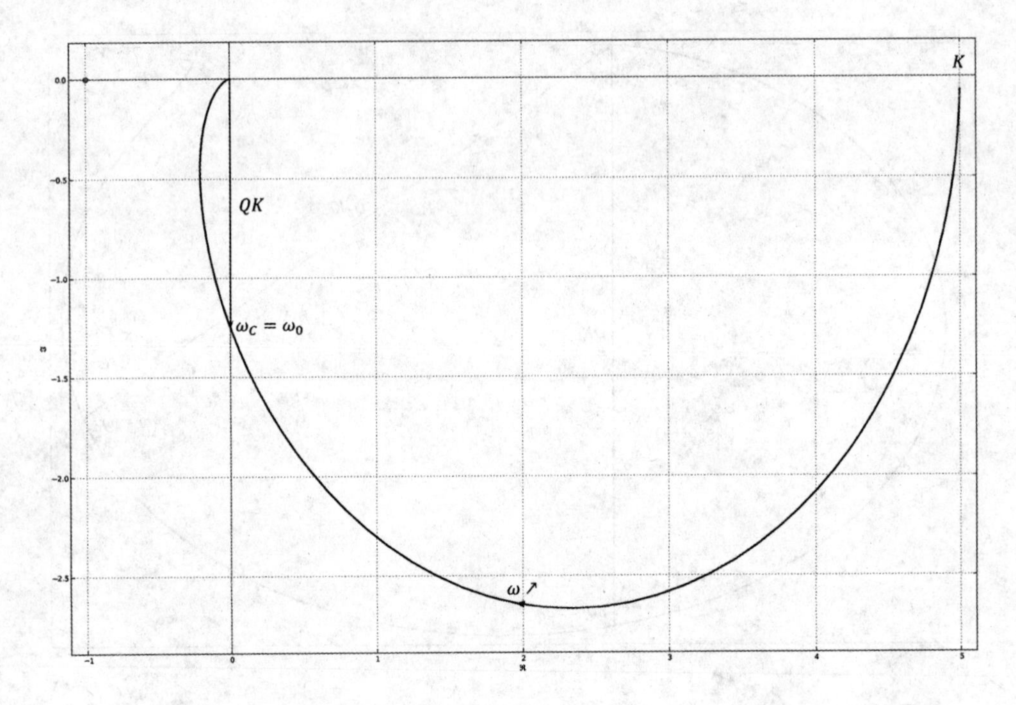

Figure 2.86 Nyquist plot for a second order system if $\xi \geqslant \frac{\sqrt{2}}{2}$

Pure Delay

$H(s) = e^{-\tau s} \Longrightarrow H(j\omega) = e^{-j\tau\omega}.$

We can hence determine that:

$$\begin{cases} A(\omega) = |H(j\omega)| = 1 \\ \varphi(\omega) = \arg[H(j\omega)] = -\tau\omega \end{cases} \Rightarrow \begin{cases} A_{\text{dB}}(\omega) = 20 \log A(\omega) = 0 \\ \varphi(\omega) = -\tau\omega \end{cases}$$

A pure delay hence does not change the value of the magnitude $(\forall\omega, o_0(\omega) = i_0)$. However, the phase is decreased proportionally to ω with a slope $-\tau$, as illustrated in Figure 2.87.

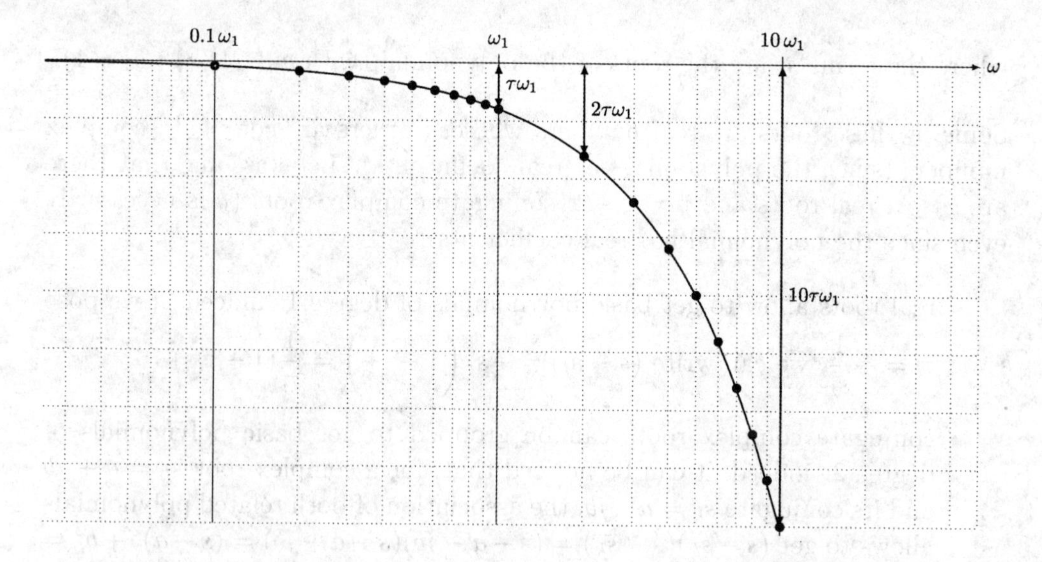

Figure 2.87 influence of a pure delay on the phase

2.9.4 Other Systems

General Form of a Transfer Function

The previous studies would be incomplete without the consideration of some delays, which can always be observed in real systems (such as the ones in the industrial science laboratory of ECPk): most of the time, these delays are due to the treatment of the information (Analog-Digital and Digital-Analog conversions) and to the transfer in the wires. As it will be seen in the second volume, these delays have a very bad influence on the stability of control systems, which will be analyzed thanks to a frequency plot: it is thus necessary to analyze the frequency influence, which is proposed in the following.

The general form of a transfer function presented in section 2.9.1 becomes:

$$H(s) = \frac{O(s)}{I(s)} = \frac{Ke^{-\tau s}}{s^\alpha} \frac{1 + b_1 s + b_2 s^2 + \cdots + b_m s^m}{1 + a_1 s + a_2 s^2 + \cdots + a_{n-\alpha} s^{n-\alpha}}$$

According to the fundamental theorem of algebra, any polynomial of degree n with real coefficients can be decomposed on the field of complex numbers $\mathbb{C}$ under the following form:

$$1 + a_1 s + a_2 s^2 + \cdots + a_n s^n = a_n (s - s_1)(s - s_2) \ldots (s - s_n)$$

where the terms s_i are the roots of the polynomial (with $a_n(-1)^n \prod_{i=1}^{n} s_i = 1$).

Some of these roots are real numbers whereas others are conjugate complex numbers (since the polynomial has real coefficients). Let's assume that there are $x \leqslant n$ real roots and $y = n - x$ conjugate complex roots (y is necessarily even since the polynomial has real coefficients):

- real roots allow to get basic polynomials of degree 1: indeed, if we pose $\tau_i = -\dfrac{1}{s_i}$, we can write $(s - s_i) = -s_i\left(1 - \dfrac{s}{s_i}\right) = \dfrac{1}{\tau_i}(1 + \tau_i s)$.
- conjugate complex roots can be grouped to get basic polynomials of degree 2: indeed, it can be noticed that, for a complex root $s_k = a + jb$ and its conjugate $\overline{s_k} = a - jb$, the association of both related polynomials allows to get $(s - s_k)(s - \overline{s_k}) = (s - a - jb)(s - a + jb) = (s - a)^2 + b^2 = s^2 + 2\xi_k \omega_{0k} s + {\omega_{0k}}^2$ by posing ${\omega_{0k}}^2 = a^2 + b^2$ and $\xi_k = -\dfrac{a}{\sqrt{a^2 + b^2}}$.

We can hence conclude that any polynomial of degree n with real coefficients can be decomposed into a product of x polynomials of degree 1 and $\dfrac{n-x}{2}$ polynomials of degree 2. The previous polynomial can hence be expressed under the following form:

$$\underbrace{(1 + \tau_1 s)(1 + \tau_2 s) \ldots (1 + \tau_x s)}_{x \text{ polynomials of degree } 1}$$
$$\times \underbrace{\left(1 + 2\xi_a \frac{s}{\omega_{0a}} + \frac{s^2}{{\omega_{0a}}^2}\right) \ldots \left(1 + 2\xi_n \frac{s}{\omega_{0_n}} + \frac{s^2}{{\omega_{0_n}}^2}\right)}_{\frac{n-x}{2} \text{ polynomials of degree } 2}$$

The general form of a transfer function can hence be written under the following form:

$$H(s) = \frac{O(s)}{I(s)} = \frac{Ke^{-\tau s}}{s^\alpha} \prod_{k=1}^{a} (1 + \tau_k s)^{\alpha_k} \prod_{l=1}^{b} \left(1 + 2\xi_l \frac{s}{\omega_{0l}} + \frac{s^2}{{\omega_{0l}}^2}\right)^{\beta_l}$$

The different terms of this expression are as follows:

- α is the **class** of the system, with $\alpha \in \{0, 1, 2\}$.
- K is the **gain** of the system.

- τ is the **delay** of the system.

- $\prod_{k=1}^{a}(1+\tau_k s)^{\alpha_k}$ is a product of first order systems: the time constant τ_k is positive and $\alpha_k \in \mathbb{Z}$.

- $\prod_{l=1}^{b}\left(1+2\xi_l\frac{s}{\omega_{0l}}+\frac{s^2}{{\omega_{0l}}^2}\right)^{\beta_l}$ is a product of second order systems: the natural frequency ω_{0l} is positive and so is the damping ratio ξ_l (ξ_l must be strictly lower than 1; otherwise, the corresponding second order system is not really a second order system but corresponds to the serial association of two first order systems), and $\beta_l \in \mathbb{Z}$.

General Methodology and Consequences

Let's consider a transfer function under the general form

$$H(s) = \frac{O(s)}{I(s)} = \frac{Ke^{-\tau s}}{s^{\alpha}}\prod_{k=1}^{a}(1+\tau_k s)^{\alpha_k}\prod_{l=1}^{b}\left(1+2\xi_l\frac{s}{\omega_{0l}}+\frac{s^2}{{\omega_{0l}}^2}\right)^{\beta_l}$$

We can hence determine that

$$H(j\omega) = \frac{Ke^{-j\tau\omega}}{j^{\alpha}\omega^{\alpha}}\prod_{k=1}^{a}(1+j\tau_k\omega)^{\alpha_k}\prod_{l=1}^{b}\left(\left(1-\frac{\omega^2}{{\omega_{0l}}^2}\right)+2j\xi_l\frac{\omega}{\omega_{0l}}\right)^{\beta_l}$$

and we get the following expressions:

- for the magnitude:

$$|H(j\omega)| = \frac{K}{\omega^{\alpha}}\prod_{k=1}^{a}(1+{\tau_k}^2\omega^2)^{\frac{\alpha_k}{2}}\prod_{l=1}^{b}\left(\left(1-\frac{\omega^2}{{\omega_{0l}}^2}\right)^2+4{\xi_l}^2\frac{\omega^2}{{\omega_{0l}}^2}\right)^{\frac{\beta_l}{2}}$$

$$A_{\mathrm{dB}}(\omega) = 20\log K - 20\,\alpha\log\omega + 10\sum_{k=1}^{a}\alpha_k\log(1+{\tau_k}^2\omega^2)$$
$$+10\sum_{l=1}^{b}\beta_l\log\left(\left(1-\frac{\omega^2}{{\omega_{0l}}^2}\right)^2+4{\xi_l}^2\frac{\omega^2}{{\omega_{0l}}^2}\right)$$

- for the phase:

$$\varphi(\omega) = -\tau\omega - \alpha\frac{\pi}{2} + \sum_{k=1}^{a} \alpha_k \arctan(\tau_k\omega) + \sum_{l=1}^{b} \beta_l \left(-\frac{\pi}{2} + \arctan \frac{1 - \dfrac{\omega^2}{{\omega_0}_l{}^2}}{2\xi_l \dfrac{\omega}{{\omega_0}_l}} \right)$$

We first need to see how the system reacts at low frequencies, i.e. when $\omega = 2\pi f \to 0$ (on the left of the Bode plots and at the beginning of the Nichols and Nyquist plots): since $s = j\omega$, it is equivalent to seeing how $H(s)$ reacts when $s \to 0$. $H(s) = \dfrac{O(s)}{I(s)} \underset{s\to 0}{\sim} \dfrac{K}{s^\alpha}$, so:

- the magnitude is equivalent to $A_{\mathrm{dB}}(\omega) \underset{s\to 0}{\sim} 20 \log K - 20\,\alpha \log\omega$, which implies that the 1st asymptot has a slope of $-20\,\alpha$ dB per decade; besides, the 1st asymptot goes through the point $(\omega = 1, A_{\mathrm{dB}} = 20 \log K)$.
- the phase is equivalent to $\varphi(\omega) \underset{s\to 0}{\sim} -\alpha\dfrac{\pi}{2}$: the phase plot begins with the value $-\alpha\dfrac{\pi}{2}$ rad.

Then:

- **First order** systems $(1 + \tau_k s)^{\alpha_k}$ have a noticeable influence for frequencies greater than $\omega_k = \dfrac{1}{\tau_k}$ and:
  - Magnitude: they add $20\,\alpha_k$ dB per decade to the previous asymptotic slope;
  - Phase: they add $\alpha_k\dfrac{\pi}{2}$ rad to the previous value of the asymptotic phase.
- **Second order** systems $\left(1 + 2\xi_l\dfrac{s}{{\omega_0}_l} + \dfrac{s^2}{{\omega_0}_l{}^2}\right)^{\beta_l}$ have a noticeable influence for frequencies greater than $\omega_l = {\omega_0}_l$ and:
  - Magnitude: they add $40\,\beta_l$ dB per decade to the previous asymptotic slope;
  - Phase: they add $\beta_l\pi$ rad to the previous value of the asymptotic phase.
- The **delay** $e^{-\tau s}$ has no influence on the magnitude, but it shifts the phase by $\tau\omega$ rad for each frequency ω.

2.10 Conclusions and Perspectives

The study of linear continuous-time time-invariant systems is a very important component of the training of a future engineer as it allows to study the controlled evolutions mainly used in industrial systems. The second volume will present the structure of such control systems and the different methods used to determine the tuning of the parameters of the controller, in order to reach all or part of the desired performances in terms of stability, damping, rapidity and precision/robustness.

As seen in this chapter, the linearization of the evolution of a SISO (Single Input, Single Output) continuous-time time-invariant system provides a linear differential equation. The transformation of this equation to the Laplace symbolic domain allows to obtain a transfer function of a given order in the mathematical form of a polynomial rational fraction. This polynomial structure can be considered as a set of first and second order systems: the wide study of these two basic systems is thus important in order to evaluate the behavior of more complex systems.

Even if most systems are neither linear nor time-invariant, the study proposed in this first volume and completed in the second volume is very important and is the foundation of a more general study of a control system. Indeed, the setting and validation of a model of a system are always done in a linearized way, even in the most complex cases. The study of the influence of non-linearities or model changes is then mainly done with suitable softwares starting from this linearized model which can then be modified, optimized, etc. to lower the difference between the measures and the simulation of the model.

Chapter 3

Kinematics of Systems of Solid Bodies

Definition 17 (Kinematics)

Kinematics is the part of mechanics dedicated to the description of the relative motion of rigid solid bodies without taking into account the causes of this motion.

The term "kinematics" comes from the ancient Greek word $\kappa\iota\nu\eta\mu\alpha$, *kinema*, which means "motion".

In this chapter, we are going to study the motion of rigid solid bodies, and in particular the relative motion of rigid solid bodies.

3.1 Mathematical Tools

3.1.1 Calculation Figures

An angle can be defined between two vectors: a plane construction can hence be drawn in the vector plane which is defined by the two vectors.

It is thus possible to define an angle between two bases which have one vector in common, the common vector being orthogonal to the plane which contains the angle to define.

For instance, let's consider the case of a 2-D problem, and let's suppose that two bases $(\overrightarrow{x_1}, \overrightarrow{y_1}, \overrightarrow{z_1})$ and $(\overrightarrow{x_2}, \overrightarrow{y_2}, \overrightarrow{z_2})$ have a common vector $\overrightarrow{z_1} = \overrightarrow{z_2}$. A plane figure, called **calculation figure**, allows to define the angle $\alpha = (\overrightarrow{x_1}, \overrightarrow{x_2}) = (\overrightarrow{y_1}, \overrightarrow{y_2})$. The calculation figure contains the vectors $\overrightarrow{x_1}, \overrightarrow{x_2}, \overrightarrow{y_1}$ and $\overrightarrow{y_2}$, and the vector $\overrightarrow{z_1} = \overrightarrow{z_2}$ is orthogonal to the calculation figure. The calculation figure which allows to define the angle α is presented in Figure 3.1. The normal vector always points towards the reader.

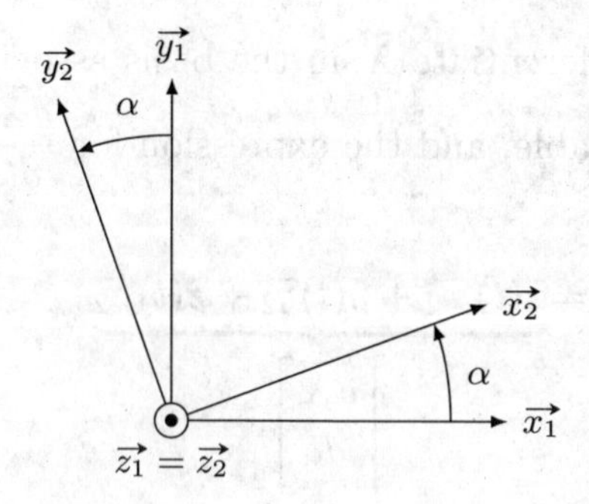

Figure 3.1 calculation figure allowing to define the angle α

3.1.2 Vector Differentiation

As many rigid solid bodies (and hence many frames) will be involved in any kinematics problem, a vector $\overrightarrow{X}$ can have many expressions depending on the basis in which it is expressed. However, when differentiating such a vector $\overrightarrow{X}$, it is important to take into account both the basis in which it is expressed and the basis in which it is differentiated.

For instance, let's consider the two frames F_1 and F_2 whose related bases are presented in Figure 3.1 and have one vector in common. Let's assume that the vector $\overrightarrow{X}$ is defined as:

$$\overrightarrow{X} = x(t)\,\overrightarrow{x_2} + y(t)\,\overrightarrow{y_2} + z(t)\,\overrightarrow{z_2}$$

If we want to differentiate $\overrightarrow{X}$, we need to know which of these elements are constant, and which are variable. However, the vectors $\overrightarrow{x_2}$, $\overrightarrow{y_2}$ and $\overrightarrow{z_2}$ are constant in F_2, but they are mobile relatively to F_1 (and hence variable), so two cases are possible:

- if we differentiate $\overrightarrow{X}$ in the basis associated with F_2, then $\overrightarrow{x_2}$, $\overrightarrow{y_2}$ and $\overrightarrow{z_2}$ are constant, so the expression for $\frac{d\overrightarrow{X}}{dt}$ can be determined as:

$$\left[\frac{d\overrightarrow{X}}{dt}\right]_2 = \dot{x}(t)\,\overrightarrow{x_2} + \dot{y}(t)\,\overrightarrow{y_2} + \dot{z}(t)\,\overrightarrow{z_2} \tag{3.1}$$

- however, if we differentiate $\overrightarrow{X}$ in the basis associated with F_1, then $\overrightarrow{x_2}$, $\overrightarrow{y_2}$ and $\overrightarrow{z_2}$ are variable, and the expression for $\frac{d\overrightarrow{X}}{dt}$ becomes:

$$\left[\frac{d\overrightarrow{X}}{dt}\right]_1 = \underbrace{\dot{x}(t)\,\overrightarrow{x_2} + \dot{y}(t)\overrightarrow{y_2} + \dot{z}(t)\,\overrightarrow{z_2}}_{\left[\frac{d\overrightarrow{X}}{dt}\right]_2} + x(t)\left[\frac{d\overrightarrow{x_2}}{dt}\right]_1 + y(t)\left[\frac{d\overrightarrow{y_2}}{dt}\right]_1 + z(t)\left[\frac{d\overrightarrow{z_2}}{dt}\right]_1 \tag{3.2}$$

The expressions (3.1) and (3.2) are obviously different, so the basis in which a vector is differentiated must necessarily be specified, and the notation $\frac{d\overrightarrow{X}}{dt}$ is **mathematically false** as it makes absolutely no sense.

Definition 18 (Differentiation of a Vector in a Basis)

The time derivative of a vector $\overrightarrow{X}$ makes sense only in a given basis B. It is noted $\left[\frac{d\overrightarrow{X}}{dt}\right]_B$.

In the remainder of this chapter, when a vector $\overrightarrow{X}$ will be differentiated in a basis B_i associated with a frame F_i, this derivative will be noted $\left[\frac{d\overrightarrow{X}}{dt}\right]_i$ instead of $\left[\frac{d\overrightarrow{X}}{dt}\right]_{B_i}$, for the sake of clarity and for its shorter notation.

The differentiation of a vector in a basis has the same properties as the differentiation of a scalar regarding:

- the differentiation of a sum:

$$\forall \vec{U}, \vec{V}, \left[\frac{d(\vec{U}+\vec{V})}{dt}\right]_B = \left[\frac{d\vec{U}}{dt}\right]_B + \left[\frac{d\vec{V}}{dt}\right]_B$$

- the differentiation of a product; as we are dealing with vectors, 3 types of product are possible:
  - a product between a scalar and a vector:

$$\forall \lambda \in \mathbb{R}, \forall \vec{U}, \left[\frac{d(\lambda\vec{U})}{dt}\right]_B = \dot{\lambda}\vec{U} + \lambda\left[\frac{d\vec{U}}{dt}\right]_B$$

  - a scalar product between two vectors:

$$\forall \vec{U}, \vec{V}, \frac{d(\vec{U}\cdot\vec{V})}{dt} = \left[\frac{d\vec{U}}{dt}\right]_B \cdot \vec{V} + \vec{U} \cdot \left[\frac{d\vec{V}}{dt}\right]_B$$

  for **any** basis B of space.
  - a vector product between two vectors:

$$\forall \vec{U}, \vec{V}, \left[\frac{d(\vec{U}\times\vec{V})}{dt}\right]_B = \left[\frac{d\vec{U}}{dt}\right]_B \times \vec{V} + \vec{U} \times \left[\frac{d\vec{V}}{dt}\right]_B$$

Let's consider the expressions (3.1) and (3.2):

$$\begin{cases} \left[\frac{d\vec{X}}{dt}\right]_2 = \dot{x}(t)\,\vec{x_2} + \dot{y}(t)\,\vec{y_2} + \dot{z}(t)\,\vec{z_2} \\ \left[\frac{d\vec{X}}{dt}\right]_1 = \dot{x}(t)\,\vec{x_2} + \dot{y}(t)\,\vec{y_2} + \dot{z}(t)\,\vec{z_2} \\ \qquad + x(t)\left[\frac{d\vec{x_2}}{dt}\right]_1 + y(t)\left[\frac{d\vec{y_2}}{dt}\right]_1 + z(t)\left[\frac{d\vec{z_2}}{dt}\right]_1 \end{cases}$$

We hence have:

$$\left[\frac{d\vec{X}}{dt}\right]_1 = \left[\frac{d\vec{X}}{dt}\right]_2 + x(t)\left[\frac{d\vec{x_2}}{dt}\right]_1 + y(t)\left[\frac{d\vec{y_2}}{dt}\right]_1 + z(t)\left[\frac{d\vec{z_2}}{dt}\right]_1$$

The vectors of the basis $(\overrightarrow{x_2}, \overrightarrow{y_2}, \overrightarrow{z_2})$ are unit vectors, so we can write:

$$\overrightarrow{x_2} \cdot \overrightarrow{x_2} = \|\overrightarrow{x_2}\|^2 = 1 \Rightarrow \frac{d}{dt}(\overrightarrow{x_2} \cdot \overrightarrow{x_2}) = \overrightarrow{x_2} \cdot \left[\frac{d\overrightarrow{x_2}}{dt}\right]_B + \left[\frac{d\overrightarrow{x_2}}{dt}\right]_B \cdot \overrightarrow{x_2} = 0$$

for any basis B of space. As the scalar product is commutative, $\overrightarrow{x_2} \cdot \left[\frac{d\overrightarrow{x_2}}{dt}\right]_B = \left[\frac{d\overrightarrow{x_2}}{dt}\right]_B \cdot \overrightarrow{x_2}$ and we hence have $\overrightarrow{x_2} \cdot \left[\frac{d\overrightarrow{x_2}}{dt}\right]_B = 0$, which implies that the vector $\left[\frac{d\overrightarrow{x_2}}{dt}\right]_B$ is orthogonal to the vector $\overrightarrow{x_2}$ and hence belongs to the vector plane whose normal vector is the vector $\overrightarrow{x_2}$. As a consequence, there exists two real scalars a_{12} and a_{13} such that $\left[\frac{d\overrightarrow{x_2}}{dt}\right]_B = a_{12}\overrightarrow{y_2} + a_{13}\overrightarrow{z_2}$. In the same way, there exists two real scalars a_{21} and a_{23} such that $\left[\frac{d\overrightarrow{y_2}}{dt}\right]_B = a_{21}\overrightarrow{x_2} + a_{23}\overrightarrow{z_2}$, and there exists two real scalars a_{31} and a_{32} such that $\left[\frac{d\overrightarrow{z_2}}{dt}\right]_B = a_{31}\overrightarrow{x_2} + a_{32}\overrightarrow{y_2}$.
Besides, the vectors of the basis $(\overrightarrow{x_2}, \overrightarrow{y_2}, \overrightarrow{z_2})$ are orthogonal. If we consider the two vectors $\overrightarrow{x_2}$ and $\overrightarrow{y_2}$, we have $\overrightarrow{x_2} \cdot \overrightarrow{y_2} = 0$. If we differentiate this expression in the basis associated with the rigid solid body (1), we get:

$$\frac{d}{dt}(\overrightarrow{x_2} \cdot \overrightarrow{y_2}) = \underbrace{\overrightarrow{x_2} \cdot \left[\frac{d\overrightarrow{y_2}}{dt}\right]_1}_{a_{21}} + \underbrace{\left[\frac{d\overrightarrow{x_2}}{dt}\right]_1 \cdot \overrightarrow{y_2}}_{a_{12}} = 0 \Leftrightarrow a_{21} + a_{12} = 0$$

In the same way, the relation $\overrightarrow{x_2} \cdot \overrightarrow{z_2} = 0$ implies that $a_{13} = a_{31} = 0$, and the relation $\overrightarrow{y_2} \cdot \overrightarrow{z_2} = 0$ implies that $a_{23} = a_{32} = 0$. We hence have:

$$\begin{cases} \left[\frac{d\overrightarrow{x_2}}{dt}\right]_1 = a_{12}\overrightarrow{y_2} + a_{13}\overrightarrow{z_2} = a_{12}\overrightarrow{y_2} - a_{31}\overrightarrow{z_2} \\ \left[\frac{d\overrightarrow{y_2}}{dt}\right]_1 = a_{21}\overrightarrow{x_2} + a_{23}\overrightarrow{z_2} = -a_{12}\overrightarrow{x_2} + a_{23}\overrightarrow{z_2} \\ \left[\frac{d\overrightarrow{z_2}}{dt}\right]_1 = a_{31}\overrightarrow{x_2} + a_{32}\overrightarrow{y_2} = a_{31}\overrightarrow{x_2} - a_{23}\overrightarrow{y_2} \end{cases}$$

This system corresponds to an antisymmetric linear transformation, and there exists a vector $\vec{\Omega}(2/1)$ such that $\begin{cases} \left[\frac{d\vec{x_2}}{dt}\right]_1 = \vec{\Omega}(2/1) \times \vec{x_2} \\ \left[\frac{d\vec{y_2}}{dt}\right]_1 = \vec{\Omega}(2/1) \times \vec{y_2} \\ \left[\frac{d\vec{z_2}}{dt}\right]_1 = \vec{\Omega}(2/1) \times \vec{z_2} \end{cases}$. If we note $\vec{\Omega}(2/1) = \alpha\vec{x_2} + \beta\vec{y_2} + \gamma\vec{z_2}$, we get:

$$\begin{cases} \left[\frac{d\vec{x_2}}{dt}\right]_1 = \vec{\Omega}(2/1)\times\vec{x_2} = (\alpha\vec{x_2} + \beta\vec{y_2} + \gamma\vec{z_2})\times\vec{x_2} = \gamma\vec{y_2} - \beta\vec{z_2} = a_{12}\vec{y_2} - a_{31}\vec{z_2} \\ \left[\frac{d\vec{y_2}}{dt}\right]_1 = \vec{\Omega}(2/1)\times\vec{y_2} = (\alpha\vec{x_2} + \beta\vec{y_2} + \gamma\vec{z_2})\times\vec{y_2} = -\gamma\vec{x_2} + \alpha\vec{z_2} = -a_{12}\vec{x_2} + a_{23}\vec{z_2} \\ \left[\frac{d\vec{z_2}}{dt}\right]_1 = \vec{\Omega}(2/1)\times\vec{z_2} = (\alpha\vec{x_2} + \beta\vec{y_2} + \gamma\vec{z_2})\times\vec{z_2} = \beta\vec{x_2} - \alpha\vec{y_2} = a_{31}\vec{x_2} - a_{23}\vec{y_2} \end{cases}$$

and we hence have $\vec{\Omega}(2/1) = a_{23}\vec{x_2} + a_{31}\vec{y_2} + a_{12}\vec{z_2}$. As a consequence:

$$\begin{aligned} \left[\frac{d\vec{X}}{dt}\right]_1 &= \left[\frac{d\vec{X}}{dt}\right]_2 + x(t)\underbrace{\left[\frac{d\vec{x_2}}{dt}\right]_1}_{\vec{\Omega}(2/1)\times\vec{x_2}} + y(t)\underbrace{\left[\frac{d\vec{y_2}}{dt}\right]_1}_{\vec{\Omega}(2/1)\times\vec{y_2}} + z(t)\underbrace{\left[\frac{d\vec{z_2}}{dt}\right]_1}_{\vec{\Omega}(2/1)\times\vec{z_2}} \\ &= \left[\frac{d\vec{X}}{dt}\right]_2 + \vec{\Omega}(2/1) \times \underbrace{(x(t)\vec{x_2} + y(t)\vec{y_2} + z(t)\vec{z_2})}_{\vec{X}} \\ &= \left[\frac{d\vec{X}}{dt}\right]_2 + \vec{\Omega}(2/1) \times \vec{X} \end{aligned}$$

The vector $\vec{\Omega}(2/1)$ is called the **angular velocity vector** of F_2 relatively to F_1.

Definition 19 (Vector Differentiation Formula)

The **vector differentiation formula** is the formula which allows to change the basis in which a vector is differentiated. Its expression is:

$$\left[\frac{d\vec{X}}{dt}\right]_1 = \left[\frac{d\vec{X}}{dt}\right]_2 + \vec{\Omega}(2/1) \times \vec{X} \tag{3.3}$$

3.1.3 Vector Triple Product

One of the properties of the vector triple product is as follows:

$$\vec{a} \times (\vec{b} \times \vec{c}) = (\vec{a} \cdot \vec{c})\vec{b} - (\vec{a} \cdot \vec{b})\vec{c}$$

This relation is equivalent to:

$$\boxed{(\vec{a} \cdot \vec{c})\vec{b} = \vec{a} \times (\vec{b} \times \vec{c}) + (\vec{a} \cdot \vec{b})\vec{c}}$$

3.1.4 Euclidean Space

Definition 20 (Euclidean Space)

A real vector space $\mathcal{E}$ is a Euclidean space if and only if it is equipped with a symmetric bilinear form $\varphi : \mathcal{E} \times \mathcal{E} \to \mathbb{R}$ which is also positive definite, and which is called a scalar product.

3.2 Setting

3.2.1 General Hypothesis of the Study: Rigid Solid Bodies

Definition 21 (Rigid Solid Body)

A **rigid solid body** is a set of material points such that the distance between any two material points does not vary over the time.

A solid body (S) is rigid if and only if: $\boxed{\forall A, B \in S, \|\overrightarrow{AB}\| = C}$, where C is a constant scalar. This notion is illustrated in Figure 3.2.
In the remainder of this chapter, we will consider that all solid bodies are rigid.

3.2.2 Association of a Reference Frame to a Solid Body

Definition 22 (Reference Frame)

A **reference frame** allows to locate an event in both space and time.

A reference frame thus is the association of:

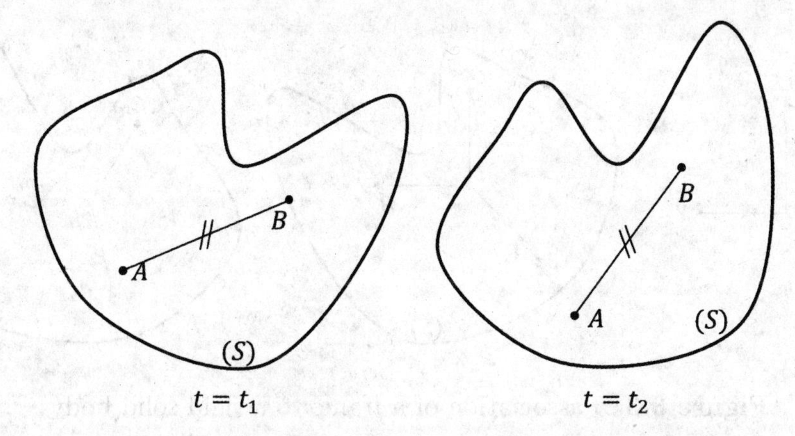

Figure 3.2 a rigid solid body

- a **space frame**: solid bodies evolve in a physical space which can be modeled by an affine 3-dimensional space that can be characterized by a direct orthonormal frame $(O; \vec{x}, \vec{y}, \vec{z})$; and

- a **time frame**: time can be considered as a set of infinitely close instants which are interrelated by means of an order relation; it can be modeled by an affine 1-dimensional space.

Time can be associated with two terms: the *date* (which characterizes a specific instant), and the *duration* (which corresponds to the time elapsed between two dates).

Let's consider a rigid solid body (S) and an orthonormal and direct basis $(\vec{x}, \vec{y}, \vec{z})$. It is possible to choose a point I of (S) to create a frame $F(I; \vec{x}, \vec{y}, \vec{z})$. At a given instant of time, it is also possible to find three points $A, B, C \in S$ such that $(\vec{x}, \vec{y}, \vec{z}) = (\overrightarrow{IA}, \overrightarrow{IB}, \overrightarrow{IC})$.
As (S) is a rigid solid body, the distance between any two of the four points I, A, B, C will remain the same over the time, and so will the relative orientation of any two of the three vectors $\overrightarrow{IA}$, $\overrightarrow{IB}$ and $\overrightarrow{IC}$. $(I; \overrightarrow{IA}, \overrightarrow{IB}, \overrightarrow{IC})$ can hence be considered as an orthonormal and direct frame which can be associated with the rigid solid body (S) and will follow its motion over the time, as illustrated in Figure 3.3. This associated frame hence is equivalent to the rigid solid body (S).
Besides, this equivalence holds at any instant of time, so **a reference frame is equivalent to a rigid solid body**.

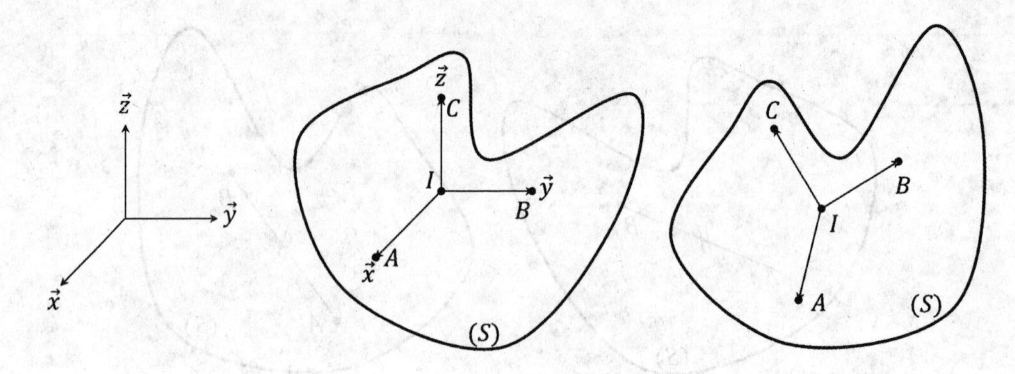

Figure 3.3 association of a frame to a rigid solid body

To study the motion of a rigid solid body relatively to another one, it will hence be sufficient to study the relative motion of the frames which are associated with these rigid solid bodies. In the remainder of this chapter, all bases (and frames) will be considered orthonormal and direct.

3.2.3 Setting of the Attitude in 3-D Problems

In the remainder of this chapter, we are going to study the motion of rigid solid bodies relatively to a frame of reference. We showed in section 3.2.2 that each rigid solid body can be assimilated to a frame (which can be the reference frame): we are thus going to study the motion of frames relatively to a frame of reference.

Let's consider a frame $F_2(O_2; \overrightarrow{x_2}, \overrightarrow{y_2}, \overrightarrow{z_2})$ which is mobile relatively to a frame of reference $F_1(O_1; \overrightarrow{x_1}, \overrightarrow{y_1}, \overrightarrow{z_1})$, as illustrated in Figure 3.4. The **attitude** corresponds to the set of the parameters which are necessary and sufficient to fully define the position and orientation of one frame relatively to another one. The attitude of the frame F_2 relatively to the frame of reference F_1 hence is defined by:

- the **position** of F_2 relatively to F_1, which is defined by the vector $\overrightarrow{O_1O_2}$ (and which corresponds to the relative position of the origins); and
- the **orientation** of F_2 relatively to F_1, which corresponds to the orientation of the basis $(\overrightarrow{x_2}, \overrightarrow{y_2}, \overrightarrow{z_2})$ relatively to the basis $(\overrightarrow{x_1}, \overrightarrow{y_1}, \overrightarrow{z_1})$.

It is necessary to determine a **setting** of this attitude, i.e. parameters which are necessary and sufficient, in order to know the motion of the frame F_2 relatively to the frame of reference F_1.

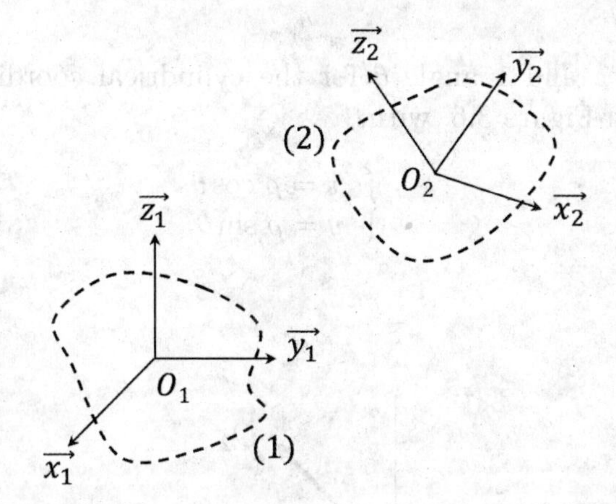

Figure 3.4 a frame F_2 which is mobile relatively to a frame of reference F_1

Setting of the Position

The expression for the vector $\overrightarrow{O_1O_2}$ can be determined in different coordinate systems (e.g. cartesian, cylindrical, or spherical). However, whatever the chosen coordinate system, 3 parameters will be needed to fully define the vector $\overrightarrow{O_1O_2}$ and only the type of these parameters will change depending on the chosen coordinate system. It can be:

- 3 lengths x, y, z for the cartesian coordinate system, as illustrated in Figure 3.5;

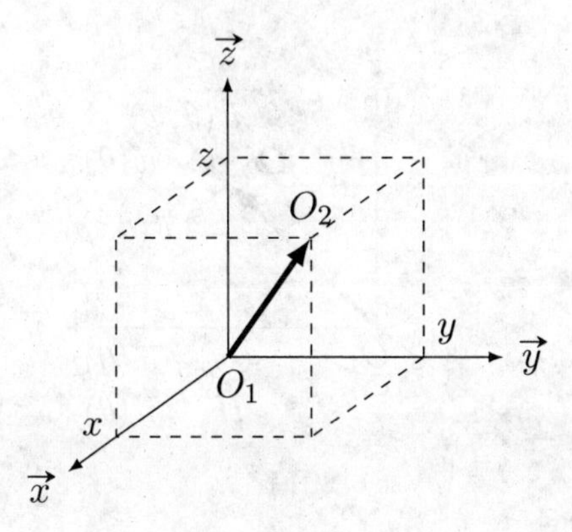

Figure 3.5 the cartesian coordinate system

- 2 lengths ρ, z and 1 angle θ for the cylindrical coordinate system, as illustrated in Figure 3.6, with:

$$\begin{cases} x = \rho\cos\theta \\ y = \rho\sin\theta \end{cases}$$

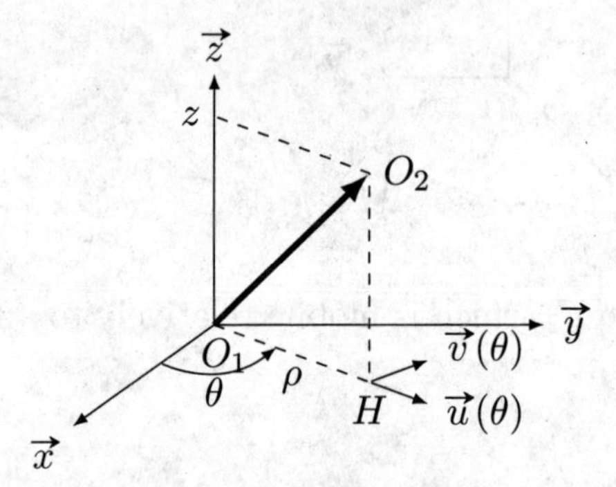

Figure 3.6 the cylindrical coordinate system

- 1 length r and 2 angles θ, φ for the spherical coordinate system, as illustrated in Figure 3.7, with:

$$\begin{cases} x = r\,\sin\theta\cos\varphi \\ y = r\,\sin\theta\sin\varphi \\ z = r\,\cos\theta \end{cases}$$

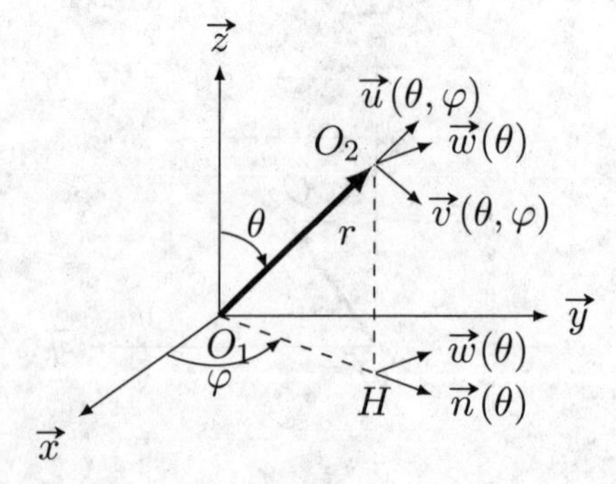

Figure 3.7 the spherical coordinate system

Setting of the Orientation

In the most general case, the bases associated with the two frames F_2 and F_1 do not have any vector in common, and the orientation of $(\overrightarrow{x_2}, \overrightarrow{y_2}, \overrightarrow{z_2})$ relatively to $(\overrightarrow{x_1}, \overrightarrow{y_1}, \overrightarrow{z_1})$ is illustrated in Figure 3.8.

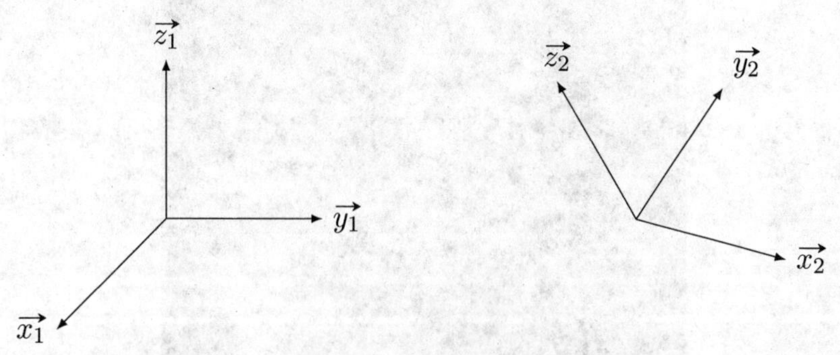

Figure 3.8 problematic of the orientation of $(\overrightarrow{x_2}, \overrightarrow{y_2}, \overrightarrow{z_2})$ relatively to $(\overrightarrow{x_1}, \overrightarrow{y_1}, \overrightarrow{z_1})$

3 angles are needed to determine a setting of the relative orientation of both bases. Some commonly used angles are the **Euler angles** which were introduced by the Swiss mathematician and physician Leonhard Paul Euler (Figure 3.9).

These 3 angles allow the setting of the orientation of a vector basis relatively to another one by means of three successive rotations, as illustrated in Figure 3.10:

- a first rotation, called **precession** (the word *precession* comes from the verb *precede*), around one of the three vectors of one basis;

- a third rotation, called **intrinsic rotation** (or **spin**), around one of the three vectors of the other basis; and

- a second rotation, called **nutation** (the word *nutation* comes from the Latin word *nutatio*, "action of nodding"), around one of the two vectors which are orthogonal to the two previously chosen vectors, this vector being called **nodal vector** (this name comes from the **line of nodes**, which is the name given to the intersection line between the two planes in which the precession and the spin occur).

Figure 3.9 Leonhard Paul Euler (1707-1783)

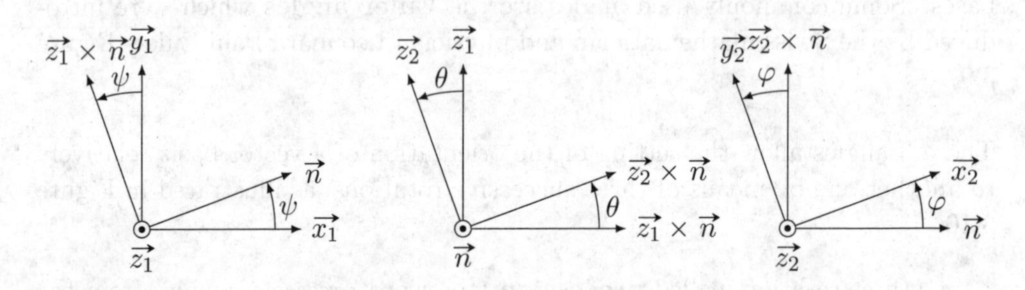

Figure 3.10 calculation figures allowing to define the Euler angles ψ, θ and φ

For instance, in the case of the bases $(\vec{x_1}, \vec{y_1}, \vec{z_1})$ and $(\vec{x_2}, \vec{y_2}, \vec{z_2})$, let's consider that the precession is around $\vec{z_1}$ and that the spin is around $\vec{z_2}$. The nodal vector $\vec{n}$ hence is orthogonal to both $\vec{z_1}$ and $\vec{z_2}$ (since the line of nodes is the intersection between the planes $(\vec{x_1}, \vec{y_1})$ and $(\vec{x_2}, \vec{y_2})$), and it is defined as:

$$\vec{n} = \frac{\vec{z_1} \times \vec{z_2}}{\|\vec{z_1} \times \vec{z_2}\|}$$

It is then possible to decompose the different rotations:

- the precession (the 1st rotation) is around $\overrightarrow{z_1}$, and the nutation (the 2nd rotation) is around $\overrightarrow{n}$, so the precession must make $\overrightarrow{x_1}$ become $\overrightarrow{n}$ (since $\overrightarrow{n}$ belongs to the vector plane $(\overrightarrow{x_1}, \overrightarrow{y_1})$ as it is orthogonal to $\overrightarrow{z_1}$), and we hence have $(\overrightarrow{x_1}, \overrightarrow{n}) = \psi$: the basis $(\overrightarrow{x_1}, \overrightarrow{y_1}, \overrightarrow{z_1})$ becomes the basis $(\overrightarrow{n}, \overrightarrow{z_1} \times \overrightarrow{n}, \overrightarrow{z_1})$ after the precession.

- the nutation (the 2nd rotation) is around $\overrightarrow{n}$, and the spin (the 3rd rotation) is around $\overrightarrow{z_2}$, so the nutation must make $\overrightarrow{z_1}$ become $\overrightarrow{z_2}$ (since $\overrightarrow{n}$ is orthogonal to both $\overrightarrow{z_1}$ and $\overrightarrow{z_2}$) and we hence have $(\overrightarrow{z_1}, \overrightarrow{z_2}) = \theta$: the basis $(\overrightarrow{n}, \overrightarrow{z_1} \times \overrightarrow{n}, \overrightarrow{z_1})$ becomes the basis $(\overrightarrow{n}, \overrightarrow{z_2} \times \overrightarrow{n}, \overrightarrow{z_2})$ after the nutation.

- the spin is the 3rd and last rotation: it is around $\overrightarrow{z_2}$ and makes the basis $(\overrightarrow{z_2}, \overrightarrow{n}, \overrightarrow{z_2} \times \overrightarrow{n})$ become the basis $(\overrightarrow{z_2}, \overrightarrow{x_2}, \overrightarrow{y_2})$, we hence have $(\overrightarrow{n}, \overrightarrow{x_2}) = \varphi$.

The different bases are presented in Figure 3.10. The three Euler angles have the same name as the rotation to which they are related: ψ is the precession angle, θ is the nutation angle, and φ is the spin angle.

Conclusion

To sum up, a maximum of 6 parameters are necessary to determine a setting of the attitude of F_2 relatively to F_1:

- 3 parameters (3 lengths, or 2 lengths and 1 angle, or 1 length and 2 angles) for the setting of their relative position; and

- 3 angles for the setting of their relative orientation.

These 6 parameters can vary over the time and thus are algebraic (they can either be positive or negative, depending on the instant).

It can be noticed that a second type of commonly used angles exist which are called **Tait-Bryan angles** (after the Scottish mathematical physicist Peter Guthrie Tait (1831-1901) and the British mathematician George Hartley Bryan (1864-1928)) or **Nautical or Cardan angles** (after the Italian mathematician and physicist Gerolamo Cardano (1501-1576)). These three angles correspond to the **yaw**, the **pitch**, and the **roll**. However, even if they are more understandable by the user, the related calculations are more complex since these three angles are defined relatively to the same frame, which implies that they are interrelated.

3.2.4 Setting of the Attitude in 2-D Problems

In the specific case of 2-D problems, the setting of the attitude is simpler. Indeed, if we consider the same frames as in section 3.2.3, the motion of the frame F_2 relatively to the frame of reference F_1 must necessarily be in a plane for the problem to be 2-dimensional, and both bases hence have one vector in common, as illustrated in Figure 3.11. Once again, both the relative position and orientation of both frames need to be determined.

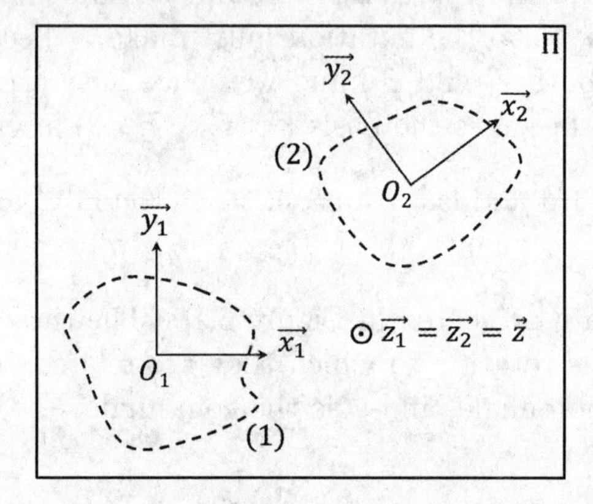

Figure 3.11 setting of the attitude in 2-D problems

Setting of the Position

The expression for the vector $\overrightarrow{O_1O_2}$ can be determined in different coordinate systems (e.g. cartesian or polar). However, whatever the chosen coordinate system, 2 parameters will be needed to fully define the vector $\overrightarrow{O_1O_2}$. Only the type of these parameters will change depending on the chosen coordinate system. It can be:

- 2 lengths for the cartesian coordinate system: $\overrightarrow{O_1O_2} = x\,\vec{x_1} + y\,\vec{y_1}$;
- 1 length and 1 angle for the polar coordinate system: $\overrightarrow{O_1O_2} = \rho\,\vec{u}$ and $(\vec{x_1}, \vec{u}) = \alpha$.

Setting of the Orientation

Since both bases have one vector in common, 1 angle is sufficient to determine a setting of the relative orientation of both bases, e.g. $\theta = (\vec{x_1}, \vec{x_2}) = (\vec{y_1}, \vec{y_2})$.

Conclusion

To sum up, a maximum of 3 parameters are necessary to determine a setting of the attitude of F_2 relatively to F_1:

- 2 parameters (2 lengths, or 1 length and 1 angle) for the setting of their relative position; and
- 1 angle for the setting of their relative orientation.

These 3 parameters can vary over the time and thus are algebraic (they are either positive or negative, depending on the instant).

3.2.5 Comment Regarding the Number of Parameters

The contact surfaces which exist between solid bodies will restrict the possible motions between them as well as the number of parameters which are necessary to define their relative attitude, as we will see in chapter 6.

3.3 General Definitions

3.3.1 Position Vector

Let's consider a point P in motion relatively to a frame of reference F.

Definition 23 (Position Vector)

A **position vector** of the point P in F is any vector $\overrightarrow{OP}$ such that the point O is immobile in the frame F (e.g. the origin of the frame F).

Definition 24 (Trajectory)

The **trajectory** of the point P is the set of all the successive positions of the mobile point P in F over the time. It is often noted $\mathcal{T}(P/F, t_1 \to t_2)$ for a trajectory between the instants $t = t_1$ and $t = t_2$.

3.3.2 Velocity Vector

The velocity vector depends on the point P considered **and** on the motion studied. If we consider a point P which belongs to a solid body (2) in motion relatively to a solid body (1), its velocity vector will be noted $\vec{V}(P, 2/1)$, or sometimes $\vec{V}(P \in 2/1)$, and it will be called the velocity vector of the point P during the motion of (2) relatively to (1).

Definition 25 (Velocity Vector)

The **velocity vector** of a point P during the motion of (2) relatively to (1) is the time derivative of a position vector of this point P in the observation basis:

$$\forall P \in 2, \vec{V}(P, 2/1) = \left[\frac{d\overrightarrow{OP}}{dt}\right]_1 \tag{3.4}$$

O being immobile in the frame associated with (1). In this formula, P has to be a point of (2), or it needs to be considered as a point of (2) during the calculation.

It can be noticed that the expression for the velocity vector of the point P will be the same, whatever the chosen position vector. For instance, let's consider a point A which is immobile in F_1. In the same way that $\overrightarrow{OP}$ is a position vector of P in F_1, $\overrightarrow{AP}$ also is a position vector of P in F_1. We can write:

$$\begin{aligned}
\left[\frac{d\overrightarrow{AP}}{dt}\right]_1 &= \left[\frac{d(\overrightarrow{AO} + \overrightarrow{OP})}{dt}\right]_1 \\
&= \underbrace{\left[\frac{d\overrightarrow{AO}}{dt}\right]_1}_{\vec{0}} + \left[\frac{d\overrightarrow{OP}}{dt}\right]_1 \\
&= \left[\frac{d\overrightarrow{OP}}{dt}\right]_1
\end{aligned}$$

The chosen position vector hence has no influence on the expression for the related velocity vector. However, most often, the initial point of the chosen position vector will be the origin of the frame.

The velocity vector of the point P during the motion of (2) relatively to (1) $\vec{V}(P, 2/1)$ is tangent to the trajectory of P in F_1 at any time, as illustrated in Figure 3.12.

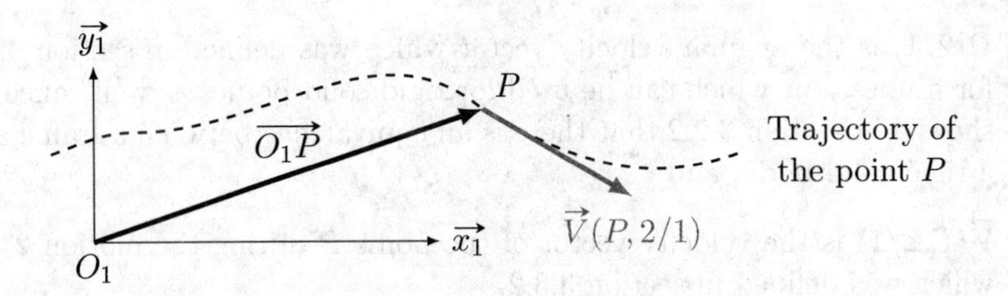

Figure 3.12 tangency between $\vec{V}(P, 2/1)$ and the trajectory of P in F_1

3.3.3 Acceleration Vector

A similar notation will be used for the acceleration vector.

Definition 26 (Acceleration Vector)

The **acceleration vector** of a point P during the motion of (2) relatively to (1) is the time derivative of the velocity vector of this point P during the same motion in the observation basis:

$$\forall P \in 2, \vec{a}(P, 2/1) = \left[\frac{d\vec{V}(P, 2/1)}{dt}\right]_1 = \left[\frac{d^2\overrightarrow{OP}}{dt^2}\right]_1 \tag{3.5}$$

O being immobile in the frame associated with (1). In this formula, P has to be a point of (2), or it needs to be considered as a point of (2) during the calculation.

3.4 The Twist (or Kinematic Screw)

3.4.1 General Form

The **twist** (also called **kinematic screw**) associated with the motion of a solid body (2) relatively to another solid body (1) can be expressed under the following form, at a point P:

$$\{\mathcal{V}(2/1)\} = \left\{ \begin{array}{c} \vec{\Omega}(2/1) \\ \vec{V}(P,2/1) \end{array} \right\}_P$$

where:

- $\vec{\Omega}(2/1)$ is the angular velocity vector which was defined in section 3.1.2 for frames, but which can be used for rigid solid bodies as well, since we showed in section 3.2.2 that there is an equivalence between a frame and a rigid solid body; and
- $\vec{V}(P,2/1)$ is the velocity vector of the point P during the motion 2 / 1 which was defined in section 3.3.2.

It can be noticed that the symbol "=" which is used between the twist $\{\mathcal{V}(2/1)\}$ and its developed form $\left\{ \begin{array}{c} \vec{\Omega}(2/1) \\ \vec{V}(P,2/1) \end{array} \right\}_P$ makes no sense. Indeed, from a mathematical point of view, a twist is a vector field whereas the developed form is a set of two vectors, so both parts are different. To be rigorous, we should use the symbols ":" (which is used in many books) or "≡" (which is more coherent from a mathematical point of view); however, it is the symbol "=" which is the most used, so it is the one which will be retained in the remainder of this book.

Another notation, called the **Plücker notation** (after the German mathematician and physicist Julius Plücker (1801-1868)), also exists. It consists in assigning 3 homogeneous coordinates to each of the 2 previously defined vectors in projective 3-space:

- $\vec{\Omega}(2/1) = p_{21}\,\vec{x} + q_{21}\,\vec{y} + r_{21}\,\vec{z}$
- $\vec{V}(P,2/1) = u_{21}^P\,\vec{x} + v_{21}^P\,\vec{y} + w_{21}^P\,\vec{z}$

and the corresponding notation for the twist then is:

$$\{\mathcal{V}(2/1)\} = \left\{ \begin{array}{cc} p_{21} & u_{21}^P \\ q_{21} & v_{21}^P \\ r_{21} & w_{21}^P \end{array} \right\}_{(P,B)}$$

Be careful: in this notation, the indication of the basis ($B = (\vec{x}, \vec{y}, \vec{z})$ here) is absolutely necessary!

The notation $\mathcal{V}$ for the twist refers to the **v**elocity vectors to which it is related.

3.4.2 Point Change Formula

Let's consider two rigid solid bodies (1) and (2) in relative motion with which two frames F_1 and F_2 are associated, and two points A and B of (2). According to the definition of the velocity vector, we have:

$$\left[\frac{d\overrightarrow{AB}}{dt}\right]_1 = \left[\frac{d(\overrightarrow{AO}+\overrightarrow{OB})}{dt}\right]_1 = \left[\frac{d\overrightarrow{OB}}{dt}\right]_1 - \left[\frac{d\overrightarrow{OA}}{dt}\right]_1 = \vec{V}(B,2/1) - \vec{V}(A,2/1)$$

O being a point of (1). Then, according to the vector differentiation formula (3.3):

$$\left[\frac{d\overrightarrow{AB}}{dt}\right]_1 = \underbrace{\left[\frac{d\overrightarrow{AB}}{dt}\right]_2}_{\vec{0}} + \vec{\Omega}(2/1) \times \overrightarrow{AB} = \vec{\Omega}(2/1) \times \overrightarrow{AB}$$

since $\left[\frac{d\overrightarrow{AB}}{dt}\right]_2 = \vec{0}$ due to the fact that both A and B belong to (2). These two relations allow to determine the point change formula for velocity vectors.

Property 1 (Point Change Formula for Velocity Vectors)

The velocity vectors of any two points of a rigid solid body respect the following formula:

$$\forall A, B \in \mathcal{E}, \vec{V}(B,2/1) = \vec{V}(A,2/1) + \vec{\Omega}(2/1) \times \overrightarrow{AB} \tag{3.6}$$

where A and B are considered as points of (2) during the calculation.

Since the velocity vector is the only vector from the twist which depends on the point considered, the point change formula also allows to change the point at which a twist is expressed. For instance, if the twist related to the motion 2 / 1 is known at the point A as

$$\{\mathcal{V}(2/1)\} = \left\{ \begin{array}{c} \vec{\Omega}(2/1) \\ \vec{V}(A,2/1) \end{array} \right\}_A$$

then the twist related to the same motion can be determined at the point B as

$$\{\mathcal{V}(2/1)\} = \left\{ \begin{array}{c} \vec{\Omega}(2/1) \\ \vec{V}(B,2/1) \end{array} \right\}_B = \left\{ \begin{array}{c} \vec{\Omega}(2/1) \\ \vec{V}(A,2/1) + \vec{\Omega}(2/1) \times \overrightarrow{AB} \end{array} \right\}_B$$

The twist related to a motion hence fully characterizes this motion. Indeed, as soon as the expression for this twist is known at a given point of a rigid solid body, then it can be determined at any other point of the rigid solid body by means of the point change formula.

3.4.3 Mathematical Properties

Scalar Invariance

Let's consider the twist related to the motion of a rigid solid body (2) relatively to a rigid solid body (1) and expressed at a point A of (2):

$$\{\mathcal{V}(2/1)\} = \left\{ \begin{array}{c} \vec{\Omega}(2/1) \\ \vec{V}(A,2/1) \end{array} \right\}_A$$

We have showed in section 3.4.2 that the twist related to the same motion can be determined at any point B of (2) as:

$$\{\mathcal{V}(2/1)\} = \left\{ \begin{array}{c} \vec{\Omega}(2/1) \\ \vec{V}(B,2/1) \end{array} \right\}_B = \left\{ \begin{array}{c} \vec{\Omega}(2/1) \\ \vec{V}(A,2/1) + \vec{\Omega}(2/1) \times \overrightarrow{AB} \end{array} \right\}_B$$

thanks to the point change formula.

Let's determine the expression for the scalar product between the two vectors of this latter twist:

$$\begin{aligned} \vec{\Omega}(2/1) \cdot \vec{V}(B,2/1) &= \vec{\Omega}(2/1) \cdot \left(\vec{V}(A,2/1) + \vec{\Omega}(2/1) \times \overrightarrow{AB} \right) \\ &= \vec{\Omega}(2/1) \cdot \vec{V}(A,2/1) + \vec{\Omega}(2/1) \cdot \left(\vec{\Omega}(2/1) \times \overrightarrow{AB} \right) \\ &= \vec{\Omega}(2/1) \cdot \vec{V}(A,2/1) + \overrightarrow{AB} \cdot \underbrace{\left(\vec{\Omega}(2/1) \times \vec{\Omega}(2/1) \right)}_{\vec{0}} \\ &= \vec{\Omega}(2/1) \cdot \vec{V}(A,2/1) \end{aligned}$$

It can be noticed that the penultimate line can be deduced from the antepenultimate one by taking into account the fact that $\vec{\Omega}(2/1) \cdot \left(\vec{\Omega}(2/1) \times \overrightarrow{AB} \right)$ is a mixed product whose value does not change when its vectors are cyclically permuted.

Property 2 (Scalar Invariance)

The scalar product between the two elements of a twist is an invariant: it does not depend on the point at which the twist is expressed:

$$\forall A, B \in \mathcal{E}, \vec{\Omega}(2/1) \cdot \vec{V}(A, 2/1) = \vec{\Omega}(2/1) \cdot \vec{V}(B, 2/1) \tag{3.7}$$

Equal Projectivity

Let's consider two points A and B of a rigid solid body (2) in motion relatively to a rigid solid body (1). According to the point change formula:

$$\vec{V}(B, 2/1) = \vec{V}(A, 2/1) + \vec{\Omega}(2/1) \times \overrightarrow{AB}$$

We hence have:

$$\begin{aligned}\vec{V}(B, 2/1) \cdot \overrightarrow{AB} &= \vec{V}(A, 2/1) \cdot \overrightarrow{AB} + \left(\vec{\Omega}(2/1) \times \overrightarrow{AB}\right) \cdot \overrightarrow{AB} \\ &= \vec{V}(A, 2/1) \cdot \overrightarrow{AB} + \underbrace{\left(\overrightarrow{AB} \times \overrightarrow{AB}\right)}_{\vec{0}} \cdot \vec{\Omega}(2/1) \\ &= \vec{V}(A, 2/1) \cdot \overrightarrow{AB}\end{aligned}$$

Property 3 (Equal Projectivity)

The velocity vectors related to the relative motion of two rigid solid bodies satisfy the following relation:

$$\forall A, B \in \mathcal{E}, \vec{V}(A, 2/1) \cdot \overrightarrow{AB} = \vec{V}(B, 2/1) \cdot \overrightarrow{AB} \tag{3.8}$$

The graphical implication of this property is presented in Figure 3.13.

Central Axis

For any twist whose angular velocity vector is not null, there exists a line from space such that any point of that line has a velocity vector which is collinear with its angular velocity vector.

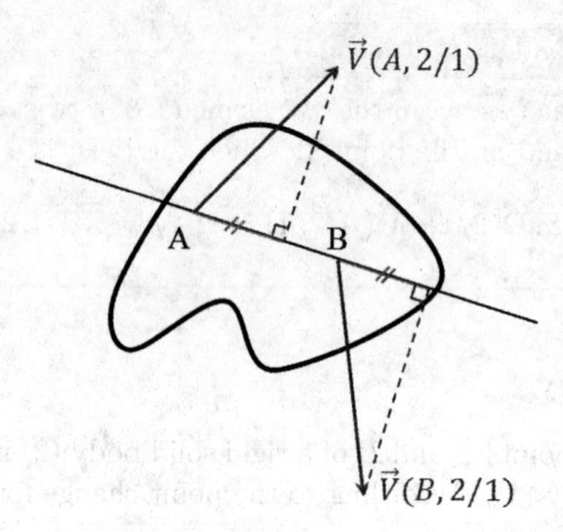

Figure 3.13 graphical implication of the equal projectivity of velocity vectors

Let's consider the twist related to the motion of a rigid solid body (2) relatively to a rigid solid body (1) and expressed at a point A of (2):

$$\{\mathcal{V}(2/1)\} = \left\{ \begin{array}{c} \vec{\Omega}(2/1) \\ \vec{V}(A,2/1) \end{array} \right\}_A$$

Let's determine the points M such that $\vec{\Omega}(2/1) \times \vec{V}(M,2/1) = \vec{0}$. According to the point change formula:

$$\vec{V}(M,2/1) = \vec{V}(A,2/1) + \vec{\Omega}(2/1) \times \overrightarrow{AM}$$

We hence have:

$$\begin{aligned}
&\vec{\Omega}(2/1) \times \vec{V}(M,2/1) = \vec{0} \\
\Leftrightarrow\, &\vec{\Omega}(2/1) \times \left(\vec{V}(A,2/1) + \vec{\Omega}(2/1) \times \overrightarrow{AM}\right) = \vec{0} \\
\Leftrightarrow\, &\vec{\Omega}(2/1) \times \vec{V}(A,2/1) + \vec{\Omega}(2/1) \times \left(\vec{\Omega}(2/1) \times \overrightarrow{AM}\right) = \vec{0} \\
\Leftrightarrow\, &\vec{\Omega}(2/1) \times \vec{V}(A,2/1) + \left(\vec{\Omega}(2/1) \cdot \overrightarrow{AM}\right)\vec{\Omega}(2/1) - \left(\vec{\Omega}(2/1) \cdot \vec{\Omega}(2/1)\right)\overrightarrow{AM} = \vec{0} \\
\Leftrightarrow\, &\left(\vec{\Omega}(2/1) \cdot \vec{\Omega}(2/1)\right)\overrightarrow{AM} = \left(\vec{\Omega}(2/1) \cdot \overrightarrow{AM}\right)\vec{\Omega}(2/1) + \vec{\Omega}(2/1) \times \vec{V}(A,2/1) \\
\Leftrightarrow\, &\overrightarrow{AM} = \frac{\vec{\Omega}(2/1) \cdot \overrightarrow{AM}}{\vec{\Omega}(2/1) \cdot \vec{\Omega}(2/1)}\vec{\Omega}(2/1) + \frac{\vec{\Omega}(2/1) \times \vec{V}(A,2/1)}{\vec{\Omega}(2/1) \cdot \vec{\Omega}(2/1)}
\end{aligned}$$

Since the vector $\overrightarrow{AM}$ is unknown, the scalar $\dfrac{\vec{\Omega}(2/1)\cdot\overrightarrow{AM}}{\vec{\Omega}(2/1)\cdot\vec{\Omega}(2/1)}$ is unknown too, and we can hence consider that $\dfrac{\vec{\Omega}(2/1)\cdot\overrightarrow{AM}}{\vec{\Omega}(2/1)\cdot\vec{\Omega}(2/1)} = \lambda \in \mathbb{R}$. We hence obtain:

$$\begin{aligned}
&\vec{\Omega}(2/1)\times\vec{V}(M,2/1)=\vec{0}\\
\Rightarrow\ &\overrightarrow{AM}=\lambda\,\vec{\Omega}(2/1)+\frac{\vec{\Omega}(2/1)\times\vec{V}(A,2/1)}{\vec{\Omega}(2/1)\cdot\vec{\Omega}(2/1)}\\
\Leftrightarrow\ &\overrightarrow{AM}=\lambda\,\vec{\Omega}(2/1)+\frac{\vec{\Omega}(2/1)\times\vec{V}(A,2/1)}{\|\vec{\Omega}(2/1)\|^2}
\end{aligned}$$

The central axis of a twist hence is a line:

- which is oriented by the vector $\vec{\Omega}(2/1)$; and
- to which belongs the point I such that $\overrightarrow{AI}=\dfrac{\vec{\Omega}(2/1)\times\vec{V}(A,2/1)}{\|\vec{\Omega}(2/1)\|^2}$.

Definition 27 (Central Axis of a Twist)

The **central axis** of a twist is the line such that the velocity vector of any point of this line and the angular velocity vector are collinear. A central axis can be defined for any twist whose angular velocity vector is not null.

It can be noticed that:

- the central axis of a twist is oriented by its angular velocity vector;
- according to the definition of the central axis of a twist, all the points of this central axis have the same velocity vector which is collinear with the angular velocity vector; and
- the velocity vector of a twist is constant along any line which is parallel to its central axis.

The central axis of a twist is also called the **instantaneous axis of screw**. The reason for this name will be illustrated in section 3.4.6.

3.4.4 Composition of Motions

Composition of Motions

The composition of motions allows to express complex motions as a composition of basic motions.

> **Definition 28** (Composition of Motions)
>
> The **composition of motions** consists in determining the general motion which corresponds to an association of intermediate motions. For instance, if we consider three rigid solid bodies (0), (1) and (2) in relative motion, the motion 2 / 0 results from the composition of the motions 2 / 1 and 1 / 0.

Composition of Velocity Vectors

Let's consider three points A, B and C which respectively belong to three rigid solid bodies (0), (1) and (2) with which three frames F_0, F_1 and F_2 are associated, as illustrated in Figure 3.14. Let's consider the motion of the point C considered as a point of (2) relatively to (0).

According to the definition of the velocity vector:

$$\vec{V}(C,2/0) = \left[\frac{d\overrightarrow{AC}}{dt}\right]_0 = \left[\frac{d(\overrightarrow{AB}+\overrightarrow{BC})}{dt}\right]_0 = \left[\frac{d\overrightarrow{AB}}{dt}\right]_0 + \left[\frac{d\overrightarrow{BC}}{dt}\right]_0$$

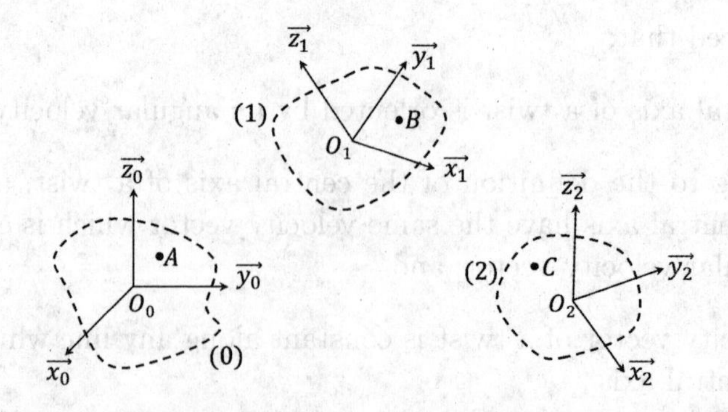

Figure 3.14 composition of motions with 3 rigid solid bodies

Since A belongs to (0), $\left[\frac{d\overrightarrow{AB}}{dt}\right]_0 = \vec{V}(B,1/0)$. As B does not belong to (0) but to (1), a relation between $\left[\frac{d\overrightarrow{BC}}{dt}\right]_0$ and $\left[\frac{d\overrightarrow{BC}}{dt}\right]_1$ can be determined by means of the vector differentiation formula as follows:

$$\begin{aligned}\left[\frac{d\overrightarrow{BC}}{dt}\right]_0 &= \left[\frac{d\overrightarrow{BC}}{dt}\right]_1 + \vec{\Omega}(1/0) \times \overrightarrow{BC} \\ &= \vec{V}(C,2/1) + \vec{\Omega}(1/0) \times \overrightarrow{BC}\end{aligned}$$

As a consequence:

$$\begin{aligned}\vec{V}(C,2/0) &= \vec{V}(C,2/1) + \vec{V}(B,1/0) + \vec{\Omega}(1/0) \times \overrightarrow{BC} \\ &= \vec{V}(C,2/1) + \vec{V}(C,1/0)\end{aligned}$$

thanks to the point change formula.

Definition 29 (Composition of Velocity Vectors)

The **composition of velocity vectors** is the relation between the velocity vectors calculated at one point and related to motions which can be composed:

$$\forall P \in \mathcal{E}, \vec{V}(P,2/0) = \vec{V}(P,2/1) + \vec{V}(P,1/0) \tag{3.9}$$

for any rigid solid bodies (0), (1) and (2).

Amongst these terms:

- the term $\vec{V}(P,2/0)$ is called **absolute velocity**;
- the term $\vec{V}(P,2/1)$ is called **relative velocity**; and
- the term $\vec{V}(P,1/0)$ is called **velocity of the moving space**.

Composition of Angular Velocity Vectors

Let's consider three rigid solid bodies (0), (1) and (2) with which three frames F_0, F_1 and F_2 are associated. The vector differentiation formula (3.3) allows us to write, for any vector $\vec{X}$:

$$\left[\frac{d\vec{X}}{dt}\right]_0 = \left[\frac{d\vec{X}}{dt}\right]_1 + \vec{\Omega}(1/0) \times \vec{X}$$
$$\left[\frac{d\vec{X}}{dt}\right]_1 = \left[\frac{d\vec{X}}{dt}\right]_2 + \vec{\Omega}(2/1) \times \vec{X}$$
$$\left[\frac{d\vec{X}}{dt}\right]_0 = \left[\frac{d\vec{X}}{dt}\right]_2 + \vec{\Omega}(2/0) \times \vec{X}$$

As a consequence, we can write:

$$\begin{aligned}
&\left[\frac{d\vec{X}}{dt}\right]_0 = \left[\frac{d\vec{X}}{dt}\right]_2 + \vec{\Omega}(2/0) \times \vec{X} \\
\Leftrightarrow &\left[\frac{d\vec{X}}{dt}\right]_1 + \vec{\Omega}(1/0) \times \vec{X} = \left[\frac{d\vec{X}}{dt}\right]_2 + \vec{\Omega}(2/0) \times \vec{X} \\
\Leftrightarrow &\left[\frac{d\vec{X}}{dt}\right]_2 + \vec{\Omega}(2/1) \times \vec{X} + \vec{\Omega}(1/0) \times \vec{X} = \left[\frac{d\vec{X}}{dt}\right]_2 + \vec{\Omega}(2/0) \times \vec{X} \\
\Leftrightarrow &\vec{\Omega}(2/1) \times \vec{X} + \vec{\Omega}(1/0) \times \vec{X} = \vec{\Omega}(2/0) \times \vec{X} \\
\Leftrightarrow &\left(\vec{\Omega}(2/0) - \vec{\Omega}(2/1) - \vec{\Omega}(1/0)\right) \times \vec{X} = \vec{0} \\
\Rightarrow &\vec{\Omega}(2/0) - \vec{\Omega}(2/1) - \vec{\Omega}(1/0) = \vec{0} \\
\Leftrightarrow &\vec{\Omega}(2/0) = \vec{\Omega}(2/1) + \vec{\Omega}(1/0)
\end{aligned}$$

since the antepenultimate relation must hold for any vector $\vec{X}$.

Definition 30 (Composition of Angular Velocity Vectors)

The **composition of angular velocity vectors** is the relation between the angular velocity vectors related to motions which can be composed:

$$\vec{\Omega}(2/0) = \vec{\Omega}(2/1) + \vec{\Omega}(1/0) \tag{3.10}$$

for any rigid solid bodies (0), (1) and (2).

Consequence for Twists

We have just seen that both the velocity vectors (according to the relation (3.9)) and the angular velocity vectors (according to the relation (3.10)) can be composed. Since these two vectors are the elements of twists, twists can hence be composed too.

Definition 31 (Composition of Twists)

The **composition of twists** is the relation between the twists related to motions which can be composed:

$$\{\mathcal{V}(2/0)\} = \{\mathcal{V}(2/1)\} + \{\mathcal{V}(1/0)\} \tag{3.11}$$

for any rigid solid bodies (0), (1) and (2). Of course, this relation is valid at any point of space.

For this relation to hold, the three twists obviously need to be expressed at the same point.

3.4.5 Particular Motions

Instantaneous Translation Motion

Definition 32 (Instantaneous Translation Motion)

A rigid solid body (2) has an **instantaneous translation motion** relatively to a rigid solid body (1) if and only if any line from (2) keeps a constant direction over the time relatively to (1).

Indeed, (2) is a rigid solid body, so for any two points A and B of (2), $\|\overrightarrow{AB}\| = C$ (where C is a scalar constant). Besides, if (2) has an instantaneous translation motion, then the direction of $\overrightarrow{AB}$ does not vary over the time relatively to (1), and we hence have:

$$
\begin{aligned}
&\left[\frac{d\overrightarrow{AB}}{dt}\right]_1 = \vec{0} \Leftrightarrow \left[\frac{d(\overrightarrow{AO}+\overrightarrow{OB})}{dt}\right]_1 = \vec{0} \\
&\Leftrightarrow \left[\frac{d\overrightarrow{OA}}{dt}\right]_1 = \left[\frac{d\overrightarrow{OB}}{dt}\right]_1 \Leftrightarrow \vec{V}(A,2/1) = \vec{V}(B,2/1) \\
&\Leftrightarrow \vec{V}(A,2/1) = \vec{V}(A,2/1) + \vec{\Omega}(2/1) \times \overrightarrow{AB} \\
&\Leftrightarrow \vec{\Omega}(2/1) \times \overrightarrow{AB} = \vec{0}
\end{aligned}
$$

where O is the origin of the frame associated with (1). Since the latter relation must hold for any two points A and B of (2), we hence necessarily have $\vec{\Omega}(2/1) = \vec{0}$.

The twist related to such a motion hence has the following form:

$$\forall P \in \mathcal{E}, \{\mathcal{V}(2/1)\} = \left\{ \begin{array}{c} \vec{0} \\ \vec{V}(P,2/1) \end{array} \right\}_P$$

where $\vec{V}(P,2/1)$ does not depend on the point P considered since:

$$\forall P,Q \in \mathcal{E}, \vec{V}(P,2/1) = \vec{V}(Q,2/1)$$

Instantaneous Rotation Motion

Definition 33 (Instantaneous Rotation Motion)

A rigid solid body (2) has an **instantaneous rotation motion** relatively to a rigid solid body (1) if and only if there exists a line from (2) which is immobile relatively to (1).

At any point P of this line, we hence have:

$$\vec{V}(P,2/1) = \vec{0} \Rightarrow \vec{\Omega}(2/1) \times \vec{V}(P,2/1) = \vec{0}$$

and P hence belongs to the central axis of the twist related to this motion.

The twist related to such a motion hence has the following form, at any point P of its central axis:

$$\{\mathcal{V}(2/1)\} = \left\{ \begin{array}{c} \vec{\Omega}(2/1) \\ \vec{0} \end{array} \right\}_P$$

The central axis is then called the **instantaneous axis of rotation** of 2 / 1. Besides, if we consider that the frames associated with (2) and (1) have one vector in common, as illustrated in Figure 3.15, and if we consider a point A of (2) such that $\overrightarrow{PA} = r\,\vec{x_2}$ where P belongs to the instantaneous axis of rotation of 2 / 1, we can write:

$$\vec{V}(A,2/1) = \vec{V}(P,2/1) + \vec{\Omega}(2/1) \times \overrightarrow{PA} = \dot{\alpha}\,\vec{z_2} \times r\,\vec{x_2} = r\dot{\alpha}\,\vec{y_2}$$

and the expression for the twist becomes:

$$\{\mathcal{V}(2/1)\} = \left\{ \begin{array}{c} \dot{\alpha}\,\vec{z_2} \\ r\dot{\alpha}\,\vec{y_2} \end{array} \right\}_A$$

where r stands for the distance of the point A to the instantaneous axis of rotation of 2 / 1. We can hence conclude that the vector $\vec{V}(A,2/1)$:

- is orthogonal to the vector $\overrightarrow{PA}$; and
- has a length which is proportional to the distance of A to the instantaneous axis of rotation of 2 / 1.

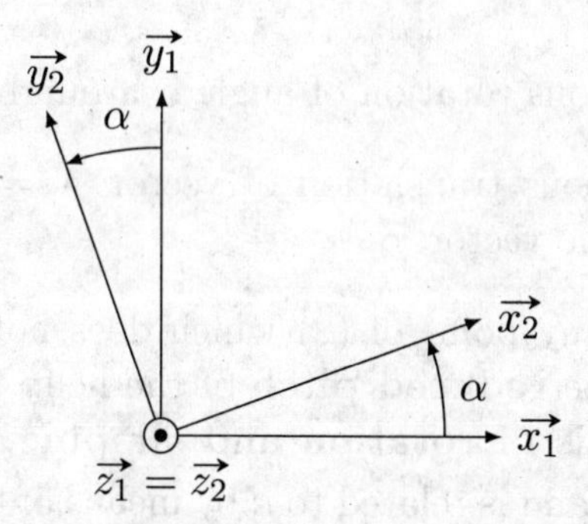

Figure 3.15 calculation figure allowing to define the angle α between the bases of the frames associated with the rigid solid bodies (2) and (1)

These 2 properties are illustrated in Figure 3.16 with different points A_0, $A_1, \cdots, A_{n-1}, A_n$ of (2) located at various distances from the point P (and hence from the instantaneous axis of rotation of 2 / 1).

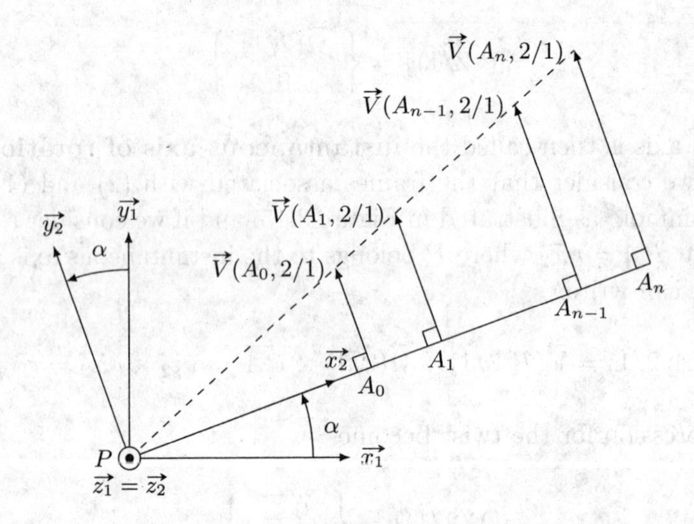

Figure 3.16 properties satisfied by the velocity vector of a point belonging to a rigid solid body having an instantaneous rotation motion

3.4.6 Helical Tangent Motion

Definition 34 (Helical Tangent Motion)

A rigid solid body (2) has an **helical tangent motion** relatively to a rigid solid body (1) if and only if the motion 2 / 1 results from the composition of:

- an instantaneous rotation of angle α around an axis Δ;
- an instantaneous translation of vector $\vec{\lambda} = k\,\alpha\,\vec{r}$, where Δ is oriented by the vector $\vec{r}$.

The trajectory of any point of (2) which does not belong to Δ is an helix. k is called the **reduced pitch** of the helix, and Δ is called the **instantaneous axis of rotation and slipping**. The **real pitch** of the helix is noted p and is related to k by means of the relation $k = \dfrac{p}{2\pi}$. It is illustrated in Figure 3.17.

Such a motion is illustrated in Figure 3.18 with $\vec{r} = \vec{z_1} = \vec{z_2}$, the frames $(O_1; \vec{x_1}, \vec{y_1}, \vec{z_1})$ and $(O_2; \vec{x_2}, \vec{y_2}, \vec{z_2})$ being respectively attached to (1) and (2), with $\overrightarrow{O_1O_2} = k\,\alpha\,\vec{z_2}$.

Figure 3.17 the real pitch of a screw

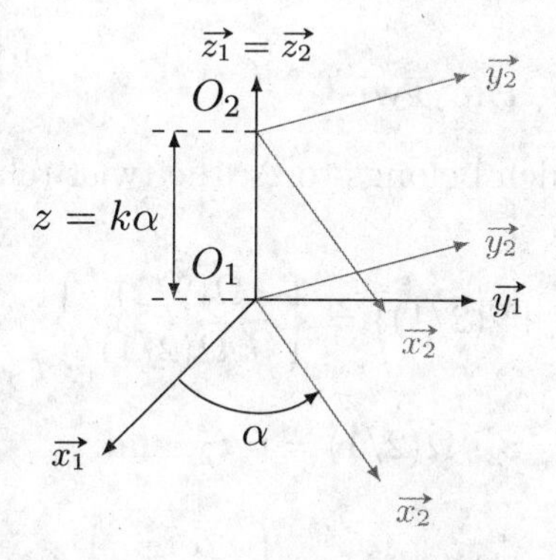

Figure 3.18 an helical tangent motion

Velocity Vector of a Point $I \in \Delta$

According to the definition of the central axis, the twist related to the motion 2 / 1 can be expressed under the following form, at the point I:

$$\{\mathcal{V}(2/1)\} = \left\{ \begin{array}{c} \vec{\Omega}(2/1) \\ k\,\vec{\Omega}(2/1) \end{array} \right\}_I$$

with $k \in \mathbb{R}$. As a consequence, all the points of the axis Δ have the same velocity vector, which has the same direction as Δ.

Velocity Vectors of Two Points $P, Q \notin \Delta$

According to the point change formula, we have:

$$\begin{aligned}\vec{V}(P,2/1) &= \vec{V}(I,2/1) + \vec{\Omega}(2/1) \times \overrightarrow{IP} \\ &= k\,\vec{\Omega}(2/1) + \vec{\Omega}(2/1) \times \overrightarrow{IP}\end{aligned}$$

The velocity vector of any point Q (considered as a point of (2) during the calculation) which does not belong to Δ can hence be expressed as the sum of two vectors:

- one vector which is equal to $\vec{V}(I,2/1)$ and is called **instantaneous translation velocity**; and
- one vector which is orthogonal to both $\vec{\Omega}(2/1)$ and $\overrightarrow{IQ}$ and is called **instantaneous rotation velocity**.

This decomposition is illustrated in Figure 3.19.

Conclusion Regarding the Twist

For any point $I \in (2)$ which belongs to Δ, the twist related to the motion 2 / 1 can be expressed as:

$$\{\mathcal{V}(2/1)\} = \left\{ \begin{array}{c} \vec{\Omega}(2/1) \\ k\,\vec{\Omega}(2/1) \end{array} \right\}_I \tag{3.12}$$

For instance, if $\vec{r} = \vec{z_1} = \vec{z_2}$, $\vec{\Omega}(2/1) = \dot{\alpha}\,\vec{z_2}$, and:

$$\{\mathcal{V}(2/1)\} = \left\{ \begin{array}{c} \dot{\alpha}\,\vec{z_2} \\ k\,\dot{\alpha}\,\vec{z_2} \end{array} \right\}_I$$

For any point P (considered as a point of (2) during the calculation) which does not belong to Δ, the twist related to the motion 2 / 1 can be expressed as:

$$\{\mathcal{V}(2/1)\} = \left\{ \begin{array}{c} \vec{\Omega}(2/1) \\ \vec{V}(I,2/1) + \vec{\Omega}(2/1) \times \overrightarrow{IP} \end{array} \right\}_P \tag{3.13}$$

For instance, if $\vec{r} = \vec{z_1} = \vec{z_2}$ and if $\overrightarrow{IP} = d\,\vec{x_2}$, $\vec{\Omega}(2/1) = \dot{\alpha}\,\vec{z_2}$, and:

$$\vec{V}(I,2/1) + \vec{\Omega}(2/1) \times \overrightarrow{IP} = k\,\dot{\alpha}\,\vec{z_2} + \dot{\alpha}\,\vec{z_2} \times d\,\vec{x_2} = d\,\dot{\alpha}\,\vec{y_2} + k\,\dot{\alpha}\,\vec{z_2}$$

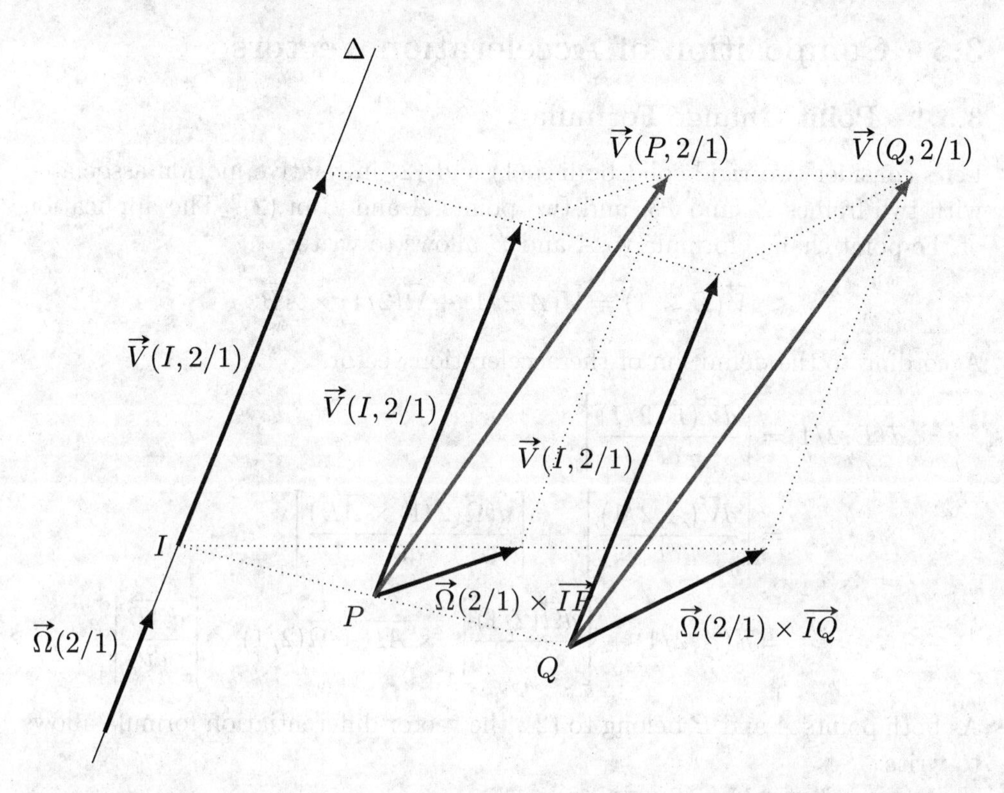

Figure 3.19 decomposition of the global velocity into an instantaneous translation velocity and an instantaneous rotation velocity

We hence have:

$$\{\mathcal{V}(2/1)\} = \left\{ \begin{array}{c} \dot\alpha\,\vec{z_2} \\ d\,\dot\alpha\,\vec{y_2} + k\,\dot\alpha\,\vec{z_2} \end{array} \right\}_P$$

Amongst the two terms of the velocity vector of P:

- the term $k\,\dot\alpha\,\vec{z_2}$ is collinear with $\vec{r}$ (and hence with $\vec{\Omega}(2/1)$) and corresponds to the **instantaneous translation velocity**; and
- the term $d\,\dot\alpha\,\vec{y_2}$ is orthogonal to $\vec{r}$ (and hence to $\vec{\Omega}(2/1)$) and to $\vec{IP}$, and hence corresponds to the **instantaneous rotation velocity**.

The global velocity results from the association of these two velocities, and so does the instant trajectory, which can be assimilated to an helix.

3.5 Composition of Acceleration Vectors

3.5.1 Point Change Formula

Let's consider two rigid solid bodies (1) and (2) in relative motion associated with two frames F_1 and F_2, and two points A and B of (2). The application of the point change formula to A and B allows to write:

$$\vec{V}(B,2/1) = \vec{V}(A,2/1) + \vec{\Omega}(2/1) \times \overrightarrow{AB}$$

According to the definition of the acceleration vector,

$$\begin{aligned}\vec{a}(B,2/1) &= \left[\frac{d\vec{V}(B,2/1)}{dt}\right]_1 \\ &= \left[\frac{d\vec{V}(A,2/1)}{dt}\right]_1 + \left[\frac{d(\vec{\Omega}(2/1) \times \overrightarrow{AB})}{dt}\right]_1 \\ &= \vec{a}(A,2/1) + \left[\frac{d\vec{\Omega}(2/1)}{dt}\right]_1 \times \overrightarrow{AB} + \vec{\Omega}(2/1) \times \left[\frac{d\overrightarrow{AB}}{dt}\right]_1\end{aligned}$$

As both points A and B belong to (2), the vector differentiation formula allows to write:

$$\left[\frac{d\overrightarrow{AB}}{dt}\right]_1 = \left[\frac{d\overrightarrow{AB}}{dt}\right]_2 + \vec{\Omega}(2/1) \times \overrightarrow{AB} = \vec{\Omega}(2/1) \times \overrightarrow{AB}$$

The point change formula for acceleration vectors hence is:

$$\vec{a}(B,2/1) = \vec{a}(A,2/1) + \left[\frac{d\vec{\Omega}(2/1)}{dt}\right]_1 \times \overrightarrow{AB} + \vec{\Omega}(2/1) \times \left(\vec{\Omega}(2/1) \times \overrightarrow{AB}\right)$$

Property 4 (Point Change Formula for Acceleration Vectors)

The acceleration vectors of any two points of a rigid solid body respect the following formula:

$$\begin{aligned}\forall A, B \in \mathcal{E}, \vec{a}(B,2/1) = \vec{a}(A,2/1) &+ \left[\frac{d\vec{\Omega}(2/1)}{dt}\right]_1 \times \overrightarrow{AB} \\ &+ \vec{\Omega}(2/1) \times \left(\vec{\Omega}(2/1) \times \overrightarrow{AB}\right) \qquad (3.14)\end{aligned}$$

where A and B are considered as points of (2) during the calculation.

3.5.2 Composition of Acceleration Vectors

Let's consider three points A, B and C which respectively belong to three rigid solid bodies (0), (1) and (2) with which three frames F_0, F_1 and F_2 are associated, as illustrated in Figure 3.14. Let's consider the motion of the point C considered as a point of (2) relatively to (0).

According to the definition of the acceleration vector:

$$\vec{a}(C,2/0) = \left[\frac{d^2\overrightarrow{AC}}{dt^2}\right]_0 = \left[\frac{d^2(\overrightarrow{AB}+\overrightarrow{BC})}{dt^2}\right]_0 = \left[\frac{d^2\overrightarrow{AB}}{dt^2}\right]_0 + \left[\frac{d^2\overrightarrow{BC}}{dt^2}\right]_0$$

Since A belongs to (0), $\left[\frac{d^2\overrightarrow{AB}}{dt^2}\right]_0 = \vec{a}(B,1/0)$. As B does not belong to (0) but to (1), a relation between $\left[\frac{d\overrightarrow{BC}}{dt}\right]_0$ and $\left[\frac{d\overrightarrow{BC}}{dt}\right]_1$ can be determined by means of the vector differentiation formula as follows:

$$\left[\frac{d\overrightarrow{BC}}{dt}\right]_0 = \left[\frac{d\overrightarrow{BC}}{dt}\right]_1 + \vec{\Omega}(1/0) \times \overrightarrow{BC}$$

Differentiating this expression in the basis associated with F_0 allows to write:

$$\begin{aligned}
\left[\frac{d^2\overrightarrow{BC}}{dt^2}\right]_0 &= \left[\frac{d}{dt}\left[\frac{d\overrightarrow{BC}}{dt}\right]_1\right]_0 + \left[\frac{d(\vec{\Omega}(1/0)\times\overrightarrow{BC})}{dt}\right]_0 \\
&= \left[\frac{d}{dt}\left[\frac{d\overrightarrow{BC}}{dt}\right]_1\right]_1 + \vec{\Omega}(1/0) \times \left[\frac{d\overrightarrow{BC}}{dt}\right]_1 \\
&\quad + \left[\frac{d\vec{\Omega}(1/0)}{dt}\right]_0 \times \overrightarrow{BC} + \vec{\Omega}(1/0) \times \left[\frac{d\overrightarrow{BC}}{dt}\right]_0 \\
&= \left[\frac{d^2\overrightarrow{BC}}{dt^2}\right]_1 + \vec{\Omega}(1/0) \times \left[\frac{d\overrightarrow{BC}}{dt}\right]_1 + \left[\frac{d\vec{\Omega}(1/0)}{dt}\right]_0 \times \overrightarrow{BC} \\
&\quad + \vec{\Omega}(1/0) \times \left(\left[\frac{d\overrightarrow{BC}}{dt}\right]_1 + \vec{\Omega}(1/0) \times \overrightarrow{BC}\right)
\end{aligned}$$

$$\left[\frac{d^2\overrightarrow{BC}}{dt^2}\right]_0 = \vec{a}(C,2/1) + 2\,\vec{\Omega}(1/0) \times \left[\frac{d\overrightarrow{BC}}{dt}\right]_1 + \left[\frac{d\vec{\Omega}(1/0)}{dt}\right]_0 \times \overrightarrow{BC}$$
$$+\vec{\Omega}(1/0) \times \left(\vec{\Omega}(1/0) \times \overrightarrow{BC}\right)$$

We can hence write:

$$\vec{a}(C,2/0) = \vec{a}(B,1/0) + \vec{a}(C,2/1) + 2\,\vec{\Omega}(1/0) \times \left[\frac{d\overrightarrow{BC}}{dt}\right]_1$$
$$+ \left[\frac{d\vec{\Omega}(1/0)}{dt}\right]_0 \times \overrightarrow{BC} + \vec{\Omega}(1/0) \times \left(\vec{\Omega}(1/0) \times \overrightarrow{BC}\right)$$
$$= \vec{a}(C,2/1) + \vec{a}(C,1/0) + 2\,\vec{\Omega}(1/0) \times \vec{V}(C,2/1)$$

by applying the point change formula for acceleration vectors since, according to the formula (3.14), we have

$$\vec{a}(C,1/0) = \vec{a}(B,1/0) + \left[\frac{d\vec{\Omega}(1/0)}{dt}\right]_0 \times \overrightarrow{BC} + \vec{\Omega}(1/0) \times \left(\vec{\Omega}(1/0) \times \overrightarrow{BC}\right)$$

Definition 35 (Composition of Acceleration Vectors)

The **composition of acceleration vectors** is the relation between the acceleration vectors calculated at one point and related to motions which can be composed:

$$\forall P \in \mathcal{E},\ \vec{a}(P,2/0) = \vec{a}(P,2/1) + \vec{a}(P,1/0) + 2\,\vec{\Omega}(1/0) \times \vec{V}(P,2/1) \tag{3.15}$$

for any rigid solid bodies (0), (1) and (2).

Amongst these terms:

- the term $\vec{a}(P,2/0)$ is called **absolute acceleration**;
- the term $\vec{a}(P,2/1)$ is called **relative acceleration**;
- the term $\vec{a}(P,1/0)$ is called **acceleration of the moving space**, and it is **not** equal to $\left[\frac{d\vec{V}(P,1/0)}{dt}\right]_0$ since this acceleration results from the association of many other vectors; and

- the additional term $2\,\vec{\Omega}(1/0) \times \vec{V}(P, 2/1)$ is called **Coriolis acceleration**. This term is due to the motion of the point P relatively to (1), (1) having an instantaneous rotation motion relatively to (0).

This complex formula is quite impossible to use as soon as there are many motions. This is the reason why, most of the time, $\vec{a}(P, 2/0)$ will be obtained by differentiating the vector $\vec{V}(P, 2/0)$ thanks to the relation $\vec{a}(P, 2/0) = \left[\dfrac{d\vec{V}(P, 2/0)}{dt}\right]_0$.

3.6 Kinematics of Contacts

3.6.1 Definition of a Contact at a Point

Definition 36 (Contact at a Point)

Two rigid solid bodies (1) and (2) are **in contact** if a basic geometric surface from (2) is in contact with a basic geometric surface from (1). Such a contact happens **at a point** if the area of the surface of contact tends to 0.

Let's consider two rigid solid bodies (1) and (2) in contact at a point P and in relative motion. We can hence define a tangent plane Π at the point P with unit normal vector $\vec{n}$, as illustrated in Figure 3.20. The velocity and angular velocity vectors which appear in Figure 3.20 have specific properties.

3.6.2 Velocity Vector

Let's consider that $\vec{b} = \vec{X}$ in the expression presented in section 3.1.3. It is hence possible to decompose the vector $\vec{X}$ into a sum of two vectors. If we choose $\vec{a} = \vec{c} = \vec{n}$, we get:

$$\begin{aligned} (\vec{n} \cdot \vec{n})\vec{X} &= \vec{n} \times (\vec{X} \times \vec{n}) + (\vec{n} \cdot \vec{X})\vec{n} \\ \Leftrightarrow \vec{X} &= \vec{n} \times (\vec{X} \times \vec{n}) + (\vec{X} \cdot \vec{n})\vec{n} \end{aligned} \tag{3.16}$$

If we apply the relation (3.16) to the case where $\vec{X} = \vec{V}(P, 2/1)$, we get:

$$\vec{V}(P, 2/1) = \underbrace{\vec{n} \times (\vec{V}(P, 2/1) \times \vec{n})}_{\text{slipping velocity}} + \underbrace{(\vec{V}(P, 2/1) \cdot \vec{n})\vec{n}}_{\text{penetration velocity}}$$

The vector which is normal to Π is called **penetration velocity**, and the vector which is included in Π is called **slipping velocity**.

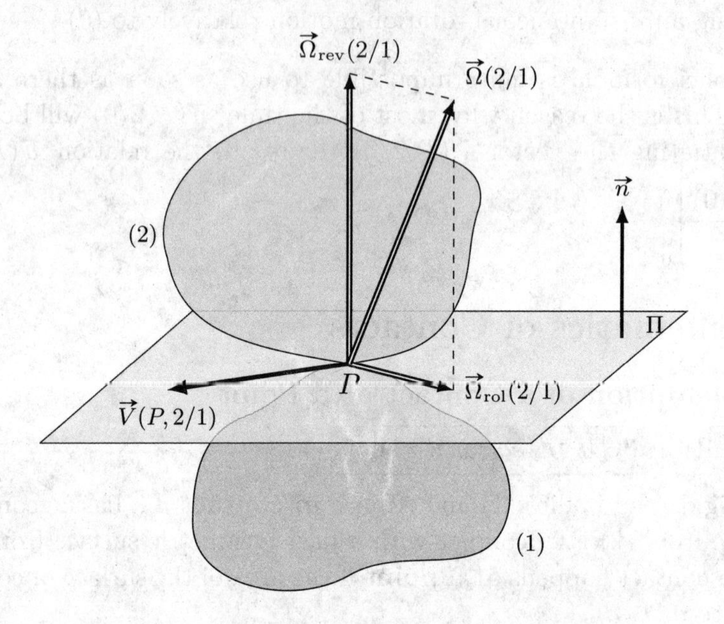

Figure 3.20 contact at a point between two rigid solid bodies (1) and (2)

If the slipping velocity is null, (2) is **rolling without slipping** relatively to (1). The revolving angular velocity (which will be defined in section 3.6.3) can be non-null, but the rolling angular velocity (which will also be defined in section 3.6.3) is influenced by this null velocity condition at the contact, which is the reason why this term "rolling without slipping" is used without specifying anything about the revolving.

3.6.3 Angular Velocity Vector

In the same way that we decomposed the velocity vector $\vec{V}(P,2/1)$, we can decompose the angular velocity vector $\vec{\Omega}(2/1)$ by means of the relation (3.16):

$$\vec{\Omega}(2/1) = \underbrace{\vec{n} \times (\vec{\Omega}(2/1) \times \vec{n})}_{\text{rolling angular velocity } \vec{\Omega}_{\mathrm{rol}}(2/1)} + \underbrace{(\vec{\Omega}(2/1) \cdot \vec{n})\vec{n}}_{\text{revolving angular velocity } \vec{\Omega}_{\mathrm{rev}}(2/1)} \qquad (3.17)$$

The vector which is normal to Π is called **revolving angular velocity** and noted $\vec{\Omega}_{\mathrm{rev}}(2/1)$, and the vector which is included in Π is called **rolling angular velocity** and noted $\vec{\Omega}_{\mathrm{rol}}(2/1)$.

3.7 Case of Plane Problems

Definition 37 (Plane Problem)

A problem is **plane** from a kinematic point of view if and only if, for any rigid solid bodies (i) and (j), a plane from (i) remains coincident with a plane from (j).

3.7.1 Consequence for Twists

Let's consider two rigid solid bodies (1) and (2) associated with two frames F_1 $(O_1; \vec{x_1}, \vec{y_1}, \vec{z_1})$ and F_2 $(O_2; \vec{x_2}, \vec{y_2}, \vec{z_2})$. Let's suppose that the planes $(O_1, \vec{x_1}, \vec{y_1})$ and $(O_2, \vec{x_2}, \vec{y_2})$ are coincident, and that $\vec{z_1} = \vec{z_2} = \vec{z}$ is hence normal to the plane in which motion occurs, as illustrated in Figure 3.21. The attitude of F_2 relatively to F_1 can be defined as follows:

- $\overrightarrow{O_1O_2} = x\,\vec{x_1} + y\,\vec{y_1}$
- $(\vec{x_1}, \vec{x_2}) = (\vec{y_1}, \vec{y_2}) = \alpha$

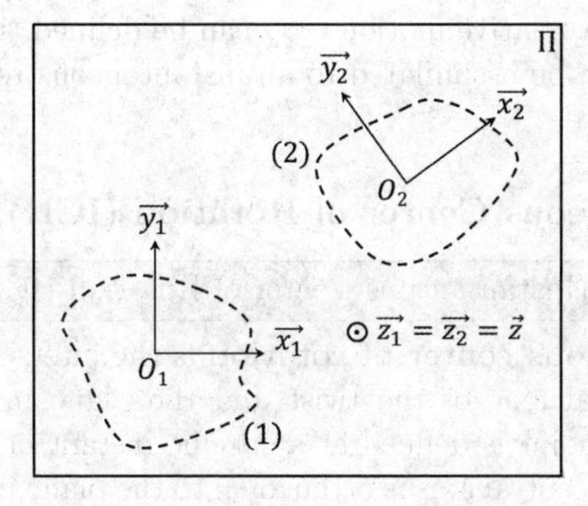

Figure 3.21 two rigid solid bodies (1) and (2) in relative plane motion

The twist $\{\mathcal{V}(2/1)\}$ hence has the following form at the point O_2:

$$\{\mathcal{V}(2/1)\} = \left\{ \begin{array}{c} \dot{\alpha}\,\vec{z} \\ \dot{x}\,\vec{x_1} + \dot{y}\,\vec{y_1} \end{array} \right\}_{O_2}$$

Besides, for any point P of the plane such that $\overrightarrow{O_1P} = x_P\,\vec{x_1} + y_P\,\vec{y_1}$:

$$\begin{aligned}\vec{V}(P,2/1) &= \vec{V}(O_2,2/1) + \vec{\Omega}(2/1) \times \overrightarrow{O_2P} \\ &= \dot{x}\,\vec{x_1} + \dot{y}\,\vec{y_1} + \dot{\alpha}\,\vec{z} \times [(x_P - x)\vec{x_1} + (y_P - y)\vec{y_1}] \\ &= [\dot{x} - (y_P - y)\dot{\alpha}]\,\vec{x_1} + [\dot{y} + (x_P - x)\dot{\alpha}]\,\vec{y_1}\end{aligned}$$

The twist expressed at any point P of the plane hence has the following expression:

$$\{\mathcal{V}(2/1)\} = \left\{ \begin{array}{c} \dot{\alpha}\,\vec{z} \\ [\dot{x} - (y_P - y)\dot{\alpha}]\,\vec{x_1} + [\dot{y} + (x_P - x)\dot{\alpha}]\,\vec{y_1} \end{array} \right\}_P$$

which allows to notice that:

- all velocity vectors are included in the plane in which motion occurs; and
- all angular velocity vectors are normal to the plane in which motion occurs (and all the central axes hence are parallel and orthogonal to this plane).

We hence have:

$$\forall P \in \mathcal{E}, \vec{V}(P,2/1) \cdot \vec{\Omega}(2/1) = 0$$

If a central axis for a relative motion i / j can be defined (i.e. if $\vec{\Omega}(i/j) \neq \vec{0}$), then this motion can be assimilated to an instantaneous rotation around this central axis.

3.7.2 Instantaneous Center of Rotation (ICR)

Definition 38 (Instantaneous Center of Rotation (ICR))

The **instantaneous center of rotation** is the intersection point between the central axis of the twist and the plane in which motion occurs. Such a point is defined at a specific instant of time, and it is unique since the central axis is orthogonal to the plane in which motion occurs.

The instantaneous center of rotation is generally abbreviated to ICR and noted $I_{i/j}$ for a motion i / j, as illustrated in Figure 3.22. As the ICR $I_{i/j}$ belongs to the plane in which motion occurs, $\vec{V}(I_{i/j}, i/j) \cdot \vec{\Omega}(i/j) = 0$. As the ICR $I_{i/j}$ belongs to the central axis of the twist, $\vec{V}(I_{i/j}, i/j) \times \vec{\Omega}(i/j) = \vec{0}$. We hence have:

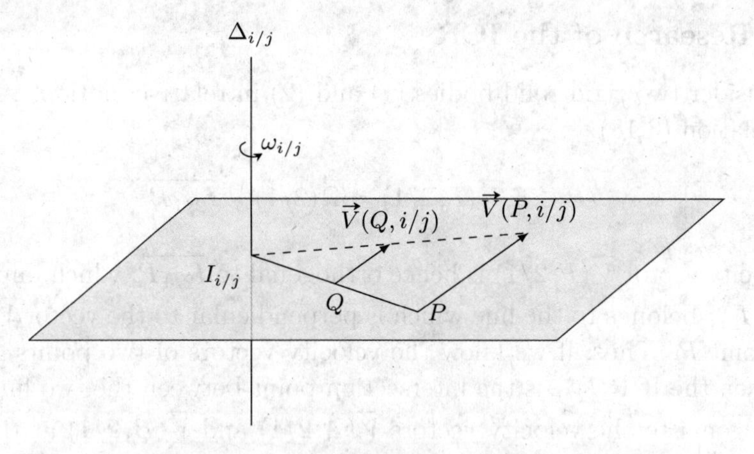

Figure 3.22 definition of the instantaneous center of rotation for a motion i/j

$$\left\{\begin{array}{l}\vec{V}(I_{i/j},i/j)\cdot\vec{\Omega}(i/j)=0\\ \vec{V}(I_{i/j},i/j)\times\vec{\Omega}(i/j)=\vec{0}\end{array}\right. \Rightarrow \vec{V}(I_{i/j},i/j)=\vec{0}$$

As a consequence, for any point $P \in \mathcal{E}$:

$$\vec{V}(P,i/j)=\vec{V}(I_{i/j},i/j)+\vec{\Omega}(i/j)\times\overrightarrow{I_{i/j}P}=\vec{\Omega}(i/j)\times\overrightarrow{I_{i/j}P} \quad (3.18)$$

We can obviously also notice that $I_{i/j} = I_{j/i}$.

3.8 Graphical Constructions

The different properties which have been presented in this chapter allow to solve some kinematics problems graphically. However, such a graphical study does not allow to obtain a *general expression* for the velocity vectors as functions of time, contrarily to an analytical study. The scheme which is used to perform a graphical study of a system represents this system in a particular position at a given instant of time t, and a graphical study hence allows to determine a *numerical value* for the velocity vectors of different points in this particular position.

3.8.1 Research of the ICR

Let's consider two rigid solid bodies (1) and (2) in relative motion. According to the relation (3.18):

$$\forall P \in \mathcal{E}, \vec{V}(P, 2/1) = \vec{\Omega}(2/1) \times \overrightarrow{I_{2/1}P}$$

The velocity vector $\vec{V}(P, 2/1)$ is hence orthogonal to $\overrightarrow{I_{2/1}P}$, which implies that the ICR $I_{2/1}$ belongs to the line which is perpendicular to the vector $\vec{V}(P, 2/1)$ at the point P. Thus, if we know the velocity vectors of two points A and B of (2), then the ICR $I_{2/1}$ is the intersection point between the two lines which are orthogonal to the velocity vectors $\vec{V}(A, 2/1)$ and $\vec{V}(B, 2/1)$ at the points A and B, as illustrated in Figure 3.23.

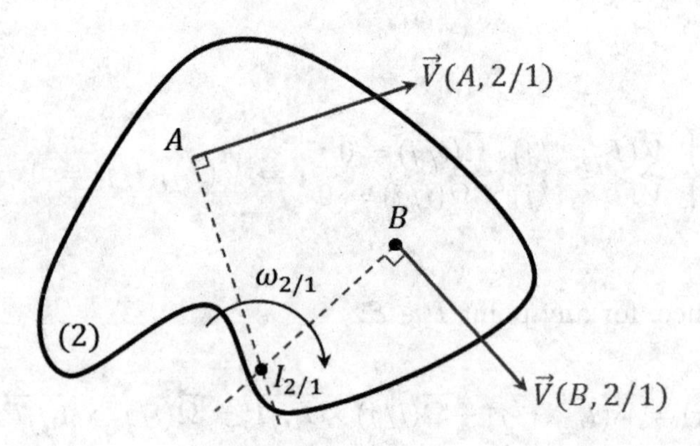

Figure 3.23 determination of the ICR in a given position

3.8.2 Plotting of Velocity Vectors

Let's consider two rigid solid bodies (1) and (2) in relative motion, and two points A and B of (2). Let's suppose that the velocity vector $\vec{V}(A, 2/1)$ is known and that we want to determine the velocity vector $\vec{V}(B, 2/1)$. A few approaches allow to do so.

Equal Projection

As the velocity vectors $\vec{V}(A, 2/1)$ and $\vec{V}(B, 2/1)$ have the same projection:

$$\vec{V}(A, 2/1) \cdot \overrightarrow{AB} = \vec{V}(B, 2/1) \cdot \overrightarrow{AB}$$

This relation allows to determine $\vec{V}(B,2/1)$ if the direction of $\vec{V}(B,2/1)$ is known. Otherwise, two such relations are needed if two velocity vectors of two points of (2) are known.

Instantaneous Center of Rotation

If the ICR $I_{2/1}$ is known, then the velocity vector $\vec{V}(B,2/1)$ can be determined by means of the relation (3.18):

$$\vec{V}(B,2/1) = \vec{\Omega}(2/1) \times \overrightarrow{I_{2/1}B}$$

We hence have:

- $\vec{V}(B,2/1) \perp \overrightarrow{I_{2/1}B}$
- $\|\vec{V}(B,2/1)\| = \|\vec{\Omega}(2/1)\| \cdot \|\overrightarrow{I_{2/1}B}\| = \dfrac{\|\overrightarrow{I_{2/1}B}\|}{\|\overrightarrow{I_{2/1}A}\|}\|\vec{V}(A,2/1)\|$

$$= \frac{\|\overrightarrow{I_{2/1}B'}\|}{\|\overrightarrow{I_{2/1}A}\|}\|\vec{V}(A,2/1)\|$$

where B' is the point of $[I_{2/1}A)$ such that $\|\overrightarrow{I_{2/1}B'}\| = \|\overrightarrow{I_{2/1}B}\|$. The graphical implication of these two properties is presented in Figure 3.24.

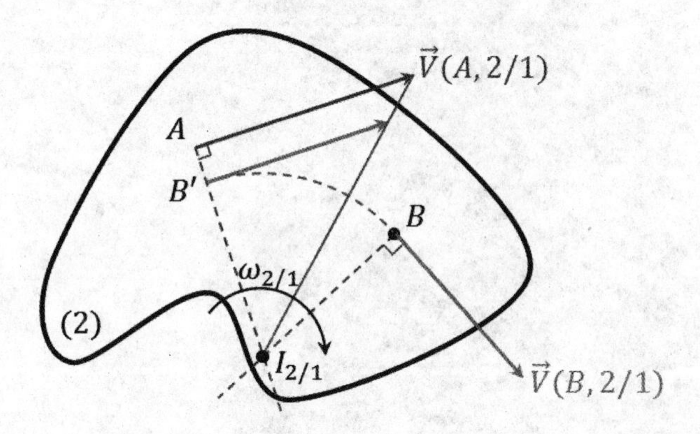

Figure 3.24 determination of a velocity vector by means of the instantaneous center of rotation

3.9 Conclusions and Perspectives

One must keep in mind that all the results presented in this chapter are valid if and only if the solid bodies which are considered are **rigid**. Indeed, it is this hypothesis which allows to study the relative motions of frames instead of the relative motions of solid bodies, and to describe motions by means of twists. Even if such a tool may look complex at first glance, it is really useful since the two vectors which constitute every twist – i.e. the angular velocity vector and the velocity vector of any point P of the Euclidean space $\mathcal{E}$ – and the relation which links them – i.e. the point change formula – allow to characterize the motion of any point of a rigid solid body relatively to another.

However, in most of the cases which will be faced by engineers, solid bodies will not be rigid but almost-rigid. In such a situation, the results presented in this chapter may still be applied to determine approximate results, and hence an approximate solution to any kinematics problem, as the deformations remain, in most cases, very low compared to the general motion. Indeed, this method can almost always be used, and there are only a few very particular cases (such as crash tests) for which it cannot be used.

Chapter 4

Modeling of Mechanical Actions

4.1 Notion of Mechanical Action (Effort and Moment)

Definition 39 (Mechanical Action)

A **mechanical action** is a cause which may leave a solid body (or a set of solid bodies) in an equilibrium state, engender its motion, or deform it.

Mechanical actions are generally divided into two groups:

- **volumic actions** or **actions at a distance**, which are exerted without solid bodies to be in contact: the action of gravitation, as well as the magnetic attraction and repulsion, are examples of actions at a distance; and
- **surfacic actions** or **contact actions**, which are exerted between two solid bodies in contact or between one solid body and one fluid.

We are going to show that such a mechanical action can be decomposed into two related terms: an *effort* and a *moment*. Let's consider the example of a tower crane which lifts a load L of mass M, as illustrated in Figure 4.1.
The effect of the mechanical action which is exerted at the point A by the load L can first be modeled by a vector $\vec{R}(L \rightarrow \text{crane})$ whose characteristics are as follows:

- its origin is the point A;

- its direction is vertical, oriented downwards; and
- its length is defined as $W = Mg$, since W here corresponds to the action of gravitation, known as the weight.

The model obtained is depicted in Figure 4.2.

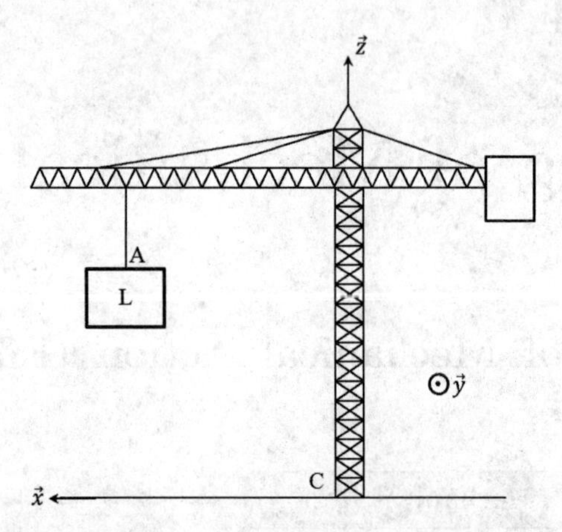

Figure 4.1 a tower crane lifting a load L

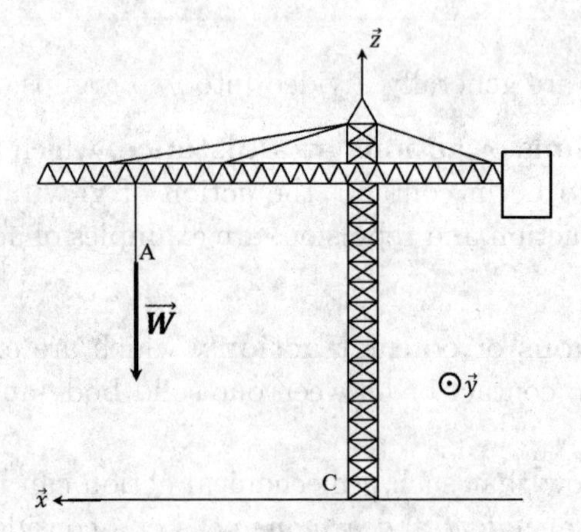

Figure 4.2 first model for the mechanical action exerted by the load

We saw in chapter 3 that the motion of a solid body (i) relatively to a solid body (j) is noted $i\ /\ j$. In this chapter, we are going to study the mechanical actions which are exerted by a solid body (i) on a solid body (j). In order to avoid confusion between the notations used in statics and in kinematics, the mechanical action exerted by a solid body (i) on a solid body (j) will be noted $i \to j$.

The vector $\vec{R}(L \to \text{crane})$ represents the **effort** exerted by the load L on the tower crane, and it will hence be noted:

$$\vec{R}(L \to \text{crane}) = \vec{W} = -Mg\,\vec{z} \tag{4.1}$$

where g is the gravitational acceleration, whose value is $g \approx 9.806\ \text{m.s}^{-2}$ in Beijing and about $g \approx 9.797\ \text{m.s}^{-2}$ in Shenzhen (the value of g decreases while going from the poles to the equator). The commonly used value for numerical applications is $g = 9.8\ \text{m.s}^{-2}$.

$\vec{R}(L \to \text{crane})$ is also called the **resultant** of the mechanical actions exerted by the load L on the crane, hence its notation.

However, to study the condition of the non-crashing of the crane, we need to know the effect of a mechanical action at any of its points. For instance, we can determine the effect of the mechanical action exerted by the load L at the point C. The effort $\vec{R}(L \to \text{crane}) = -Mg\,\vec{z}$ is still exerted at the point C, but it is also quite obvious that the mechanical action exerted at the point A causes a rotation effect at the point C which may make the tower crane fall on the left, and the greater the distance between the points A and C will be, the more important this effect will be. This rotation effect can also be represented by a vector, which is called the **moment** at the point C of the mechanical action exerted by the load L on the crane, and which is noted $\vec{\mathcal{M}}(C, L \to \text{crane})$ and **defined** as:

$$\vec{\mathcal{M}}(C, L \to \text{crane}) = \vec{CA} \times \vec{R}(L \to \text{crane}) = \vec{R}(L \to \text{crane}) \times \vec{AC} \tag{4.2}$$

4.2 The Wrench (or Screw of Mechanical Actions)

4.2.1 Point Change Formula

Let's consider a solid body (1) which exerts a mechanical action on a solid body (2) at a point P, and two points A and B of (2). According to the relation (4.2), the moment at the point A of the mechanical action exerted by the solid body (1) on the solid body (2) is defined as:

$$\vec{\mathcal{M}}(A, 1 \to 2) = \vec{AP} \times \vec{R}(1 \to 2),$$

where $\vec{R}(1 \to 2)$ is the resultant of the mechanical action exerted by the solid body (1) on the solid body (2).

According to the relation (4.2), the moment at the point B of the same mechanical action is defined as:

$$\begin{aligned}\vec{\mathcal{M}}(B,1 \to 2) &= \overrightarrow{BP} \times \vec{R}(1 \to 2) \\ &= \left(\overrightarrow{BA} + \overrightarrow{AP}\right) \times \vec{R}(1 \to 2) \\ &= \overrightarrow{AP} \times \vec{R}(1 \to 2) + \overrightarrow{BA} \times \vec{R}(1 \to 2) \\ &= \vec{\mathcal{M}}(A,1 \to 2) + \vec{R}(1 \to 2) \times \overrightarrow{AB}\end{aligned}$$

Property 5 (Point Change Formula)

The moments of the mechanical action exerted by a solid body (1) on a solid body (2) at any two points of the Euclidean space $\mathcal{E}$ respect the following formula:

$$\forall A, B \in \mathcal{E}, \vec{\mathcal{M}}(B,1 \to 2) = \vec{\mathcal{M}}(A,1 \to 2) + \vec{R}(1 \to 2) \times \overrightarrow{AB} \quad (4.3)$$

4.2.2 General Form

The screw of the mechanical actions exerted by a solid body (1) on a solid body (2) can be expressed under the following form, at any point P:

$$\boxed{\{\mathcal{F}(1 \to 2)\} = \left\{ \begin{array}{c} \vec{R}(1 \to 2) \\ \vec{\mathcal{M}}(P,1 \to 2) \end{array} \right\}_P}$$

where:

- $\vec{R}(1 \to 2)$ is the resultant of the mechanical actions exerted by (1) on (2); and
- $\vec{\mathcal{M}}(P,1 \to 2)$ is the moment at the point P of the mechanical actions exerted by (1) on (2).

The screw of the mechanical actions exerted by a solid body (1) on a solid body (2) fully characterizes these mechanical actions. Indeed, as soon as the expression for this screw of mechanical actions (also called **wrench**) is known at a given point of a solid body, then it can be determined at any other point of the solid body by means of the point change formula (4.3).

In the same way that a Plücker notation exists for twists, a Plücker notation also exists for wrenches. The coordinates which are assigned are as follows:

- $\vec{R}(1 \to 2) = X_{12}\,\vec{x} + Y_{12}\,\vec{y} + Z_{12}\,\vec{z}$
- $\vec{\mathcal{M}}(P, 1 \to 2) = L_{12}^{P}\,\vec{x} + M_{12}^{P}\,\vec{y} + N_{12}^{P}\,\vec{z}$

and the corresponding notation for the wrench is as follows:

$$\boxed{\{\mathcal{F}(1 \to 2)\} = \left\{ \begin{array}{cc} X_{12} & L_{12}^{P} \\ Y_{12} & M_{12}^{P} \\ Z_{12} & N_{12}^{P} \end{array} \right\}_{(P,B)}}$$

4.2.3 Mathematical Properties

Scalar Invariance

Let's consider the screw of the mechanical actions exerted by a solid body (1) on a solid body (2) and expressed at a point A of (2):

$$\{\mathcal{F}(1 \to 2)\} = \left\{ \begin{array}{c} \vec{R}(1 \to 2) \\ \vec{\mathcal{M}}(A, 1 \to 2) \end{array} \right\}_{A}$$

It can be noticed that the symbol "=" which is used between the screw $\{\mathcal{F}(1 \to 2)\}$ and its developed form $\left\{ \begin{array}{c} \vec{R}(1 \to 2) \\ \vec{\mathcal{M}}(A, 1 \to 2) \end{array} \right\}_{A}$ makes no sense. Indeed, from a mathematical point of view, a screw is a vector field whereas the developed form is a set of two vectors, so both parts are different. To be rigorous, we should use the symbols ":" (which is used in many books) or "≡"(which is more coherent from a mathematical point of view); however, it is the symbol "=" which is the most used, so it is the one which will be retained in the remainder of this book.

We have showed in section 4.2.2 that the screw of the same mechanical actions can be determined at any point B of (2) as:

$$\{\mathcal{F}(1 \to 2)\} = \left\{ \begin{array}{c} \vec{R}(1 \to 2) \\ \vec{\mathcal{M}}(B, 1 \to 2) \end{array} \right\}_{B} = \left\{ \begin{array}{c} \vec{R}(1 \to 2) \\ \vec{\mathcal{M}}(A, 1 \to 2) + \vec{R}(1 \to 2) \times \overrightarrow{AB} \end{array} \right\}_{B}$$

thanks to the point change formula (4.3).

Let's determine the expression for the scalar product between the two vectors of this latter wrench:

$$\begin{aligned}
&\vec{R}(1 \to 2) \cdot \overrightarrow{\mathcal{M}}(B, 1 \to 2) \\
&= \vec{R}(1 \to 2) \cdot \left(\overrightarrow{\mathcal{M}}(A, 1 \to 2) + \vec{R}(1 \to 2) \times \overrightarrow{AB}\right) \\
&= \vec{R}(1 \to 2) \cdot \overrightarrow{\mathcal{M}}(A, 1 \to 2) + \vec{R}(1 \to 2) \cdot \left(\vec{R}(1 \to 2) \times \overrightarrow{AB}\right) \\
&= \vec{R}(1 \to 2) \cdot \overrightarrow{\mathcal{M}}(A, 1 \to 2) + \overrightarrow{AB} \cdot \underbrace{\left(\vec{R}(1 \to 2) \times \vec{R}(1 \to 2)\right)}_{\vec{0}} \\
&= \vec{R}(1 \to 2) \cdot \overrightarrow{\mathcal{M}}(A, 1 \to 2)
\end{aligned}$$

Property 6 (Scalar Invariance)

The scalar product between the two elements of a wrench is an invariant: it does not depend on the point at which the wrench is expressed:

$$\forall A, B \in \mathcal{E}, \vec{R}(1 \to 2) \cdot \overrightarrow{\mathcal{M}}(A, 1 \to 2) = \vec{R}(1 \to 2) \cdot \overrightarrow{\mathcal{M}}(B, 1 \to 2) \quad (4.4)$$

Central Axis

For any wrench whose resultant vector is not null, there exists a line from space such that the moment vector at any point of that line is collinear with its resultant vector, which means that

$$\exists P \in \mathcal{E}, \vec{R}(1 \to 2) \times \overrightarrow{\mathcal{M}}(P, 1 \to 2) = \vec{0}$$

Let's consider the screw of the mechanical actions exerted by a solid body (1) on a solid body (2) and expressed at a point A:

$$\{\mathcal{F}(1 \to 2)\} = \left\{ \begin{array}{c} \vec{R}(1 \to 2) \\ \overrightarrow{\mathcal{M}}(A, 1 \to 2) \end{array} \right\}_A$$

Let's determine the points P such that $\vec{R}(1 \to 2) \times \overrightarrow{\mathcal{M}}(P, 1 \to 2) = \vec{0}$. According to the point change formula (4.3):

$$\overrightarrow{\mathcal{M}}(P, 1 \to 2) = \overrightarrow{\mathcal{M}}(A, 1 \to 2) + \vec{R}(1 \to 2) \times \overrightarrow{AP}$$

We hence have:

$$
\begin{aligned}
&\vec{R}(1 \to 2) \times \vec{\mathcal{M}}(P, 1 \to 2) = \vec{0} \\
&\Leftrightarrow \vec{R}(1 \to 2) \times \left(\vec{\mathcal{M}}(A, 1 \to 2) + \vec{R}(1 \to 2) \times \overrightarrow{AP}\right) = \vec{0} \\
&\Leftrightarrow \vec{R}(1 \to 2) \times \vec{\mathcal{M}}(A, 1 \to 2) + \vec{R}(1 \to 2) \times \left(\vec{R}(1 \to 2) \times \overrightarrow{AP}\right) = \vec{0} \\
&\Leftrightarrow \vec{R}(1 \to 2) \times \vec{\mathcal{M}}(A, 1 \to 2) + \left(\vec{R}(1 \to 2) \cdot \overrightarrow{AP}\right) \vec{R}(1 \to 2) \\
&\quad - \left(\vec{R}(1 \to 2) \cdot \vec{R}(1 \to 2)\right) \overrightarrow{AP} = \vec{0} \\
&\Leftrightarrow \|\vec{R}(1 \to 2)\|^2 \overrightarrow{AP} = \left(\vec{R}(1 \to 2) \cdot \overrightarrow{AP}\right) \vec{R}(1 \to 2) \\
&\quad + \vec{R}(1 \to 2) \times \vec{\mathcal{M}}(A, 1 \to 2) \\
&\Leftrightarrow \overrightarrow{AP} = \frac{\vec{R}(1 \to 2) \cdot \overrightarrow{AP}}{\|\vec{R}(1 \to 2)\|^2} \vec{R}(1 \to 2) + \frac{\vec{R}(1 \to 2) \times \vec{\mathcal{M}}(A, 1 \to 2)}{\|\vec{R}(1 \to 2)\|^2}
\end{aligned}
$$

Since the vector $\overrightarrow{AP}$ is unknown, the scalar $\dfrac{\vec{R}(1 \to 2) \cdot \overrightarrow{AP}}{\|\vec{R}(1 \to 2)\|^2}$ is unknown too, and we can hence consider that $\dfrac{\vec{R}(1 \to 2) \cdot \overrightarrow{AP}}{\|\vec{R}(1 \to 2)\|^2} = \lambda \in \mathbb{R}$. We hence obtain:

$$
\begin{aligned}
&\vec{R}(1 \to 2) \times \vec{\mathcal{M}}(P, 1 \to 2) = \vec{0} \\
&\Rightarrow \overrightarrow{AP} = \lambda \vec{R}(1 \to 2) + \frac{\vec{R}(1 \to 2) \times \vec{\mathcal{M}}(A, 1 \to 2)}{\|\vec{R}(1 \to 2)\|^2}
\end{aligned}
$$

The central axis of a wrench hence is a line:

- which is oriented by the vector $\vec{R}(1 \to 2)$; and
- to which belongs the point I such that $\overrightarrow{AI} = \dfrac{\vec{R}(1 \to 2) \times \vec{\mathcal{M}}(A, 1 \to 2)}{\|\vec{R}(1 \to 2)\|^2}$.

Definition 40 (Central Axis of a Wrench)

The **central axis** of a wrench is the line such that the moment vector at any point of this line and the resultant vector are collinear. A central axis can be defined for any wrench whose resultant vector is not null.

It can be noticed that:

- the central axis of a wrench is oriented by its resultant vector;

- according to the definition of the central axis of a wrench, all the points of this central axis have the same moment vector which is collinear with the resultant vector; and
- the moment vector of a wrench is constant along any line which is parallel to its central axis.

4.2.4 Mechanical Actions in the Links Between Two Solid Bodies

In order to study the mechanical actions which exist in systems, we need to model the mechanical actions which exist in the links between any two solid bodies of such systems.

Let's consider two solid bodies (1) and (2) which are in contact at a point P and in relative motion with respect to a frame $F(O; \vec{x}, \vec{y}, \vec{z})$. This motion can be fully characterized by the related twist:

$$\{\mathcal{V}(2/1)\} = \left\{ \begin{array}{c} \vec{\Omega}(2/1) \\ \vec{V}(P,2/1) \end{array} \right\}_P$$

Let's consider the screw of the mechanical actions exerted by (1) on (2) at the point P,

$$\{\mathcal{F}(1 \to 2)\} = \left\{ \begin{array}{c} \vec{R}(1 \to 2) \\ \vec{\mathcal{M}}(P,1 \to 2) \end{array} \right\}_P$$

In the second volume, we will see that the power of the reciprocal actions exerted between (1) and (2) is noted $P(1 \leftrightarrow 2)$ and defined as:

$$\begin{aligned} P(1 \leftrightarrow 2) &= \{\mathcal{F}(1 \to 2)\} \otimes \{\mathcal{V}(2/1)\} \\ &= \left\{ \begin{array}{c} \vec{R}(1 \to 2) \\ \vec{\mathcal{M}}(P,1 \to 2) \end{array} \right\}_P \otimes \left\{ \begin{array}{c} \vec{\Omega}(2/1) \\ \vec{V}(P,2/1) \end{array} \right\}_P \\ &= \vec{R}(1 \to 2) \cdot \vec{V}(P,2/1) + \vec{\Omega}(2/1) \cdot \vec{\mathcal{M}}(P,1 \to 2) \end{aligned}$$

If we assume that there is no friction between (1) and (2), and that the contact between (1) and (2) is perfect, then the power of the reciprocal actions exerted between (1) and (2) is null, so:

$$\vec{R}(1 \to 2) \cdot \vec{V}(P,2/1) + \vec{\Omega}(2/1) \cdot \vec{\mathcal{M}}(P,1 \to 2) = 0$$

Since this relation must hold for any motion and at any point P, then we necessarily have:

$$\vec{R}(1 \to 2) \cdot \vec{V}(P,2/1) = \vec{\Omega}(2/1) \cdot \vec{\mathcal{M}}(P,1 \to 2) = 0$$

This latter relation allows us to determine the wrench $\{\mathcal{F}(1 \to 2)\}$:

- let's note $\vec{\Omega}(2/1) = p_{21}\vec{x} + q_{21}\vec{y} + r_{21}\vec{z}$ and $\overrightarrow{\mathcal{M}}(P, 1 \to 2) = L^P_{12}\vec{x} + M^P_{12}\vec{y} + N^P_{12}\vec{z}$. The vector $\overrightarrow{\mathcal{M}}(P, 1 \to 2)$ can be determined as follows:
  - if the vector $\vec{\Omega}(2/1)$ has a coordinate which is not null (e.g. $p_{21} \neq 0$, which means that a rotation motion is possible around the axis of direction $\vec{x}$), then the corresponding coordinate of the vector $\overrightarrow{\mathcal{M}}(P, 1 \to 2)$ needs to be null ($L^P_{12} = 0$, which means that no moment can be exerted around the axis of direction $\vec{x}$) for the product of both coordinates ($p_{21}L^P_{12}$) to be null in the scalar product $\vec{\Omega}(2/1) \cdot \overrightarrow{\mathcal{M}}(P, 1 \to 2)$.
  - conversely, if the vector $\vec{\Omega}(2/1)$ has a coordinate which is null (e.g. $p_{21} = 0$, which means that no rotation motion is possible around the axis of direction $\vec{x}$), then the corresponding coordinate of the vector $\overrightarrow{\mathcal{M}}(P, 1 \to 2)$ can take any value ($L^P_{12} \neq 0$, which means that a moment can be exerted around the axis of direction $\vec{x}$) and the product of both coordinates ($p_{21}L^P_{12}$) will remain null in the scalar product $\vec{\Omega}(2/1) \cdot \overrightarrow{\mathcal{M}}(P, 1 \to 2)$.
- let's note $\vec{V}(P, 2/1) = u^P_{21}\vec{x} + v^P_{21}\vec{y} + w^P_{21}\vec{z}$ and $\vec{R}(1 \to 2) = X_{12}\vec{x} + Y_{12}\vec{y} + Z_{12}\vec{z}$. The vector $\vec{R}(1 \to 2)$ can be determined as follows:
  - if the vector $\vec{V}(P, 2/1)$ has a coordinate which is not null (e.g. $u^P_{21} \neq 0$, which means that a translation motion is possible along the direction $\vec{x}$), then the corresponding coordinate of the vector $\vec{R}(1 \to 2)$ needs to be null ($X_{12} = 0$, which means that no effort can be exerted along the direction $\vec{x}$) for the product of both coordinates ($u^P_{21}X_{12}$) to be null in the scalar product $\vec{R}(1 \to 2) \cdot \vec{V}(P, 2/1)$.
  - conversely, if the vector $\vec{V}(P, 2/1)$ has a coordinate which is null (e.g. $u^P_{21} = 0$, which means that no translation motion is possible along the direction $\vec{x}$), then the corresponding coordinate of the vector $\vec{R}(1 \to 2)$ can take any value ($X_{12} \neq 0$, which means that an effort can be exerted along the direction $\vec{x}$) and the product of both coordinates ($u^P_{21}X_{12}$) will remain null in the scalar product $\vec{R}(1 \to 2) \cdot \vec{V}(P, 2/1)$.

For instance, let's consider that the twist of the motion 2 / 1 is:

$$\{\mathcal{V}(2/1)\} = \left\{ \begin{array}{c} \vec{\Omega}(2/1) = p_{21}\,\vec{x} + q_{21}\,\vec{y} + r_{21}\,\vec{z} \\ \vec{V}(P, 2/1) = v^P_{21}\,\vec{y} + w^P_{21}\,\vec{z} \end{array} \right\}_P$$

According to this twist:

- since p_{21}, q_{21} and r_{21} are not null, then a rotation motion is possible around the axes of directions $\vec{x}$, $\vec{y}$ and $\vec{z}$; for the scalar product $\vec{\Omega}(2/1) \cdot \overrightarrow{\mathcal{M}}(P, 1 \to 2)$ to be null, the 3 coordinates of the vector $\overrightarrow{\mathcal{M}}(P, 1 \to 2)$ hence need to be null, so we have $\overrightarrow{\mathcal{M}}(P, 1 \to 2) = \vec{0}$; and
- since u_{21}^P is null and v_{21}^P and w_{21}^P are not null, then a translation motion is possible along the directions $\vec{y}$ and $\vec{z}$; for the scalar product $\vec{R}(1 \to 2) \cdot \vec{V}(P, 2/1)$ to be null, the coordinates of the vector $\vec{R}(1 \to 2)$ with respect to both $\vec{y}$ and $\vec{z}$ need to be null, but its coordinate with respect to $\vec{x}$ can take any value: we hence have $\vec{R}(1 \to 2) = X_{12}\,\vec{x}$.

As a consequence, the screw of the mechanical actions exerted by (1) on (2) can be determined as:

$$\{\mathcal{F}(1 \to 2)\} = \left\{ \begin{array}{c} \vec{R}(1 \to 2) = X_{12}\,\vec{x} \\ \overrightarrow{\mathcal{M}}(P, 1 \to 2) = \vec{0} \end{array} \right\}_P$$

The screw of the mechanical actions exerted by any solid body (1) on any solid body (2) can hence be deduced directly from the twist of the motion 2 / 1 by using the properties defined above.

4.2.5 Particular Forms

The general form of the screw of the mechanical actions exerted by a solid body (1) on a solid body (2) is, at a point P:

$$\{\mathcal{F}(1 \to 2)\} = \left\{ \begin{array}{c} \vec{R}(1 \to 2) \\ \overrightarrow{\mathcal{M}}(P, 1 \to 2) \end{array} \right\}_P$$

Two particular forms of this wrench exist:

- if $\vec{R}(1 \to 2) = \vec{0}$, then

$$\forall P \in \mathcal{E}, \{\mathcal{F}(1 \to 2)\} = \left\{ \begin{array}{c} \vec{0} \\ \overrightarrow{\mathcal{M}}(P, 1 \to 2) = \overrightarrow{\mathcal{M}}(1 \to 2) \end{array} \right\}_P$$

  since the moment vector does not depend on the point considered, according to the point change formula (4.3).

- if $\left\{ \begin{array}{l} \vec{R}(1 \to 2) \neq \vec{0} \\ \vec{R}(1 \to 2) \cdot \overrightarrow{\mathcal{M}}(P, 1 \to 2) = 0 \end{array} \right.$, then the wrench is called a **slider**.

  A slider has the following form, at any point Q of its central axis:

$$\{\mathcal{F}(1 \to 2)\} = \left\{ \begin{array}{c} \vec{R}(1 \to 2) \\ \vec{0} \end{array} \right\}_Q$$

where Q satisfies the relation $\overrightarrow{\mathcal{M}}(P, 1 \to 2) + \vec{R}(1 \to 2) \times \overrightarrow{PQ} = \vec{0}$.

It can be noticed that any basic effort can be modeled by a slider. Indeed, the moment of an effort $\vec{R}(1 \to 2)$ is null at its application point P ($\overrightarrow{\mathcal{M}}(P, 1 \to 2) = \vec{0}$), which implies that this application point P belongs to the central axis of the slider, and this central axis hence is the line which is oriented by $\vec{R}(1 \to 2)$ and goes through the point P.

4.3 Modeling of Friction

4.3.1 Property of Friction Models

When surfaces in contact move relative to each other, the friction between the two surfaces converts kinetic energy into heat. This conversion hence results in a loss of energy, which can be modeled by a negative power.

All friction models hence consider that the power of the reciprocal actions exerted between any two solid bodies (1) and (2) at a contact point P,

$$P(1 \leftrightarrow 2) = \vec{R}(1 \to 2) \cdot \vec{V}(P, 2/1) + \vec{\Omega}(2/1) \cdot \overrightarrow{\mathcal{M}}(P, 1 \to 2)$$

is strictly negative. As a consequence:

$$\vec{R}(1 \to 2) \cdot \vec{V}(P, 2/1) + \vec{\Omega}(2/1) \cdot \overrightarrow{\mathcal{M}}(P, 1 \to 2) < 0$$

4.3.2 Coulomb Model of Friction

The Coulomb model of friction owes its name to the French physicist Charles-Augustin Coulomb (1736-1806) (Figure 4.3). It allows to express, under a very simplified form, the intensity of the friction forces which exist between two solid bodies. It is probably the simplest (and yet coherent) model of friction, which is the reason why it is very often used at first, despite its relatively bad modeling of reality since there is a difference of at least 10% to 20% between the measures and the model ... and even more in case of local deformations.

Let's consider two solid bodies (1) and (2). If these solid bodies are not rigid, because of local deformations, the real contact between both solid bodies can take place according to a surface. However, if we consider that both solid bodies are rigid, the area of this surface can then be considered as tending to 0, and the surface of contact can be assimilated to a point. If we consider that (1) and (2) are rigid, then they will be in contact at a point that we will call P, as considered in section 3.6.1. Let's note $\vec{n}(2/1)$ the unit normal vector to the plane Π which is tangent to both (1) and (2) at the point P, as illustrated in Figure 4.4, and let's consider that there is friction during the relative motion 2 / 1.

Figure 4.3 Charles-Augustin Coulomb (1736-1806)

4.3.3 Coulomb Model of Friction for the Resultant

If there were no friction, the resultant of the contact action exerted by (1) on (2) $\vec{R}(1 \to 2)$ would be collinear with $\vec{n}(2/1)$. If there is friction between (1) and (2), as it is the case here, this resultant still has a normal component $\vec{N}(1 \to 2)$, but the friction can be modeled by an additional tangential component $\vec{T}(1 \to 2)$. The resultant $\vec{R}(1 \to 2)$ hence has the following expression:

$$\vec{R}(1 \to 2) = \vec{N}(1 \to 2) + \vec{T}(1 \to 2) \tag{4.5}$$

Besides, for (1) and (2) to remain in contact, $\vec{N}(1 \to 2)$ needs to have the same orientation as $\vec{n}(2/1)$. The relation (4.5) can hence be simplified to:

$$\vec{R}(1 \to 2) = \underbrace{\|\vec{N}(1 \to 2)\|\vec{n}(2/1)}_{\text{normal contact effort}} + \underbrace{\vec{T}(1 \to 2)}_{\text{tangential contact effort}}$$

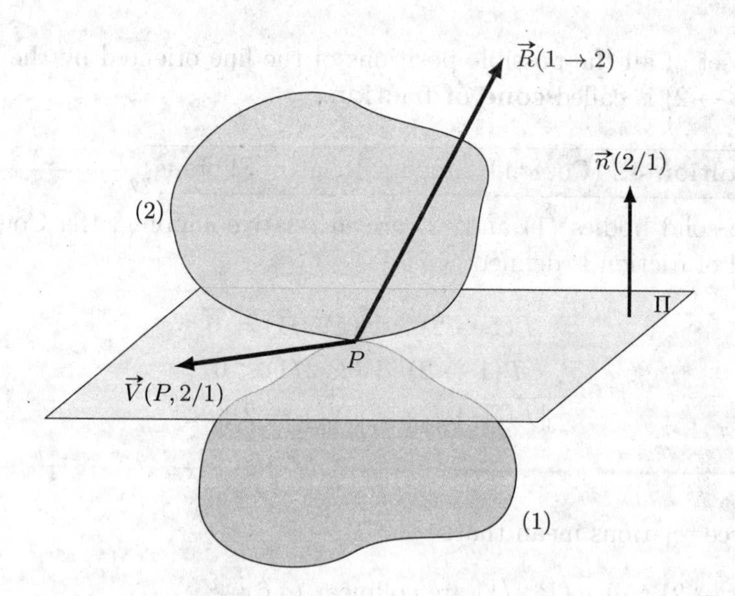

Figure 4.4 contact with friction between two rigid solid bodies (1) and (2)

The twist of the motion 2 / 1 has the following expression at the point P:

$$\{\mathcal{V}(2/1)\} = \left\{ \begin{array}{c} \vec{\Omega}(2/1) \\ \vec{V}(P, 2/1) \end{array} \right\}_P$$

We saw in section 3.6.2 that the velocity vector $\vec{V}(P, 2/1)$ can be decomposed into the sum of a normal component (the *penetration velocity*) and a tangential component (the *slipping velocity*). The penetration velocity needs to be null for (1) and (2) to remain in contact if both (1) and (2) are rigid solid bodies. The vector $\vec{V}(P, 2/1)$ hence belongs to the tangent plane Π, and it is equal to the slipping velocity vector. The Coulomb model of friction distinguishes two cases, depending on whether the vector $\vec{V}(P, 2/1)$ is null or not.

1st case: $\vec{V}(P, 2/1) \neq \vec{0}$, there is relative slipping between (1) **and** (2). In this case, the resultant $\vec{R}(1 \to 2)$ is inclined at an angle φ with respect to the normal vector $\vec{n}(2/1)$, so that its tangential component $\vec{T}(1 \to 2)$ is opposed to the relative slipping of (2) with respect to (1), and hence to $\vec{V}(P, 2/1)$:

- the angle φ is called **angle of friction**;
- the coefficient tan φ is called **coefficient of kinetic friction** (and noted μ_k or f) between (1) and (2); and

- the set of all the possible positions of the line oriented by the resultant $\vec{R}(1 \to 2)$ is called **cone of friction**.

Definition 41 (Coulomb Model - Relative Slipping)

If two solid bodies (1) and (2) are in relative slipping, the Coulomb model of friction is defined as:

$$\vec{T}(1 \to 2) \times \vec{V}(P, 2/1) = \vec{0} \tag{4.6}$$

$$\vec{T}(1 \to 2) \cdot \vec{V}(P, 2/1) < 0 \tag{4.7}$$

$$\|\vec{T}(1 \to 2)\| = \|\vec{N}(1 \to 2)\|\mu_k \tag{4.8}$$

These three relations mean that:

- $\vec{T}(1 \to 2)$ and $\vec{V}(P, 2/1)$ are collinear (4.6);
- the tangential component of the resultant is opposed to the relative slipping of (2) with respect to (1) (4.7), which corresponds to a dissipation of energy at the contact;
- the tangential component of the resultant is proportional to its normal component (4.8);

and they are illustrated in Figure 4.5.

Thanks to this model, if the slipping velocity vector $\vec{V}(P, 2/1)$ is known, it is then possible to determine the direction of the tangential component $\vec{T}(1 \to 2)$ (which is collinear with $\vec{V}(P, 2/1)$ and opposed to it). Since the normal component of the resultant is proportional to its tangential component, the problem only has one unknown, i.e. the length of either the tangential component $\vec{T}(1 \to 2)$ or the normal component $\vec{N}(1 \to 2)$. The normal component $\vec{N}(1 \to 2)$ will be retained as a reference because it is practically easy to measure.

2nd case: $\vec{V}(P, 2/1) = \vec{0}$, there is no relative slipping between (1) **and** (2), **and there hence is adherence between** (1) **and** (2).
In this case, the resultant $\vec{R}(1 \to 2)$ is inclined at an angle $\alpha < \varphi_s$ with respect to the normal vector $\vec{n}(2/1)$, as illustrated in Figure 4.6. The coefficient $\tan \varphi_s$ is called **coefficient of static friction** (and noted μ_s or f_s), and its value generally is *a bit greater* than the value of the coefficient of kinetic friction (about 10 to 25% greater in most cases, even though it can be 100% greater

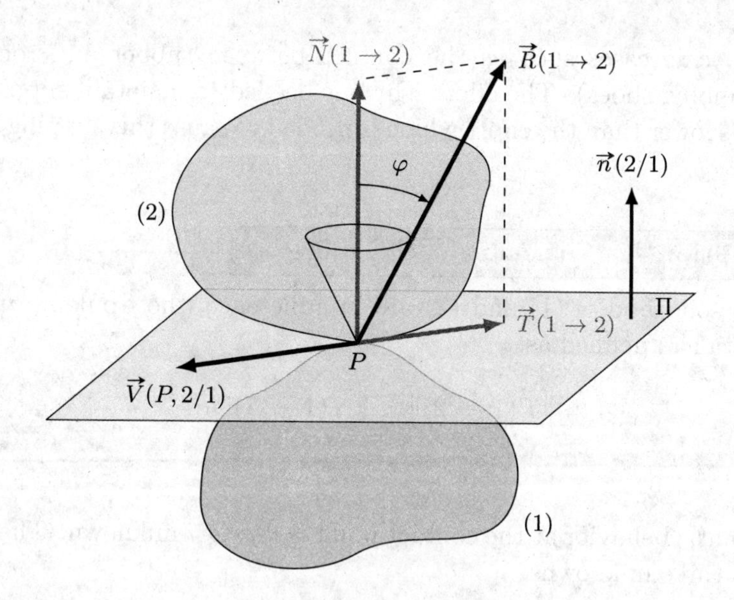

Figure 4.5 Coulomb model of friction in case of relative slipping between two rigid solid bodies (1) and (2)

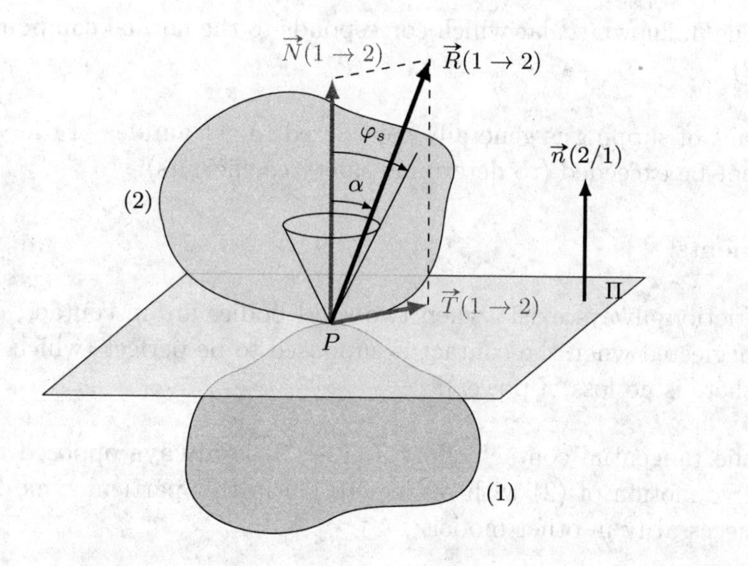

Figure 4.6 Coulomb model of friction in case of adherence between two rigid solid bodies (1) and (2)

in very specific cases such as the contact between rubber and rock in the case of climbing shoes). The effort which is needed to maintain slipping hence generally is lower than the effort which is needed to cause this slipping (because $\varphi < \varphi_s$).

Definition 42 (Coulomb Model - Adherence)

If two solid bodies (1) and (2) are in adherence, the Coulomb model of friction is defined as:

$$\|\vec{T}(1 \rightarrow 2)\| < \|\vec{N}(1 \rightarrow 2)\|\mu_s \tag{4.9}$$

The kinematic behavior at the contact point is *a priori* unknown. The problem hence has three unknowns:

- two unknown scalars which correspond to the two components of the vector $\vec{T}(1 \rightarrow 2)$ in the tangent plane Π; and
- one unknown scalar which corresponds to the normal component $\vec{N}(1 \rightarrow 2)$.

The limit of slipping is generally considered to get an idea of the value which must not be exceeded (to determine safety coefficients).

Comments:

- friction always exists when two solid bodies are in contact, even if it is neglected when the contact is supposed to be perfect (which means that there is no loss of power);
- the tangential contact effort $\vec{T}(1 \rightarrow 2)$ is **always** opposed to the relative motion of (2) with respect to (1) in this particular model, but not necessarily in other models;
- the coefficients of kinetic (μ_k) and static (μ_s) friction are adimensional terms whose value depends on the materials (main criterion), rugosity, lubrication and temperature; for instance, $\mu_k = 0.05$ for lubricated friction between steel and copper, $\mu_k = 0.2$ for dry friction between steel and steel, or $\mu_k \approx 0.7$ for dry friction between a tire and asphalt.

4.3.4 Coulomb Model of Friction for the Moment

We saw in section 3.6.3 that the angular velocity vector of the motion 2/1 $\vec{\Omega}(2/1)$ can be decomposed into the sum of two components:

$$\vec{\Omega}(2/1) = \underbrace{\vec{n}(2/1) \times (\vec{\Omega}(2/1) \times \vec{n}(2/1))}_{\text{rolling angular velocity}} + \underbrace{(\vec{\Omega}(2/1) \cdot \vec{n}(2/1))\vec{n}(2/1)}_{\text{revolving angular velocity}} \quad (4.10)$$

When the solid body (2) rolls on the solid body (1), a rolling resistance occurs which is opposed to the rotation of (2) around the instantaneous axis of rotation. This instantaneous axis of rotation is contained in the tangent plane Π and oriented by the tangential component of the angular velocity vector $\vec{\Omega}(2/1)$, that we will note $\vec{\Omega}_{\text{rol}}(2/1)$. This rolling resistance is modeled by a **rolling resistance moment vector** which is noted $\vec{\mathcal{M}}_{\text{rol}}(P, 1 \to 2)$ and illustrated in Figure 4.7.

Definition 43 (Coulomb Model - Rolling Resistance Moment)

If a solid body (2) is rolling on a solid body (1) ($\vec{\Omega}_{\text{rol}}(2/1) \neq \vec{0}$), the Coulomb model of friction is defined as:

$$\vec{\mathcal{M}}_{\text{rol}}(P, 1 \to 2) \times \vec{\Omega}_{\text{rol}}(2/1) = \vec{0} \quad (4.11)$$

$$\vec{\mathcal{M}}_{\text{rol}}(P, 1 \to 2) \cdot \vec{\Omega}_{\text{rol}}(2/1) < 0 \quad (4.12)$$

$$\|\vec{\mathcal{M}}_{\text{rol}}(P, 1 \to 2)\| = \|\vec{N}(1 \to 2)\|\eta_k \quad (4.13)$$

If there is no rolling between (1) and (2) ($\vec{\Omega}_{\text{rol}}(2/1) = \vec{0}$), we only have:

$$\|\vec{\mathcal{M}}_{\text{rol}}(P, 1 \to 2)\| < \|\vec{N}(1 \to 2)\|\eta_s \quad (4.14)$$

η_k is the **coefficient of kinetic rolling resistance** and η_s is the **coefficient of static rolling resistance**. Both have the dimension of a length. In the same way that we have $\mu_k < \mu_s$, we have $\eta_k < \eta_s$ (η_s is generally about 10 to 25% greater than η_k in most cases).

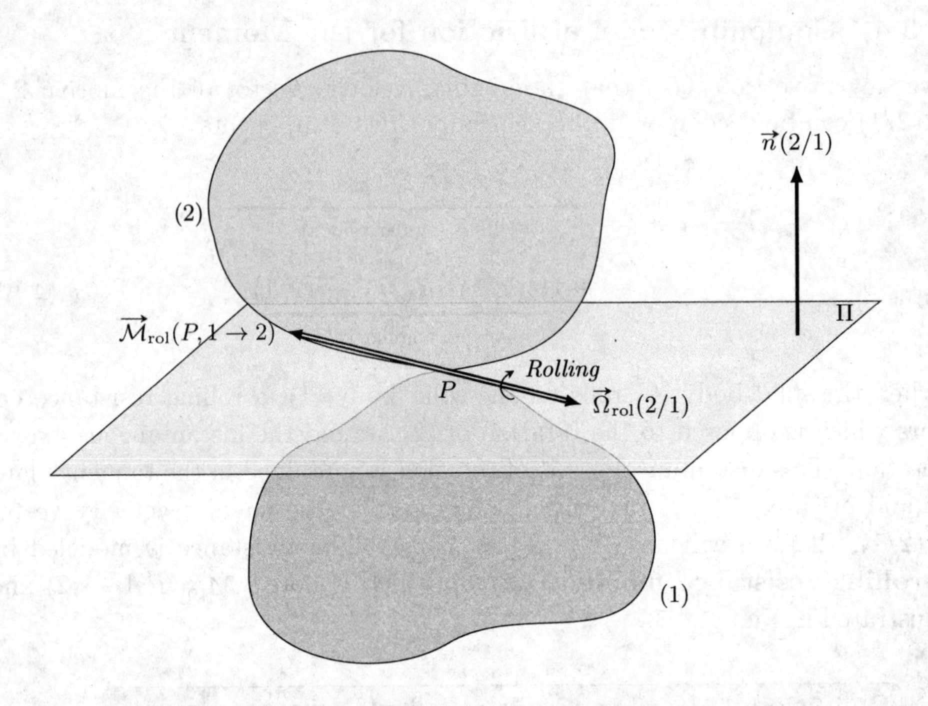

Figure 4.7 Coulomb model of friction for the rolling resistance moment between two rigid solid bodies (1) and (2)

In the same way, when the solid body (2) revolves relatively to the solid body (1), a revolving resistance occurs which is opposed to the rotation of (2) around the instantaneous axis of rotation. This instantaneous axis of rotation is normal to the tangent plane Π and oriented by the normal component of the angular velocity vector $\vec{\Omega}(2/1)$, that we will note $\vec{\Omega}_{\text{rev}}(2/1)$. This revolving resistance is modeled by a **revolving resistance moment vector** which is noted $\overrightarrow{\mathcal{M}}_{\text{rev}}(P, 1 \to 2)$ and illustrated in Figure 4.8.

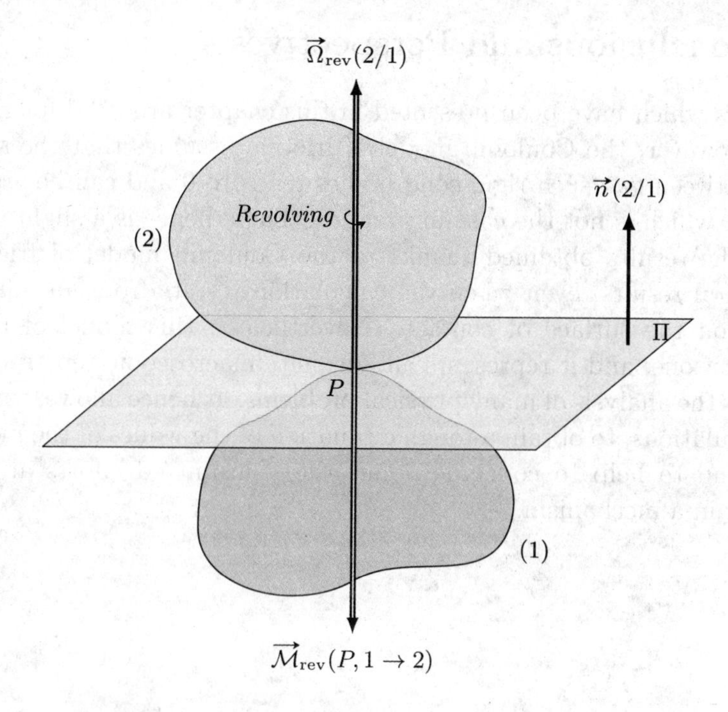

Figure 4.8 Coulomb model of friction for the revolving resistance moment between two rigid solid bodies (1) and (2)

Definition 44 (Coulomb Model - Revolving Resistance Moment)

If a solid body (2) is revolving relatively to a solid body (1) ($\vec{\Omega}_{\text{rev}}(2/1) \neq \vec{0}$), the Coulomb model of friction is defined as:

$$\vec{\mathcal{M}}_{\text{rev}}(P, 1 \to 2) \times \vec{\Omega}_{\text{rev}}(2/1) = \vec{0} \tag{4.15}$$

$$\vec{\mathcal{M}}_{\text{rev}}(P, 1 \to 2) \cdot \vec{\Omega}_{\text{rev}}(2/1) < 0 \tag{4.16}$$

$$\|\vec{\mathcal{M}}_{\text{rev}}(P, 1 \to 2)\| = \|\vec{N}(1 \to 2)\|\delta_k \tag{4.17}$$

If there is no revolving between (1) and (2) ($\vec{\Omega}_{\text{rev}}(2/1) = \vec{0}$), we only have:

$$\|\vec{\mathcal{M}}_{\text{rev}}(P, 1 \to 2)\| < \|\vec{N}(1 \to 2)\|\delta_s \tag{4.18}$$

δ_k is the **coefficient of kinetic revolving resistance** and δ_s is the **coefficient of static revolving resistance**. Both have the dimension of a length. In the same way that we have $\mu_k < \mu_s$, we have $\delta_k < \delta_s$ (δ_s is generally about 10 to 25% greater than δ_k in most cases).

4.4 Conclusions and Perspectives

The results which have been presented in this chapter are valid for rigid solid bodies. However, the Coulomb model of friction assumes that the surface of contact between two such rigid solid bodies tends to 0 and can be assimilated to a point, which is not the case in practice. There hence is a slight difference between the results obtained thanks to the Coulomb model of friction and experimental results, even when the calculation results from an infinite sum of points on the surface of contact. Nevertheless, this model of friction is the simplest one, and it represents an adequate macroscopic representation of friction for the analysis of many physical problems. It hence allows, without too many calculations, to obtain a rough evaluation of the values of the mechanical actions, and to help to conclude about the transmitted efforts at a simple contact or in a mechanism.

Chapter 5

Fundamental Principle of Equilibrium for Sets of Rigid Solid Bodies Sollicited by Mechanical Actions

5.1 Fundamental Principle of Equilibrium

The knowledge of the mechanical actions which exist in the links between two solid bodies can be useful to evaluate the power transmitted by a mechanism, to determine the dimensions of a component, or to calculate the deformation of a component. The Fundamental Principle of Equilibrium and the related vector theorems will allow us to determine these mechanical actions in the case where a system is in equilibrium.

Definition 45 (Statics)

Statics is the part of mechanics which is dedicated to the study of the equilibrium of solid bodies relatively to a reference frame.

Definition 46 (Equilibrium of a System of Solid Bodies)

A system of solid bodies (S) is in **equilibrium** relatively to a reference frame F if the motion of (S) relatively to F is null, which means that any point of (S) has a fixed position relatively to the reference frame F.

If (S) is a rigid solid body, then this particular motion can be characterized by a null twist.

5.1.1 Definition

Galilean Reference Frame

Definition 47 (Galilean Reference Frame)

A **Galilean reference frame** (or inertial reference frame) is a reference frame in which a solid body on which the resultant of mechanical actions is null has a uniform linear translation motion.

Many approximations of a Galilean reference frame exist, each of which is valid for a specific type of study:

- in the **heliocentric reference frame** (also called Copernic's reference frame), the origin of the frame is the center of the Sun ($\eta\lambda\iota o\varsigma$, *helios* means Sun, in ancient Greek) and its directions are defined by three stars; this type of reference frame allows, amongst others, to study the motion of planets in the universe, or the motion of interplanetary spacecrafts.
- in the **geocentric reference frame**, the origin of the frame is the center of the Earth ($\gamma\eta$, *ge* means Earth, in ancient Greek) and its directions are defined by three stars; this type of reference frame allows to study bodies which remain at the neighborhood of the Earth, such as satellites and airliners, or to carry out long duration terrestrial experiments (such as the Foucault pendulum experiment).

Apart from these two particular cases, and for most laboratory applications, a reference frame which is immobile relatively to the Earth (called a **terrestrial reference frame**) will be a good approximation of a Galilean reference frame.

Internal and External Efforts

Definition 48 (Isolated System)

Isolating a system of solid bodies consists in virtually separating it from the elements which are around it, and in modeling the relation between the external elements and the isolated system by means of a given number of wrenches.

If the isolated system is noted (Σ), the set of the elements which do not belong to it is noted $(\bar{\Sigma})$.

Definition 49 (Internal and External Efforts)

Two types of mechanical actions can be exerted on an isolated system of solid bodies (Σ):

- mechanical actions which are exerted by a part of the isolated system on another part of the isolated system: these are internal mechanical actions, or **internal efforts**; and
- mechanical actions which are exerted by the outside of the isolated system on the isolated system: these are external mechanical actions, or **external efforts**.

Screw of External Mechanical Actions

Definition 50 (Screw of External Mechanical Actions)

The screw of the mechanical actions which are external to a system of solid bodies (Σ) is noted $\{\mathcal{F}(\bar{\Sigma} \to \Sigma)\}$.

Fundamental Principle of Equilibrium

Theorem 4 (Fundamental Principle of Equilibrium)

If a system of solid bodies (Σ) is in equilibrium relatively to a Galilean reference frame, then the screw of the mechanical actions which are external to (Σ) is null:

$$\{\mathcal{F}(\bar{\Sigma} \to \Sigma)\} = \{0\} \tag{5.1}$$

It can be noticed that:

- the screw of external mechanical actions $\{\mathcal{F}(\bar{\Sigma} \to \Sigma)\}$ results from the sum of the screws of each external mechanical action, all these wrenches hence need to be expressed at the same point;
- the equilibrium of a system is a state which can be identified before applying the Fundamental Principle of Equilibrium: it is because this system is in equilibrium that the Fundamental Principle of Equilibrium holds, and the reverse is false; and
- the relation $\{\mathcal{F}(\bar{\Sigma} \to \Sigma)\} = \{0\}$ *does not* mean that the system is in equilibrium since this relation can be verified in other cases, e.g. if the system translates at a constant velocity.

5.1.2 Vector Theorems

The expression for the screw of the mechanical actions which are external to a system of solid bodies (Σ) is, at a point P:

$$\{\mathcal{F}(\bar{\Sigma} \to \Sigma)\} = \left\{ \begin{array}{c} \vec{R}(\bar{\Sigma} \to \Sigma) \\ \vec{\mathcal{M}}(P, \bar{\Sigma} \to \Sigma) \end{array} \right\}_P$$

If (Σ) is in equilibrium, then the Fundamental Principle of Equilibrium (5.1) implies:

$$\left\{ \begin{array}{c} \vec{R}(\bar{\Sigma} \to \Sigma) \\ \vec{\mathcal{M}}(P, \bar{\Sigma} \to \Sigma) \end{array} \right\}_P = \{0\} = \left\{ \begin{array}{c} \vec{0} \\ \vec{0} \end{array} \right\}_P$$

This relation between wrenches implies two vector theorems.

Theorem 5 (Vector Theorems)

For any system of solid bodies (Σ) which is in equilibrium relatively to a Galilean reference frame:

- the resultant vector of the screw of the mechanical actions which are external to (Σ) is null:

$$\vec{R}(\bar{\Sigma} \to \Sigma) = \vec{0} \tag{5.2}$$

- the moment vector of the screw of the mechanical actions which are external to (Σ) is null at any point of the Euclidean space $\mathcal{E}$:

$$\forall P \in \mathcal{E}, \overrightarrow{\mathcal{M}}(P, \bar{\Sigma} \to \Sigma) = \vec{0} \tag{5.3}$$

The application of the Fundamental Principle of Equilibrium to a system (Σ) will hence allow to write 6 independent scalar equations (3 scalar equations per vector of the screw of external mechanical actions) in 3-D problems, and only 3 independent scalar equations in 2-D problems (as we will see in section 5.2).

5.1.3 Law of Reciprocal Actions

Theorem 6 (Law of Reciprocal Actions)

If a solid body (i) exerts on a solid body (j) a mechanical action characterized by the wrench $\{\mathcal{F}(i \to j)\}$, then the solid body (j) exerts on the solid body (i) an opposed mechanical action which is characterized by the wrench:

$$\{\mathcal{F}(j \to i)\} = -\{\mathcal{F}(i \to j)\} \tag{5.4}$$

This law is also called the **law of mutual actions**.

Proof: let's consider two solid bodies (i) and (j), and let's note $(\Sigma) = \{i, j\}$. The Fundamental Principle of Equilibrium applied to (i) allows to write:

$$\{\mathcal{F}(\bar{i} \to i)\} = \{0\} \Leftrightarrow \{\mathcal{F}(j \to i)\} + \{\mathcal{F}(\bar{\Sigma} \to i)\} = \{0\}$$

The Fundamental Principle of Equilibrium applied to (j) allows to write:

$$\{\mathcal{F}(\bar{j} \to j)\} = \{0\} \Leftrightarrow \{\mathcal{F}(i \to j)\} + \{\mathcal{F}(\bar{\Sigma} \to j)\} = \{0\}$$

Finally, the Fundamental Principle of Equilibrium applied to (Σ) allows to write:

$$\{\mathcal{F}(\bar{\Sigma} \to \Sigma)\} = \{0\}$$

We hence have:

$$\begin{aligned}
&\{\mathcal{F}(j \to i)\}+\{\mathcal{F}(\bar{\Sigma} \to i)\}+\{\mathcal{F}(i \to j)\}+\{\mathcal{F}(\bar{\Sigma} \to j)\}-\{\mathcal{F}(\bar{\Sigma} \to \Sigma)\}=\{0\} \\
\Leftrightarrow &\{\mathcal{F}(j \to i)\}+\{\mathcal{F}(i \to j)\}+\underbrace{\{\mathcal{F}(\bar{\Sigma} \to i)\}+\{\mathcal{F}(\bar{\Sigma} \to j)\}}_{\{\mathcal{F}(\bar{\Sigma} \to \Sigma)\}}-\{\mathcal{F}(\bar{\Sigma} \to \Sigma)\}=\{0\} \\
\Leftrightarrow &\{\mathcal{F}(j \to i)\} + \{\mathcal{F}(i \to j)\} + \{\mathcal{F}(\bar{\Sigma} \to \Sigma)\} - \{\mathcal{F}(\bar{\Sigma} \to \Sigma)\} = \{0\} \\
\Leftrightarrow &\{\mathcal{F}(j \to i)\} + \{\mathcal{F}(i \to j)\} = \{0\} \\
\Leftrightarrow &\{\mathcal{F}(j \to i)\} = -\{\mathcal{F}(i \to j)\}
\end{aligned}$$

□

5.1.4 Solid Bodies in Equilibrium under the Actions of Two or Three Sliders

Case of Two Sliders

Theorem 7

If a system (Σ) is in equilibrium in a Galilean reference frame under the actions of two sliders, then these two sliders are directly opposed.

This theorem can be generalized to any equilibrium which involves two elements of the environment of the system if and only if the external mechanical actions can be modeled by sliders. Its proof is obvious from the Fundamental Principle of Equilibrium.

It can be noticed that, if a system (Σ) is in equilibrium under the actions of two elements (1) and (2) which can be modeled by the two sliders $\{\mathcal{F}(1 \to \Sigma)\} = \left\{ \begin{array}{c} \vec{R}(1 \to \Sigma) \\ \vec{0} \end{array} \right\}_A$ and $\{\mathcal{F}(2 \to \Sigma)\} = \left\{ \begin{array}{c} \vec{R}(2 \to \Sigma) \\ \vec{0} \end{array} \right\}_B$:

- the vector theorem (5.2) allows to write that $\vec{R}(1 \to \Sigma) + \vec{R}(2 \to \Sigma) = \vec{0}$: $\vec{R}(1 \to \Sigma)$ and $\vec{R}(2 \to \Sigma)$ hence have the same length and they are opposite, which is a necessary condition, but this condition is not sufficient since the system can still have an instantaneous rotation motion; and

- the vector theorem (5.3) allows to write that either $\vec{R}(1 \to \Sigma) \times \overrightarrow{AB} = \vec{0}$ or $\vec{R}(2 \to \Sigma) \times \overrightarrow{BA} = \vec{0}$: $\vec{R}(1 \to \Sigma)$ and $\vec{R}(2 \to \Sigma)$ hence are collinear with the line (AB), which is a necessary condition, but this condition is not sufficient since the system can still have an instantaneous translation motion along the line (AB).

These two conditions imply that $\vec{R}(1 \to \Sigma)$ and $\vec{R}(2 \to \Sigma)$ have the same length and direction, and that they are opposite.

Case of Three Sliders

Theorem 8

Let's consider that the actions of three sliders are exerted on a system (Σ). If (Σ) is in equilibrium in a Galilean reference frame, then the central axes of the three sliders are coplanar, and they are either parallel or concurrent.

Proof: let's consider a system (Σ) in equilibrium under the mechanical actions of three solid bodies $(1), (2), (3)$ exerted at three points A, B and C and modeled by three sliders. We have:

$$\{\mathcal{F}(1 \to \Sigma)\} = \left\{ \begin{array}{c} \vec{R}(1 \to \Sigma) \\ \vec{0} \end{array} \right\}_A$$

$$\{\mathcal{F}(2 \to \Sigma)\} = \left\{ \begin{array}{c} \vec{R}(2 \to \Sigma) \\ \vec{0} \end{array} \right\}_B$$

$$\{\mathcal{F}(3 \to \Sigma)\} = \left\{ \begin{array}{c} \vec{R}(3 \to \Sigma) \\ \vec{0} \end{array} \right\}_C$$

Let's respectively note (Δ_1), (Δ_2) and (Δ_3) the central axes of the three sliders $\{\mathcal{F}(1 \to \Sigma)\}$, $\{\mathcal{F}(2 \to \Sigma)\}$ and $\{\mathcal{F}(3 \to \Sigma)\}$. According to the vector theorem (5.2):

$$\vec{R}(1 \to \Sigma) + \vec{R}(2 \to \Sigma) + \vec{R}(3 \to \Sigma) = \vec{0} \tag{5.5}$$

The three resultants hence are coplanar, and so are the central axes of the related sliders. Two cases are possible:

1. if two resultants are collinear, then they are collinear with the third one according to the relation (5.5), and the central axes of the three sliders are parallel;

2. otherwise, the central axes of the sliders related to the three resultants are not parallel; let's note I the intersection point of (Δ_1) and (Δ_2); the vector theorem (5.3) expressed at the point I allows to write:

$$\begin{aligned}&\vec{\mathcal{M}}(I,1\to\Sigma)+\vec{\mathcal{M}}(I,2\to\Sigma)+\vec{\mathcal{M}}(I,3\to\Sigma)=\vec{0}\\ \Leftrightarrow\ &\vec{R}(1\to\Sigma)\times\vec{AI}+\vec{R}(2\to\Sigma)\times\vec{BI}+\vec{R}(3\to\Sigma)\times\vec{CI}=\vec{0}\end{aligned} \tag{5.6}$$

   $I\in(\Delta_1)$, $A\in(\Delta_1)$ since $\{\mathcal{F}(1\to\Sigma)\}$ is a slider, and $\vec{R}(1\to\Sigma)$ is collinear with (Δ_1) for the same reason, so $\vec{R}(1\to\Sigma)\times\vec{AI}=\vec{0}$.
   $I\in(\Delta_2)$, $B\in(\Delta_2)$ since $\{\mathcal{F}(2\to\Sigma)\}$ is a slider, and $\vec{R}(2\to\Sigma)$ is collinear with (Δ_2) for the same reason, so $\vec{R}(2\to\Sigma)\times\vec{BI}=\vec{0}$.
   According to the relation (5.6), we hence have $\vec{R}(3\to\Sigma)\times\vec{CI}=\vec{0}$. As $\{\mathcal{F}(3\to\Sigma)\}$ is a slider, $C\in(\Delta_3)$ and $\vec{R}(3\to\Sigma)$ is collinear with (Δ_3), which implies that $I\in(\Delta_3)$. The three central axes are hence concurrent at the point I.

□

5.1.5 Study Strategy

Most statics problems will consist in determining the mechanical actions exerted by solid bodies and/or external elements on the solid bodies of a system. The following strategy can be considered to determine such mechanical actions:

1. identify whether some sets of solid bodies are in equilibrium under the actions of two or three mechanical sliders, and
   (a) identify the length of the unknown resultants by means of the vector theorem (5.2); and/or
   (b) identify the direction of the unknown resultants by means of the vector theorem (5.3).
2. choose the vector theorems to use for:
   (a) the actions exerted by the built not to appear; and
   (b) the actions in liaisons with a few degrees of freedom (this notion will be presented in section 6.1.1) not to appear.

The two vector theorems (5.2) and (5.3) must be used *separately* and *wisely* to obtain as few scalar equations as possible. This point takes a (very) long time to acquire, but it is absolutely fundamental to the engineer as it cannot be possible to calculate all the components if only 2 or 3 of them are necessary. The practical works on real industrial systems help a lot to understand this strategy of isolation and of use of a theorem, since the sensors can measure efforts at some given points.

5.2 Case of Plane Problems

The case of plane problems corresponds to the one which was presented in section 3.7 in kinematics. It hence does not make sense regarding efforts. As a consequence, the mechanical actions which are not in accordance with the definition of a plane problem from a static point of view will be *hidden*: this will make the study simpler, and such a simpler study will generally be sufficient on initial examination. The wrenches will hence not be written under their full form, and the components of these wrenches which are problematic and are willingly hidden will be represented by black squares: these components can be null or not, they are just not taken into account.

5.2.1 Consequence for Wrenches

If the problem considered is plane, then:

- only the components of the resultant vectors which are included in the plane in which motion occurs are taken into account, and the other components are hidden; and
- only the components of the moment vectors which are normal to the plane in which motion occurs are taken into account, and the other components are hidden.

For instance, let's consider the mechanical action exerted by a solid body (1) on a solid body (2) and which can be modeled by the wrench:

$$\{\mathcal{F}(1 \to 2)\} = \left\{ \begin{array}{c} \vec{R}(1 \to 2) = Y_{12}\,\vec{y} + Z_{12}\,\vec{z} \\ \vec{\mathcal{M}}(P, 1 \to 2) = M_{12}^{P}\,\vec{y} + N_{12}^{P}\,\vec{z} \end{array} \right\}_{P}$$

The Plücker notation of this wrench is:

$$\{\mathcal{F}(1 \to 2)\} = \left\{ \begin{array}{cc} 0 & 0 \\ Y_{12} & M_{12}^{P} \\ Z_{12} & N_{12}^{P} \end{array} \right\}_{(P,\vec{x},\vec{y},\vec{z})}$$

If the problem is plane and if the plane in which motion occurs is the plane $(\vec{x}, \vec{y})$, the previous wrench can be simplified to the following form:

$$\{\mathcal{F}(1 \to 2)\} = \left\{ \begin{array}{cc} 0 & \blacksquare \\ Y_{12} & \blacksquare \\ \blacksquare & N_{12}^{P} \end{array} \right\}_{(P,\vec{x},\vec{y},\vec{z})}$$

The black squares hide components which can be null or not: as a consequence, it is necessary to stay very precise and not to conclude too quickly if such a model is used.

As a consequence, as soon as it has been identified that a problem is plane, all wrenches can be simplified in the previously defined way. Besides, in the case of plane problems, the application of the Fundamental Principle of Equilibrium allows to obtain 3 scalar equations instead of 6 in the case of 3-dimensional problems.

5.2.2 Graphical Constructions

If a problem is plane, then it may be possible to determine the searched mechanical actions graphically for a given position:

- if a system of solid bodies is in equilibrium under two mechanical actions which can be modeled by sliders, then these two mechanical actions have the same direction (thanks to the vector theorem (5.3)) and norm and they are opposite (thanks to the vector theorem (5.2)). If one of these mechanical actions is known, then the second one can be determined. If both mechanical actions are unknown, then their direction can be determined thanks to the vector theorem (5.3).

- if a system of solid bodies is in equilibrium under three mechanical actions which can be modeled by coplanar sliders, if one of these mechanical actions is known and if the direction of another one is known, then the two unknown mechanical actions can be fully determined:

  - since the central axes of the three sliders are concurrent, then the direction of the third mechanical action can be determined;
  - since the sum of the three resultants is null, according to the vector theorem (5.2), then the characteristics of the two unknown mechanical actions can be determined.

5.3 Overcenter Devices

5.3.1 Definition

Definition 51 (Overcenter Device)

An **overcenter device** is used to provide a mechanical stop and prevent a solid body from moving whatever the intensity of the external mechanical actions which are exerted. This mechanical stop is provided by means of the adherence phenomenon and/or by the special positioning which allows two mechanical sliders to always be in direct opposition.

It can be noticed that such a mechanical stop cannot be provided if there is no friction between the solid bodies involved.

5.3.2 Example

Let's consider the example of the overcenter device in Figure 5.1. This device is made of a plaque (1) on which two metal rods (2) and (3) are fixed. A hook which is articulated on the axis (4) lifts a load L whose action is modeled by the effort $\vec{F}$. This axis is mobile along the horizontal groove in (1). The system $(\Sigma) = \{1,2,3\}$ is in contact with an immobile metal rod (0) at the points A and B.
Let's suppose that the system $(\Sigma) = \{1,2,3\}$ is in equilibrium under the actions of the solid bodies (0) and (4). Let's also assume that the action of gravitation on the different solid bodies is neglectable with respect to the other mechanical actions, and that the coefficient of static friction is known at the points A and B and is $\mu_s = 0.18$.

The mechanical actions which are external to (Σ) are as follows:

- the mechanical action of the metal rod (0) at the point A, which can be modeled by the slider $\left\{\mathcal{F}(0 \overset{\mathcal{L}_A}{\rightarrow} \Sigma)\right\} = \left\{ \begin{array}{c} \vec{A}(0 \rightarrow \Sigma) \\ \vec{0} \end{array} \right\}_A$ with $\vec{A}(0 \rightarrow \Sigma) = X_A \, \vec{x} + Y_A \, \vec{y}$;
- the mechanical action of the metal rod (0) at the point B, which can be modeled by the slider $\left\{\mathcal{F}(0 \overset{\mathcal{L}_B}{\rightarrow} \Sigma)\right\} = \left\{ \begin{array}{c} \vec{B}(0 \rightarrow \Sigma) \\ \vec{0} \end{array} \right\}_B$ with $\vec{B}(0 \rightarrow \Sigma) = X_B \, \vec{x} + Y_B \, \vec{y}$; and

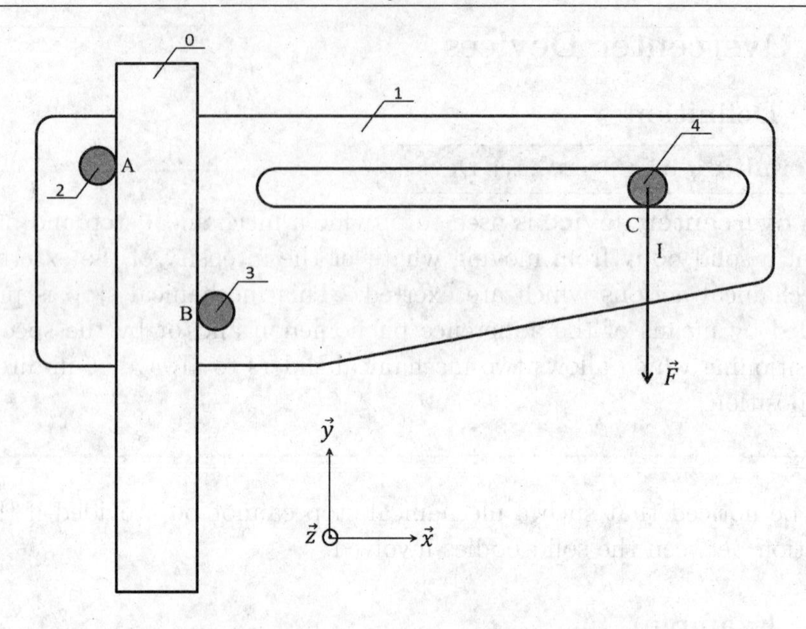

Figure 5.1 an overcenter device

- the mechanical action of the axis (4) at the point C, which can be modeled by the slider $\{\mathcal{F}(4 \to \Sigma)\} = \left\{ \begin{array}{c} \vec{F} \\ \vec{0} \end{array} \right\}_C$ with $\vec{F} = -F\,\vec{y}$.

$\vec{F}$ is known, but $\vec{A}(0 \to \Sigma)$ and $\vec{B}(0 \to \Sigma)$ are unknown. The coefficient of static friction is known at the points A and B, so the cones of friction are also known at the points A and B. If there is adherence, then the two resultant vectors $\vec{A}(0 \to \Sigma)$ and $\vec{B}(0 \to \Sigma)$ are *inside* the cones of friction. Besides, according to the Coulomb model of friction, the tangential component of the resultant vectors is opposed to motion: if there is relative slipping, (Σ) will move downwards, so the tangential component of the resultant vectors is directed upwards.

Let's now assume that the resultant vector at the point A is on the cone of friction, and that we are at the limit of slipping at the point A. We can hence determine the direction of the resultant vector at the point B, since the system (Σ) is in equilibrium under the actions of three sliders whose central axes are hence concurrent at a point I. As we know the direction of $\vec{F}$ (since $\vec{F}$ is known) and the direction of $\vec{A}(0 \to \Sigma)$ (since it is on the cone of friction at the point A), we can determine the intersection point I and the direction of $\overrightarrow{B}(0 \to \Sigma)$ (which is the line (BI)), as illustrated in Figure 5.2.

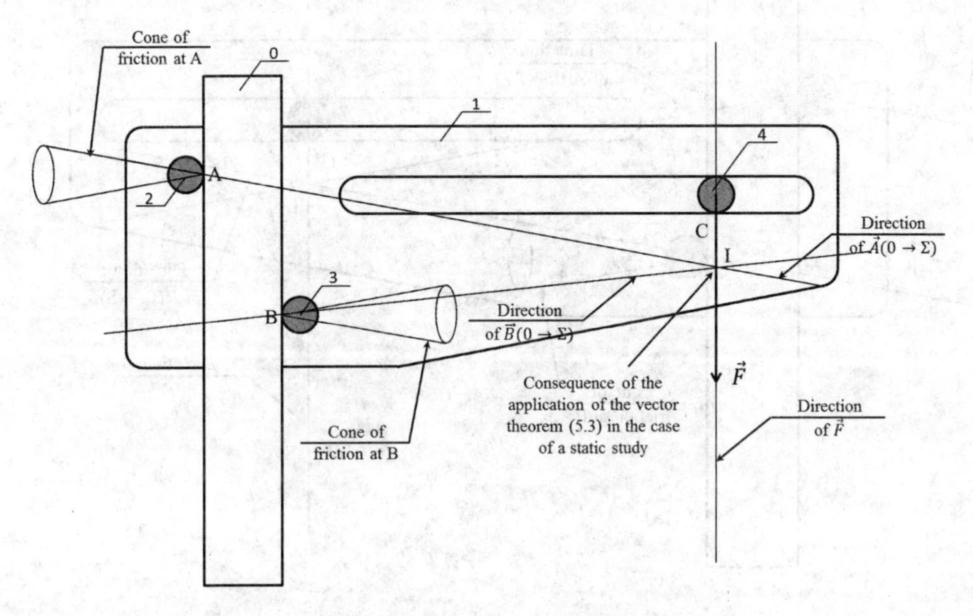

Figure 5.2 graphical determination of the direction of the resultant vectors at the points A and B

It can be noticed that $\overrightarrow{B}(0 \to \Sigma)$ is inside the cone of friction at the point B: (Σ) hence is in equilibrium, and this equilibrium does not depend on the intensity of the mechanical action $\overrightarrow{F}$.

If the load is moved on the left, the equilibrium will be strict when the direction of $\overrightarrow{F}$ goes through the point C', as illustrated in Figure 5.3. The two resultant vectors $\overrightarrow{A}(0 \to \Sigma)$ and $\overrightarrow{B}(0 \to \Sigma)$ will hence be on the cones of friction, and there will be slipping if the direction of $\overrightarrow{F}$ is on the left of C': the equilibrium of (Σ) would then be impossible, since the resultant vector $\overrightarrow{B}(0 \to \Sigma)$ would need to be outside the cone of friction, which is impossible with this Coulomb model of friction.

5.4 Conclusions and Perspectives

This study of the conditions of equilibrium of a system of rigid solid bodies under a set of external efforts is the very first step of the studies which can

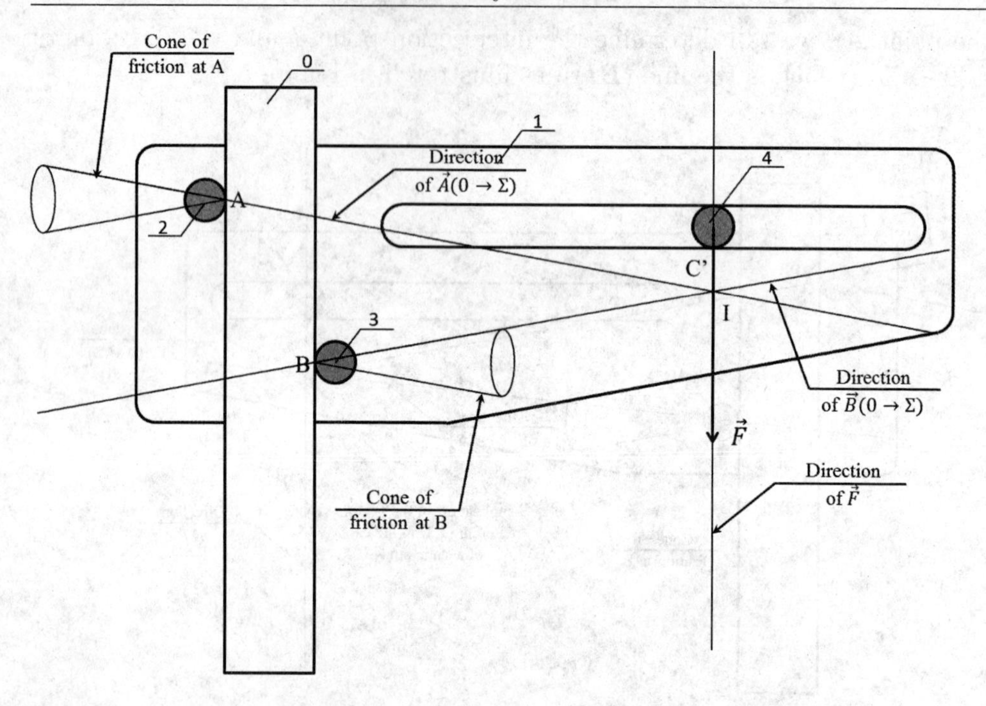

Figure 5.3 position of strict equilibrium

be carried out on a mechanical system in order to analyze and/or improve its design. Indeed, the second volume will present the dynamic and energetic laws which can be used to study the conditions of motion of a system of rigid solid bodies under a set of external efforts. Even if the study presented in this chapter can be considered as a particular case of the dynamic and energetic study, it is very important that it is mastered by an engineer as, most of the time, the evolutions of systems are either slow or at constant speed, two particular cases where the static point of view can be used to estimate the values of the different parameters. This study thus gives the engineer an idea of the mathematical relations between the parameters without too many calculations: this last point is very important as it can be very helpful to adjust a computer-aided mechanical simulation.

Chapter 6

Chains of Solid Bodies

This chapter is dedicated to the modeling of the mechanical links which exist between the pieces of a mechanism. This modeling of the links is done by means of normalized symbols called liaisons which will allow us to consider some mechanisms as chains of rigid solid bodies, and thus to determine input/output relations for these models of mechanisms by means of geometric and kinematic analyses.

6.1 Definition of a Kinematic Pair between Two Solid Bodies

6.1.1 Notion of Liaison Model

Geometry of the Contact between Two Solid Bodies

We saw in section 3.6.1 that two solid bodies (1) and (2) are in contact if a basic geometric surface from (1) is in contact with a basic geometric surface from (2). The three basic geometric surfaces are the plane, the cylinder of revolution and the sphere.

Two solid bodies can hence be in contact according to a **point** (contact between a sphere and a plane) or a **line** (contact between a sphere and a cylinder or between a cylinder and a plane) if the area of the surface of contact tends to 0. Otherwise, they are in contact according to a **surface** (contact between two spheres, between two cylinders, or between two planes).

When two solid bodies are in contact according to a line, two cases are possible:

- the line can be **straight**, in which case all normal vectors along this line are collinear; this case occurs when a cylinder is in contact with a plane, as illustrated in Figure 6.1.

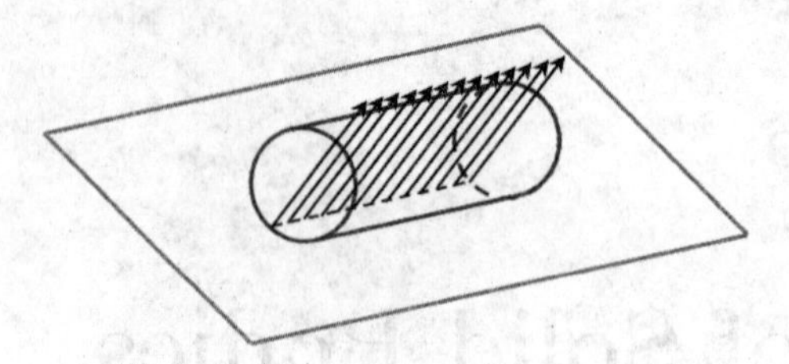

Figure 6.1 two solid bodies in contact according to a straight line

- the line can be **circular**, in which case all normal vectors are coplanar and concurrent; this case occurs when a sphere is in contact with a cylinder, as illustrated in Figure 6.2.

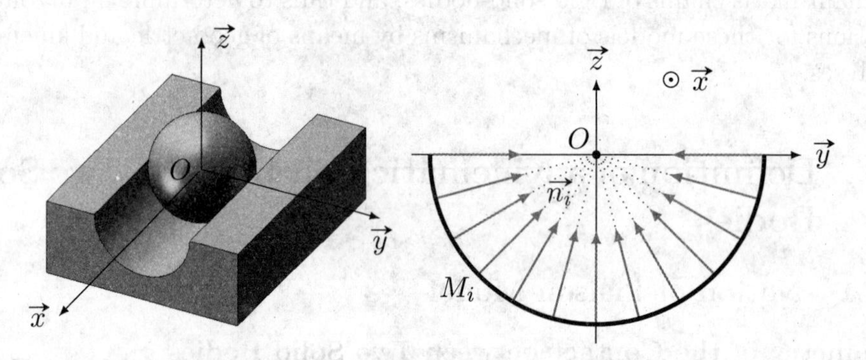

Figure 6.2 two solid bodies in contact according to a circular line

Degrees of Freedom

We saw in chapter 3 that the motion of a rigid solid body (2) relatively to a rigid solid body (1) can be fully characterized by the twist related to the motion 2 / 1:

$$\{\mathcal{V}(2/1)\} = \left\{ \begin{array}{c} \vec{\Omega}(2/1) = p_{21}\ \vec{x} + q_{21}\ \vec{y} + r_{21}\ \vec{z} \\ \vec{V}(A, 2/1) = u_{21}^A\ \vec{x} + v_{21}^A\ \vec{y} + w_{21}^A\ \vec{z} \end{array} \right\}_A$$

This twist hence is fully characterized by its 6 components:

- the 3 components p_{21}, q_{21}, r_{21} of the angular velocity vector $\vec{\Omega}(2/1)$; and

- the 3 components $u_{21}^A, v_{21}^A, w_{21}^A$ of the velocity vector of the point A, $\vec{V}(A,2/1)$.

Definition 52 (Degree of Freedom)

A **degree of freedom** is one of the 6 **independent** basic motions which are possible between two solid bodies (1) and (2).

These degrees of freedom can be of two types: **rotational** degrees of freedom (which correspond to p_{21}, q_{21} and r_{21}), and **translational** degrees of freedom (which correspond to u_{21}^A, v_{21}^A and w_{21}^A).

Amongst these 6 degrees of freedom:

- some will be suppressed because of the nature of the geometric contact between both solid bodies (for instance, a plane contact suppresses one translational degree of freedom and two rotational degrees of freedom); and

- the remaining ones will correspond to **independent** relative motions allowed by the contact between both solid bodies.

Liaisons

Definition 53 (Liaison)

A **liaison** is the relation which exists between two solid bodies in contact according to at least one common geometric structure (a point, a line or a surface).

The geometric and kinematic specificities of a liaison between two solid bodies depend on the shape of the surface of contact between these two solid bodies.

Basic Liaisons

Definition 54 (Basic Liaison)

A **basic liaison** between two solid bodies (1) and (2) is a liaison obtained from the contact between a **spherical** surface of (1) and a basic geometric structure of (2).

This association of basic geometric surfaces allows to obtain the three following liaisons:

- the **spherical liaison** (contact between two spherical surfaces);
- the **sphere/cylinder liaison**; and
- the **sphere/plane liaison**.

Composed Liaisons

Definition 55 (Composed Liaison)

A **composed liaison** between two solid bodies (1) and (2) is a liaison obtained:

- from the association of non-spherical basic geometric structures;
- from coherent associations of basic liaisons (as we will see in section 6.4.2).

The other associations of basic geometric surfaces allow to obtain the three following liaisons:

- the **revolute liaison** (contact between two cylindric basic geometric surfaces);
- the **cylinder/plane liaison**; and
- the **plane liaison** (contact between two plane basic geometric surfaces).

From the Real Link to the Liaison Model

In the case of real mechanisms, the surfaces of contact between pieces are **never** perfect.

The real links which exist between the pieces of mechanisms have imperfections:

- there are micro-geometric defects (characterized by the *rugosity* of the surfaces) and macro-geometric defects (characterized by the shape defects of the surfaces);
- there are clearances between the surfaces (which are necessary for the mechanism to work) which prevent them from being in perfect coincidence;
- there are deformations due to some efforts;
- there is friction which causes wear;
- etc.

Thus, finally, the chosen liaison model results from the consideration of all or any of these imperfections.

6.1.2 Liaison Model: Standard Kinematic Pairs

As real links have imperfections, it is difficult to fully know their behavior. We are hence going to model these real links by means of **liaisons** that we will call **kinematic pairs**, because these liaisons will be classified according to the basic **motions** which they enable between **pairs** of solid bodies.

This modeling will imply the following consequences, which are sometimes wrongly considered as "hypotheses":

- both solid bodies are in contact according to a point, a line (which can be straight or circular), or a basic geometric surface;
- the surfaces of both solid bodies are geometrically perfect;
- there is no clearance in the liaison which exists between both solid bodies; and
- the contact is bilateral (i.e. there is no loss of contact at any time).

6.1.3 3-D Standard Kinematic Pairs

The 3-dimensional standard kinematic pairs are presented in Table 6.1. On the one hand, the twist which corresponds to each kinematic pair can be easily determined by analyzing the surfaces according to which two solid bodies (1) and (2) are in contact. On the other hand, the screw of the mechanical actions exerted by (1) on (2) can be deduced from the twist of the motion 2 / 1 by considering that the contact between (1) and (2) is "energetically" perfect (there is no loss of contact), as we did in section 4.2.4. The different 3-dimensional standard kinematic pairs can be described as follows:

- the **sphere/plane pair** allows 3 rotation motions and 2 translation motions: it hence has 5 degrees of freedom. It is geometrically characterized by its point of contact and by the normal vector at the contact. For instance, if we consider a sphere/plane pair whose point of contact is A and whose normal vector at the contact is $\vec{z}$ (as illustrated in Table 6.1), then 3 rotation motions are possible around the axes $(A, \vec{x})$, $(A, \vec{y})$ and $(A, \vec{z})$, and 2 translation motions are possible along the directions $\vec{x}$ and $\vec{y}$. The related twist and wrench thus are as follows:

$$\{\mathcal{V}(2/1)\} = \left\{ \begin{array}{c} p_{21}\,\vec{x} + q_{21}\,\vec{y} + r_{21}\,\vec{z} \\ u_{21}^{A}\,\vec{x} + v_{21}^{A}\,\vec{y} \end{array} \right\}_A \quad \text{and} \quad \{\mathcal{F}(1 \to 2)\} = \left\{ \begin{array}{c} Z_{12}\,\vec{z} \\ \vec{0} \end{array} \right\}_A$$

  at any point of the axis $(A, \vec{z})$ and in any basis which contains the direction $\vec{z}$.

- the **line contact pair** allows 2 rotation motions and 2 translation motions: it hence has 4 degrees of freedom. It is geometrically characterized by its line of contact and by the tangent plane at the contact. For instance, if we consider a line contact pair whose line of contact is $(A, \vec{x})$ and whose tangent plane at the contact is $(A, \vec{x}, \vec{y})$ (as illustrated in Table 6.1), then 2 rotation motions are possible around the axes $(A, \vec{x})$ and $(A, \vec{z})$, and 2 translation motions are possible along the directions $\vec{x}$ and $\vec{y}$. The related twist and wrench thus are as follows:

$$\{\mathcal{V}(2/1)\} = \left\{ \begin{array}{c} p_{21}\,\vec{x} + r_{21}\,\vec{z} \\ u_{21}^{A}\,\vec{x} + v_{21}^{A}\,\vec{y} \end{array} \right\}_A \quad \text{and} \quad \{\mathcal{F}(1 \to 2)\} = \left\{ \begin{array}{c} Z_{12}\,\vec{z} \\ M_{12}^{A}\,\vec{y} \end{array} \right\}_A$$

  at any point of the plane $(A, \vec{x}, \vec{y})$.

Table 6.1 3-D standard kinematic pairs

Kinematic pair	3-D representation	2-D representation	Degrees of freedom
Sphere/plane pair	$\vec{z}$ 1 A $\vec{y}$ $\vec{x}$ 2	$\vec{z}$ 1 A $\vec{y}$ $\vec{x}$ $\odot$ 2	5
Line contact pair	$\vec{z}$ 1 A $\vec{y}$ $\vec{x}$ 2	$\vec{z}$ 1 $\vec{x}$ A $\odot\vec{y}$ 2 | $\vec{z}$ $\vec{x}$ $\odot$ 1 A $\vec{y}$ 2	4
Sphere/cylinder pair	$\vec{z}$ 1 A $\vec{x}$ 2 $\vec{y}$	$\odot\vec{y}$ $\vec{z}$ 1 A $\vec{x}$ 2 | $\vec{z}$ $\odot\vec{x}$ 1 A $\vec{y}$ 2	4
Planar joint	$\vec{z}$ A 1 $\vec{x}$ 2 $\vec{y}$	$\vec{z}$ A 1 $\vec{y}$ $\odot\vec{x}$ 2	3
Spherical joint	2 $\vec{z}$ 1 A $\vec{x}$ $\vec{y}$	2 $\vec{z}$ 1 A $\odot\vec{x}$ $\vec{y}$	3
Cardan joint	2 $\vec{z}$ 1 A $\vec{x}$ $\vec{y}$	2 $\vec{z}$ 1 A $\vec{y}$ $\odot\vec{x}$	2
Cylindrical joint	$\vec{z}$ 2 A 1 $\vec{x}$ $\vec{y}$	$\odot\vec{y}$ $\vec{z}$ 2 $\vec{x}$ A 1 | $\odot\vec{x}$ $\vec{z}$ 2 1 A $\vec{y}$	2

Continued

Kinematic pair	3-D representation	2-D representation	Degrees of freedom
Screw pair	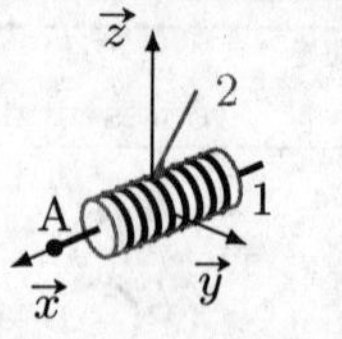 $\vec{z}$ 2 A 1 $\vec{x}$ $\vec{y}$	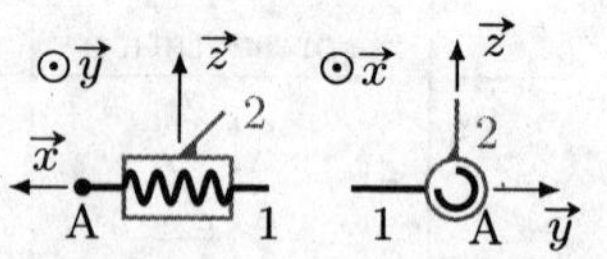 $\odot\vec{y}$ $\vec{z}$ 2 $\vec{x}$ A 1 $\odot\vec{x}$ $\vec{z}$ 2 1 A $\vec{y}$	1
Prismatic joint	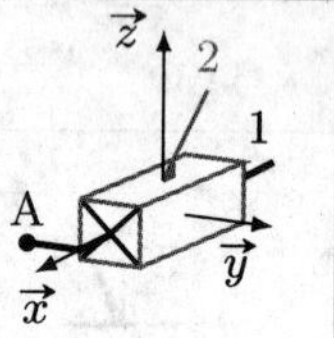 $\vec{z}$ 2 1 A $\vec{x}$ $\vec{y}$	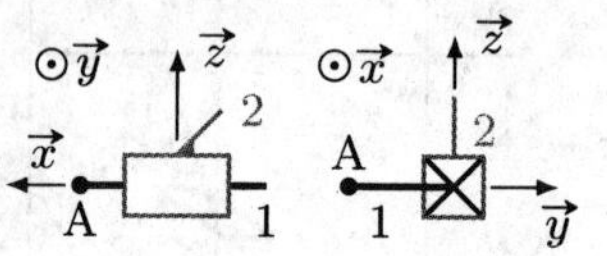 $\odot\vec{y}$ $\vec{z}$ 2 $\vec{x}$ A 1 $\odot\vec{x}$ $\vec{z}$ 2 A 1 $\vec{y}$	1
Revolute pair	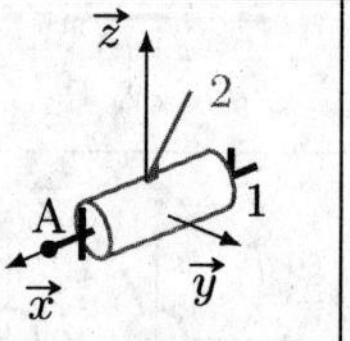 $\vec{z}$ 2 A 1 $\vec{x}$ $\vec{y}$	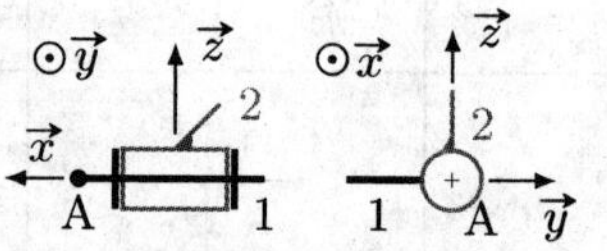 $\odot\vec{y}$ $\vec{z}$ 2 $\vec{x}$ A 1 $\odot\vec{x}$ $\vec{z}$ 2 1 A $\vec{y}$	1
Rigid joint	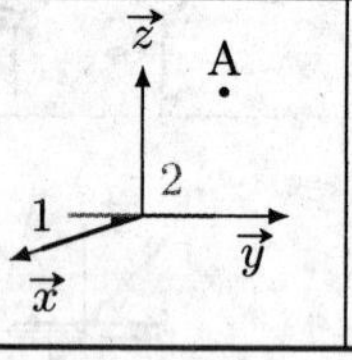 $\vec{z}$ A 2 1 $\vec{x}$ $\vec{y}$	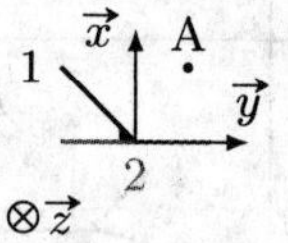 $\vec{x}$ A 1 $\vec{y}$ 2 $\otimes\vec{z}$	0

- the **sphere/cylinder pair** allows 3 rotation motions and 1 translation motion: it hence has 4 degrees of freedom. It is geometrically characterized by the center of the sphere and by the direction of the translation. For instance, if we consider a sphere/cylinder pair with a sphere of center A and a direction of translation $\vec{x}$ (as illustrated in Table 6.1), then 3 rotation motions are possible around the axes $(A, \vec{x})$, $(A, \vec{y})$ and $(A, \vec{z})$, and 1 translation motion is possible along the direction $\vec{x}$. The related twist and wrench thus are as follows:

$$\{\mathcal{V}(2/1)\} = \left\{ \begin{array}{c} p_{21}\,\vec{x} + q_{21}\,\vec{y} + r_{21}\,\vec{z} \\ u_{21}^{A}\,\vec{x} \end{array} \right\}_A$$

$$\text{and } \{\mathcal{F}(1 \to 2)\} = \left\{ \begin{array}{c} Y_{12}\,\vec{y} + Z_{12}\,\vec{z} \\ \vec{0} \end{array} \right\}_A$$

at the center A of the sphere and in any basis which contains the direction $\vec{x}$.

- the **planar joint** (also called **E-pair**) allows 1 rotation motion and 2 translation motions: it hence has 3 degrees of freedom. It is geometrically characterized by the normal vector at the contact. For instance, if we consider a planar joint whose normal vector at the contact is $\vec{z}$ (as illustrated in Table 6.1), then 1 rotation motion is possible around the axis $(A, \vec{z})$, and 2 translation motions are possible along the directions $\vec{x}$ and $\vec{y}$. The related twist and wrench thus are as follows:

$$\{\mathcal{V}(2/1)\} = \left\{ \begin{array}{c} r_{21}\,\vec{z} \\ u_{21}^{A}\,\vec{x} + v_{21}^{A}\,\vec{y} \end{array} \right\}_{A} \text{ and } \{\mathcal{F}(1 \to 2)\} = \left\{ \begin{array}{c} Z_{12}\,\vec{z} \\ L_{12}^{A}\,\vec{x} + M_{12}^{A}\,\vec{y} \end{array} \right\}_{A}$$

at any point of space and in any basis which contains the direction $\vec{z}$.

- the **spherical joint** (also called **ball joint** or **S-pair**) allows 3 rotation motions: it hence has 3 degrees of freedom. It is geometrically characterized by its center. For instance, if we consider a spherical joint of center A (as illustrated in Table 6.1), then 3 rotation motions are possible around the axes $(A, \vec{x})$, $(A, \vec{y})$ and $(A, \vec{z})$. The related twist and wrench thus are as follows:

$$\{\mathcal{V}(2/1)\} = \left\{ \begin{array}{c} p_{21}\,\vec{x} + q_{21}\,\vec{y} + r_{21}\,\vec{z} \\ \vec{0} \end{array} \right\}_{A}$$

$$\text{and } \{\mathcal{F}(1 \to 2)\} = \left\{ \begin{array}{c} X_{12}\,\vec{x} + Y_{12}\,\vec{y} + Z_{12}\,\vec{z} \\ \vec{0} \end{array} \right\}_{A}$$

at the center A of the spherical joint and in any basis.

- the **Cardan joint** (also called **U-pair**) is a spherical joint in which one rotation motion is made impossible by a rod and which allows 2 rotation motions: it hence has 2 degrees of freedom. It is geometrically characterized by its center, by the plane of the groove, and by the direction of the rod. For instance, if we consider a Cardan joint of center A, whose groove is in the plane $(A, \vec{y}, \vec{z})$, and whose rod is oriented by the vector $\vec{z}$ (as illustrated in Table 6.1), then 2 rotation motions are possible around the axes $(A, \vec{x})$ and $(A, \vec{z})$. The related twist and wrench thus are as follows:

$$\{\mathcal{V}(2/1)\} = \left\{ \begin{array}{c} p_{21}\,\vec{x} + r_{21}\,\vec{z} \\ \vec{0} \end{array} \right\}_{A}$$

$$\text{and } \{\mathcal{F}(1 \to 2)\} = \left\{ \begin{array}{c} X_{12}\,\vec{x} + Y_{12}\,\vec{y} + Z_{12}\,\vec{z} \\ M_{12}^{A}\,\vec{y} \end{array} \right\}_{A}$$

at the center A of the Cardan joint.

- the **cylindrical joint** (also called **C-pair**) allows 1 rotation motion and 1 translation motion: it hence has 2 degrees of freedom. It is geometrically characterized by its axis. For instance, if we consider a cylindrical joint of axis $(A, \vec{x})$ (as illustrated in Table 6.1), then 1 rotation motion is possible around the axis $(A, \vec{x})$ and 1 translation motion is possible along the direction $\vec{x}$. The related twist and wrench thus are as follows:

$$\{\mathcal{V}(2/1)\} = \left\{ \begin{array}{c} p_{21}\,\vec{x} \\ u_{21}^{A}\,\vec{x} \end{array} \right\}_A \quad \text{and} \quad \{\mathcal{F}(1 \to 2)\} = \left\{ \begin{array}{c} Y_{12}\,\vec{y} + Z_{12}\,\vec{z} \\ M_{12}^{A}\,\vec{y} + N_{12}^{A}\,\vec{z} \end{array} \right\}_A$$

  at any point of the axis $(A, \vec{x})$ and in any basis which contains the direction $\vec{x}$.

- the **screw pair** (also called **H-pair**) allows a relative screw motion, i.e. 1 rotation motion and 1 related translation motion: it hence has 1 degree of freedom (since both motions are not independent). It is geometrically characterized by its axis. For instance, if we consider a screw pair of axis $(A, \vec{x})$ (as illustrated in Table 6.1), then 1 rotation motion is possible around the axis $(A, \vec{x})$ and 1 related translation motion is possible along the direction $\vec{x}$. The related twist and wrench thus are as follows:

$$\{\mathcal{V}(2/1)\} = \left\{ \begin{array}{c} p_{21}\,\vec{x} \\ k\,p_{21}\,\vec{x} \end{array} \right\}_A \quad \text{and} \quad \{\mathcal{F}(1 \to 2)\} = \left\{ \begin{array}{c} X_{12}\,\vec{x} + Y_{12}\,\vec{y} + Z_{12}\,\vec{z} \\ L_{12}^{A}\,\vec{x} + M_{12}^{A}\,\vec{y} + N_{12}^{A}\,\vec{z} \end{array} \right\}_A$$

  where k is the reduced pitch of the helix related to the relative motion, as explained in section 3.4.6, and where X_{12} and L_{12}^{A} must respect:

$$L_{12}^{A} + k\,X_{12} = 0,$$

  for the power of the reciprocal actions exerted between both solid bodies to be null. As a consequence, we hence have:

$$\{\mathcal{V}(2/1)\} = \left\{ \begin{array}{c} p_{21}\,\vec{x} \\ k\,p_{21}\,\vec{x} \end{array} \right\}_A$$

$$\text{and } \{\mathcal{F}(1 \to 2)\} = \left\{ \begin{array}{c} X_{12}\,\vec{x} + Y_{12}\,\vec{y} + Z_{12}\,\vec{z} \\ -k\,X_{12}\,\vec{x} + M_{12}^{A}\,\vec{y} + N_{12}^{A}\,\vec{z} \end{array} \right\}_A$$

  at any point of the axis $(A, \vec{x})$ and in any basis which contains the direction $\vec{x}$.

Even if the representation in Table 6.1 corresponds to the latest standard, some representations from a former standard can still be found in the literature. These former representations are presented in Figures 6.3 and 6.4 for left-hand and right-hand helices.

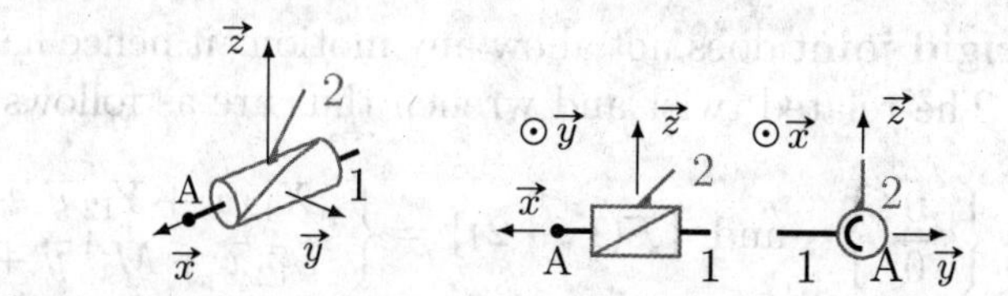

Figure 6.3 former representation of a screw pair with a left-hand helix

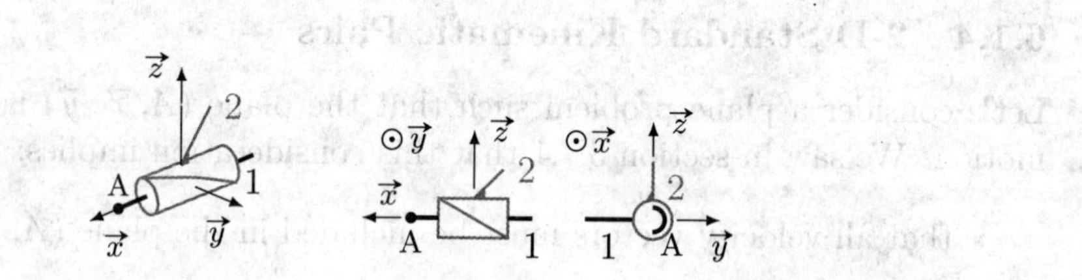

Figure 6.4 former representation of a screw pair with a right-hand helix

- the **prismatic joint** (also called **slider** or **P-pair**) allows 1 translation motion: it hence has 1 degree of freedom. It is geometrically characterized by its direction. For instance, if we consider a prismatic joint of direction $\vec{x}$ (as illustrated in Table 6.1), then 1 translation motion is possible along the direction $\vec{x}$. The related twist and wrench thus are as follows:

$$\{\mathcal{V}(2/1)\} = \left\{ \begin{array}{c} \vec{0} \\ u_{21}^{A}\,\vec{x} \end{array} \right\}_A \text{ and } \{\mathcal{F}(1 \to 2)\} = \left\{ \begin{array}{c} Y_{12}\,\vec{y} + Z_{12}\,\vec{z} \\ L_{12}^{A}\,\vec{x} + M_{12}^{A}\,\vec{y} + N_{12}^{A}\,\vec{z} \end{array} \right\}_A$$

at any point of space and in any basis which contains the direction $\vec{x}$.

- the **revolute pair** (also called **hinged joint** or **R-pair**) allows 1 rotation motion: it hence has 1 degree of freedom. It is geometrically characterized by its axis. For instance, if we consider a revolute pair of axis $(A, \vec{x})$ (as illustrated in Table 6.1), then 1 rotation motion is possible around the axis $(A, \vec{x})$. The related twist and wrench thus are as follows:

$$\{\mathcal{V}(2/1)\} = \left\{ \begin{array}{c} p_{21}\,\vec{x} \\ \vec{0} \end{array} \right\}_A \text{ and } \{\mathcal{F}(1 \to 2)\} = \left\{ \begin{array}{c} X_{12}\,\vec{x} + Y_{12}\,\vec{y} + Z_{12}\,\vec{z} \\ M_{12}^{A}\,\vec{y} + N_{12}^{A}\,\vec{z} \end{array} \right\}_A$$

at any point of the axis $(A, \vec{x})$ and in any basis which contains the direction $\vec{x}$.

- finally, the **rigid joint** does not allow any motion: it hence has no degree of freedom. The related twist and wrench thus are as follows:

$$\{\mathcal{V}(2/1)\} = \left\{ \begin{array}{c} \vec{0} \\ \vec{0} \end{array} \right\}_A \quad \text{and} \quad \{\mathcal{F}(1 \to 2)\} = \left\{ \begin{array}{c} X_{12}\vec{x} + Y_{12}\vec{y} + Z_{12}\vec{z} \\ L_{12}^A\vec{x} + M_{12}^A\vec{y} + N_{12}^A\vec{z} \end{array} \right\}_A$$

  at any point of space and in any basis.

6.1.4 2-D Standard Kinematic Pairs

Let's consider a plane problem such that the plane $(A, \vec{x}, \vec{y})$ is the plane of motion. We saw in section 3.7.1 that this consideration implies:

- that all velocity vectors must be included in the plane $(A, \vec{x}, \vec{y})$; and
- that all angular velocity vectors must be normal to the plane $(A, \vec{x}, \vec{y})$.

All velocity vectors can hence have only two components according to the directions $\vec{x}$ and $\vec{y}$, and all angular velocity vectors can have only one component according to an axis of direction $\vec{z}$. As a consequence, 8 forms of twists can be obtained by composing these 3 possible components:

1. $\{\mathcal{V}(2/1)\} = \left\{ \begin{array}{c} \vec{0} \\ \vec{0} \end{array} \right\}_A$, which corresponds to a rigid joint;

2. $\{\mathcal{V}(2/1)\} = \left\{ \begin{array}{c} r_{21}\vec{z} \\ \vec{0} \end{array} \right\}_P$ $\forall P \in (A, \vec{z})$, which corresponds to a R-pair whose axis is normal to the plane of motion;

3. $\{\mathcal{V}(2/1)\} = \left\{ \begin{array}{c} \vec{0} \\ u_{21}^P\vec{x} \end{array} \right\}_P$ $\forall P \in \mathcal{E}$, which corresponds to a P-pair whose direction is included in the plane of motion;

4. $\{\mathcal{V}(2/1)\} = \left\{ \begin{array}{c} \vec{0} \\ v_{21}^P\vec{y} \end{array} \right\}_P$ $\forall P \in \mathcal{E}$, which corresponds to a P-pair whose direction is included in the plane of motion;

5. $\{\mathcal{V}(2/1)\} = \left\{ \begin{array}{c} r_{21}\vec{z} \\ u_{21}^A\vec{x} \end{array} \right\}_A$, which corresponds to a line contact pair whose line of contact is $(A, \vec{z})$ and whose tangent plane at the contact is $(A, \vec{x}, \vec{z})$;

6. $\{\mathcal{V}(2/1)\} = \left\{ \begin{array}{c} r_{21}\vec{z} \\ v_{21}^{A}\vec{y} \end{array} \right\}_A$, which corresponds to a line contact pair whose line of contact is $(A, \vec{z})$ and whose tangent plane at the contact is $(A, \vec{y}, \vec{z})$;

7. $\{\mathcal{V}(2/1)\} = \left\{ \begin{array}{c} \vec{0} \\ u_{21}^{P}\vec{x} + v_{21}^{P}\vec{y} \end{array} \right\}_P$ $\forall P \in \mathcal{E}$, which does not correspond to any normalized kinematic pair;

8. $\{\mathcal{V}(2/1)\} = \left\{ \begin{array}{c} r_{21}\vec{z} \\ u_{21}^{A}\vec{x} + v_{21}^{A}\vec{y} \end{array} \right\}_A$, which corresponds to a E-pair whose normal vector is $\vec{z}$, but such a kinematic pair cannot be represented in the plane $(A, \vec{x}, \vec{y})$, and will hence not be retained.

Only four 2-dimensional standard kinematic pairs hence exist:

1. the **sliding pair**, whose line of contact is normal to the plane of motion and whose tangent plane at the contact is orthogonal to the plane of motion;

2. the **plane prismatic joint**, whose direction is included in the plane of motion;

3. the **articulating pair**, whose axis is normal to the plane of motion; and

4. the **plane rigid joint**.

These four 2-dimensional standard kinematic pairs are presented in Table 6.2.

6.2 Link Graph

6.2.1 Structure

Definition 56 (Equivalence Class)

An **equivalence class** is a set of pieces which have no relative motion, and which hence have the same motion at any time.

Table 6.2 2-D standard kinematic pairs

Kinematic pair	Representation	Degrees of freedom
Sliding pair	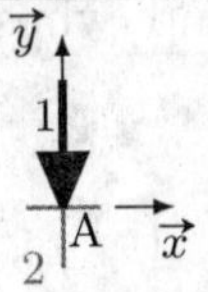 $\vec{y}$ 1 A 2 $\vec{x}$	2
Plane prismatic joint	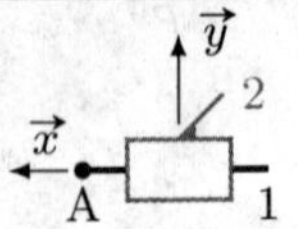 $\vec{y}$ 2 $\vec{x}$ A 1	1
Articulating pair	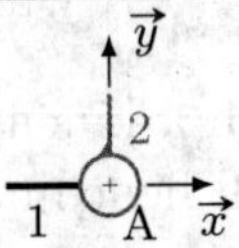 $\vec{y}$ 2 1 A $\vec{x}$	1
Plane rigid joint	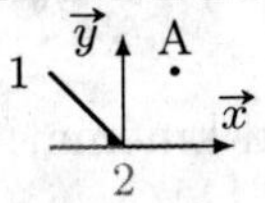 $\vec{y}$ A 1 $\vec{x}$ 2	0

Definition 57 (Link Graph)

A **link graph** is a graphical representation of a mechanism which is based on the description of the links which exist between its different equivalence classes.

Link graphs allow to identify the structure of a mechanism by identifying its equivalence classes and the links which exist between them.

Graphically, a link graph is a set of nodes and arcs:

- each node represents an equivalence class: this equivalence class is represented by a number written in a circle; and
- each arc links two nodes and represents a kinematic pair between two equivalence classes: this kinematic pair needs to be described as thoroughly as possible by means of its geometric characteristics (center, axis or direction).

An example of link graph is depicted in Figure 6.5.

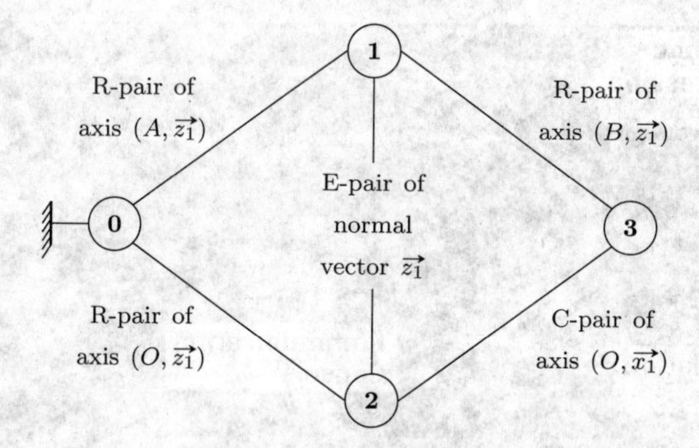

Figure 6.5 an example of link graph

6.2.2 Implementation

The construction of a link graph consists in 3 steps:

1. the equivalence classes which are in contact need to be identified;

2. the relative motions which exist between these equivalence classes need to be identified or even defined from observations; and

3. a kinematic pair needs to be chosen to model these relative motions.

Let's illustrate this construction process on the example of a real sailing boat autopilot (SBA) depicted in Figure 6.6 and used in the industrial science laboratory of ECPk. This system is used to maintain the direction of a sailing boat constant, and to compensate (or offset) the effects of the sea on the boat. The compass measures in real time the difference between the direction of the boat and the magnetic North, and the calculation unit analyzes the difference between the command and the real direction of the boat: after treating the signal from the angle sensor, the calculation unit controls the hydraulic transmission system to cause the rotation of the rudder arm.

Figure 6.7 shows the geometric diagram of the SBA system without the angle sensor. Four equivalence classes can be identified:

- (0): fix body;
- (1): rudder arm;
- (2): cylinder rod;

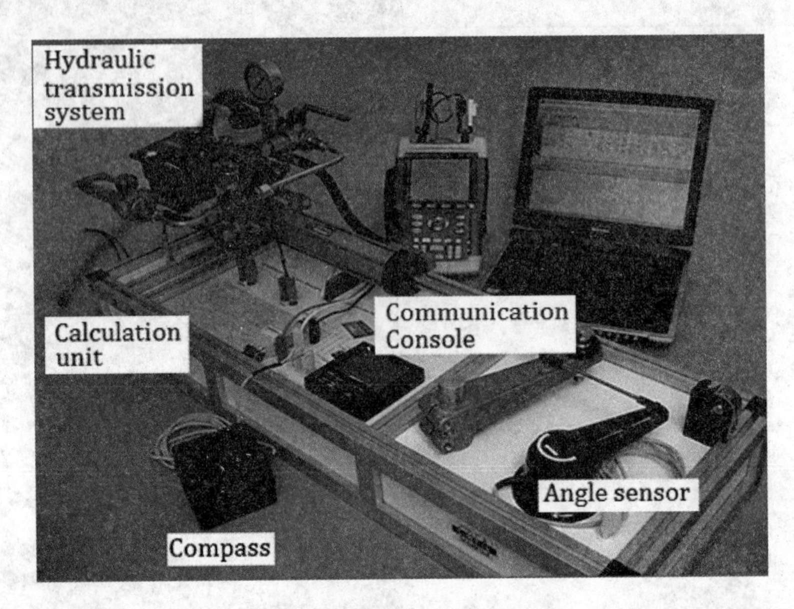

Figure 6.6 the sailing boat autopilot system in the pedagogical structure used in the industrial science laboratory

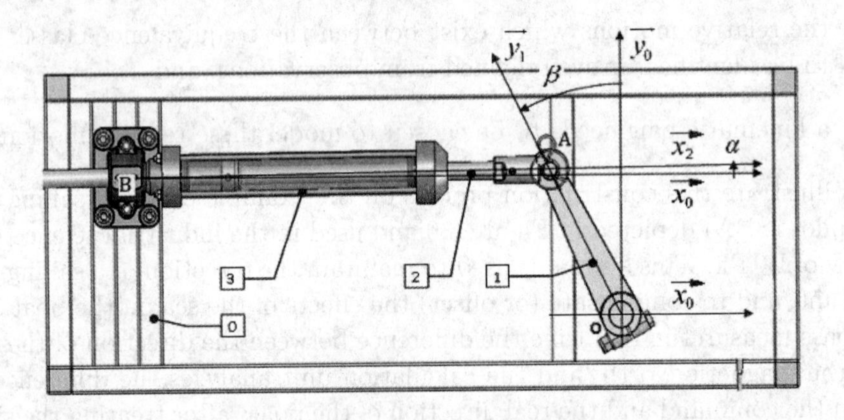

Figure 6.7 geometric diagram of the SBA system

- (3): cylinder chamber.

The links related to the motions 1/0 and 2/1 are modeled by R-pairs, the one related to the motion 3/0 by a S-pair, and the one related to the motion 3/2 by a C-pair: we can thus determine the link graph of the model of the SBA system shown in Figure 6.8.

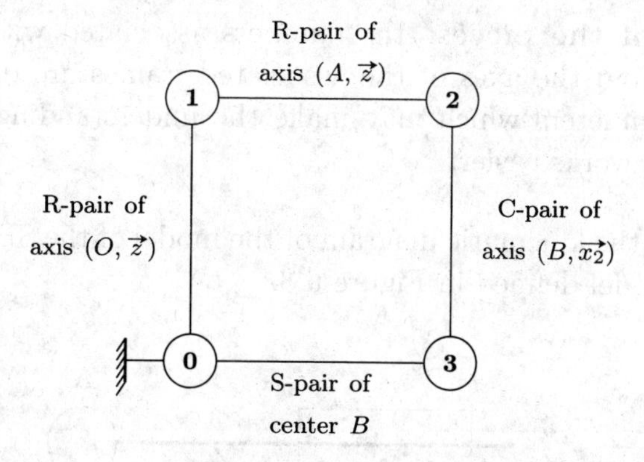

Figure 6.8 link graph of the model of the SBA system

6.3 Kinematic Diagram

6.3.1 General Principle

Definition 58 (Kinematic Diagram)

A **kinematic diagram** is a simplified geometric representation of the pieces of a mechanism and of the liaisons which exist between them by means of standard kinematic pairs.

6.3.2 Implementation and Construction

The following steps can be followed to build a kinematic diagram of a mechanism:

1. **setting of the geometry** of the mechanism: the different points and axes of the mechanism are drawn on the diagram;

2. **drawing of the symbols**: the symbol of each standard kinematic pair is drawn on the diagram while respecting its geometric characteristics (center, axis or direction);

3. **drawing of the links** between symbols: these symbols are linked together by means of lines in order to represent the subsets of pieces which have no relative motion; it is advised to use different colors for these different subsets, and to indicate the numbers associated with their pieces;

4. **spotting of the pieces**: the numbers associated with the pieces are indicated, and the axes of the associated frames are drawn, as well as all the parameters which may make the understanding of the way the mechanism works easier.

Figure 6.9 shows the kinematic diagram of the model of the SBA system, based on the chosen model defined in Figure 6.8.

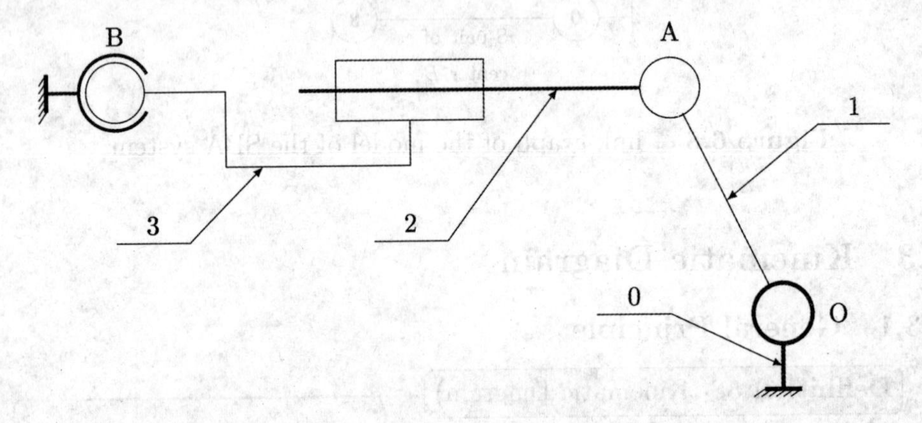

Figure 6.9 kinematic diagram of the model of the SBA system

6.4 Kinematically Equivalent Pairs

Since standard kinematic pairs are kinematic models whose characteristics are fully defined, it is possible to combine them to determine the kinematic pairs which are equivalent to associations of kinematic pairs, in order to simplify them and make the motions which are possible between pieces more explicit. When these serial and parallel associations of kinematic pairs are replaced by their equivalent kinematic pairs in a kinematic diagram, it is called a **minimal kinematic diagram**.

Definition 59 (Equivalent Pair)

A kinematic pair between two solid bodies (1) and (2) is **equivalent** to the set of the kinematic pairs which exist between these two solid bodies if it allows the same relative motions between them.

6.4.1 Serial Pairs

Let's consider a serial association of two known kinematic pairs (which are defined by their twists $\{\mathcal{V}(2/1)\}$ and $\{\mathcal{V}(3/2)\}$) between 3 solid bodies such as the one presented in Figure 6.10.

Figure 6.10 a serial association of two kinematic pairs

It is possible to determine the equivalent kinematic pair which exists between the solid bodies (1) and (3) by determining the six unknown kinematic scalars thanks to the resolution of the following equation between twists (which is called the composition of motions):

$$\{\mathcal{V}(3/1)\} = \{\mathcal{V}(3/2)\} + \{\mathcal{V}(2/1)\} \tag{6.1}$$

where the 6 unknowns are the 3 components of each one of the 2 vectors of the twist $\{\mathcal{V}(3/1)\}$.

For instance, let's consider the example of Figure 6.10 with a prismatic joint of direction $\vec{x}$ and a revolute pair of axis $(B, \vec{x})$. This association of kinematic pairs is illustrated in Figure 6.11.

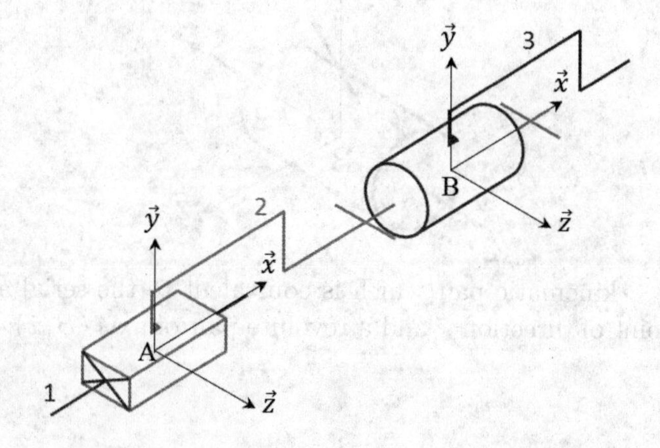

Figure 6.11 serial association of a prismatic joint of direction $\vec{x}$ and a revolute pair of axis $(B, \vec{x})$

The twists related to both kinematic pairs are as follows:

$$\{\mathcal{V}(2/1)\} = \left\{ \begin{array}{c} \vec{0} \\ u_{21}^{A}\,\vec{x} \end{array} \right\}_{A} \text{ and } \{\mathcal{V}(3/2)\} = \left\{ \begin{array}{c} p_{32}\,\vec{x} \\ \vec{0} \end{array} \right\}_{B}$$

These two twists can be summed only if they are expressed at the same point. However, the angular velocity vector of $\{\mathcal{V}(2/1)\}$ is null, and its velocity vector does hence not depend on the point at which the twist is expressed ($u_{21}^{A} = u_{21}$). We hence have:

$$\begin{aligned} \{\mathcal{V}(3/1)\} &= \{\mathcal{V}(3/2)\} + \{\mathcal{V}(2/1)\} \\ &= \left\{ \begin{array}{c} p_{32}\,\vec{x} \\ \vec{0} \end{array} \right\}_{B} + \left\{ \begin{array}{c} \vec{0} \\ u_{21}\,\vec{x} \end{array} \right\}_{B} \\ &= \left\{ \begin{array}{c} p_{32}\,\vec{x} \\ u_{21}\,\vec{x} \end{array} \right\}_{B} \end{aligned}$$

The twist $\{\mathcal{V}(3/1)\}$ is the twist of a cylindrical joint whose axis is $(B, \vec{x})$. The serial association of a prismatic joint of direction $\vec{x}$ and a revolute pair of axis $(B, \vec{x})$ hence is a cylindrical joint of axis $(B, \vec{x})$, as illustrated in Figure 6.12.

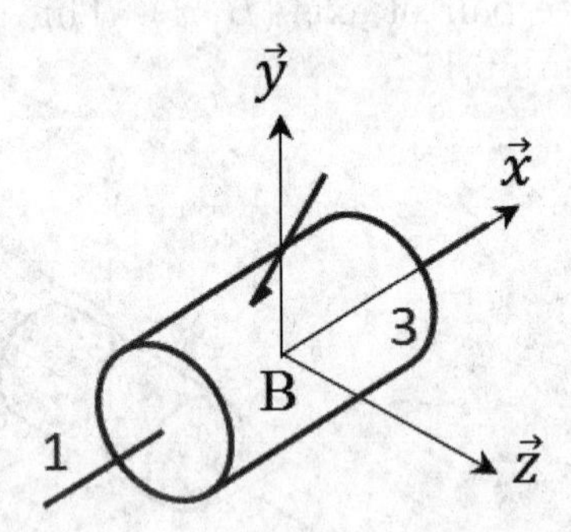

Figure 6.12 kinematic pair which is equivalent to the serial association of a prismatic joint of direction $\vec{x}$ and a revolute pair of axis $(B, \vec{x})$: a cylindrical joint

This approach can be generalized to the case of the serial association of n kinematic pairs.

Property 7 (Serial Association of n Kinematic Pairs)

The kinematic pair which is equivalent to the serial association of n kinematic pairs can be identified from its twist $\{\mathcal{V}(S_n/S_0)\}$, which is defined as the sum of the twists $\{\mathcal{V}(S_{i+1}/S_i)\}$ of these n kinematic pairs:

$$\{\mathcal{V}(S_n/S_0)\} = \sum_{i=1}^{n} \{\mathcal{V}(S_i/S_{i-1})\} \tag{6.2}$$

This relation is called the **composition of motions**, and it means that a motion is possible if it is allowed by **at least** one intermediate liaison.

6.4.2 Parallel Pairs

Let's consider a parallel association of two known kinematic pairs P_1 and P_2 (which are defined by their twists $\{\mathcal{V}_{P_1}(2/1)\}$ and $\{\mathcal{V}_{P_2}(2/1)\}$) between two solid bodies (1) and (2) such as the one presented in Figure 6.13.

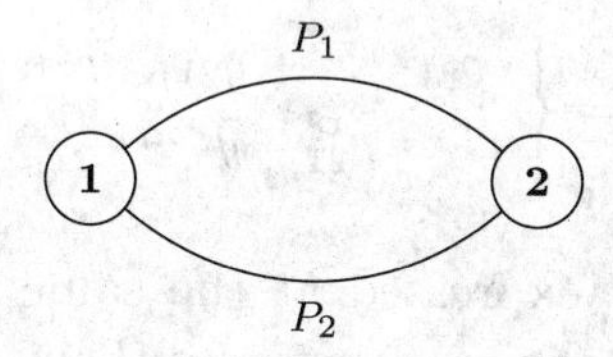

Figure 6.13 a parallel association of two kinematic pairs

It is possible to determine the equivalent kinematic pair which exists between the solid bodies (1) and (2) by determining the six unknown kinematic scalars thanks to the resolution of the following system of equations between twists:

$$\begin{cases} \{\mathcal{V}_{P_{\text{eq}}}(2/1)\} = \{\mathcal{V}_{P_1}(2/1)\} \\ \{\mathcal{V}_{P_{\text{eq}}}(2/1)\} = \{\mathcal{V}_{P_2}(2/1)\} \end{cases} \tag{6.3}$$

where the 6 unknowns are the 3 components of each one of the 2 vectors of the twist $\{\mathcal{V}_{P_{\text{eq}}}(2/1)\}$. The system (6.3) results from the fact that the twist of the equivalent kinematic pair P_{eq} must be compatible with the twists of the two kinematic pairs P_1 and P_2, i.e. the motions which are allowed by P_{eq} are the motions which are allowed by P_1 **and** P_2.

For instance, let's consider the example of Figure 6.13 with two sphere/plane pairs. A possible association of these kinematic pairs is illustrated in Figure 6.14, in the case where the two plane surfaces which are associated with the solid body (1) are coincident.

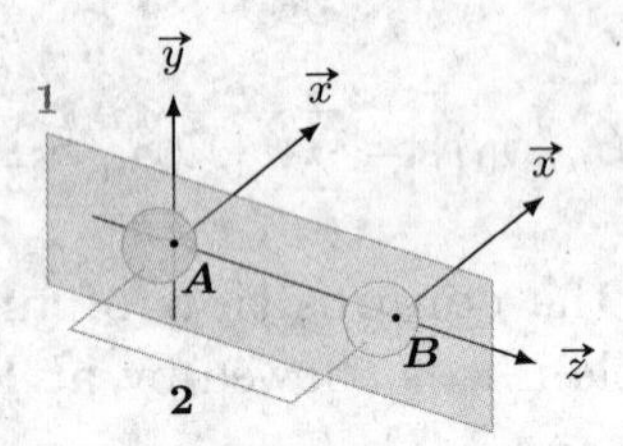

Figure 6.14 parallel association of two sphere/plane pairs with coincident planes

The twists related to both kinematic pairs are as follows:

$$\{\mathcal{V}_{P_1}(2/1)\} = \begin{Bmatrix} p_{21_{P_1}}\vec{x} + q_{21_{P_1}}\vec{y} + r_{21_{P_1}}\vec{z} \\ v^A_{21_{P_1}}\vec{y} + w^A_{21_{P_1}}\vec{z} \end{Bmatrix}_A$$

$$\{\mathcal{V}_{P_2}(2/1)\} = \begin{Bmatrix} p_{21_{P_2}}\vec{x} + q_{21_{P_2}}\vec{y} + r_{21_{P_2}}\vec{z} \\ v^B_{21_{P_2}}\vec{y} + w^B_{21_{P_2}}\vec{z} \end{Bmatrix}_B$$

We need both twists to be expressed at the same point to solve the system (6.3). For instance, we can choose to express them at the point B. If we note $\overrightarrow{AB} = c\,\vec{z}$, then we have:

$$\begin{aligned} \vec{V}_{P_1}(B,2/1) &= \vec{V}_{P_1}(A,2/1) + \vec{\Omega}_{P_1}(2/1) \times \overrightarrow{AB} \\ &= v^A_{21_{P_1}}\vec{y} + w^A_{21_{P_1}}\vec{z} + (p_{21_{P_1}}\vec{x} + q_{21_{P_1}}\vec{y} + r_{21_{P_1}}\vec{z}) \times c\,\vec{z} \\ &= q_{21_{P_1}}c\vec{x} + (v^A_{21_{P_1}} - p_{21_{P_1}}c)\vec{y} + w^A_{21_{P_1}}\vec{z} \end{aligned}$$

The system (6.3) hence is equivalent to:

$$\begin{cases} \{\mathcal{V}_{P_{eq}}(2/1)\} = \begin{Bmatrix} p_{21_{P_1}}\vec{x} + q_{21_{P_1}}\vec{y} + r_{21_{P_1}}\vec{z} \\ q_{21_{P_1}}c\vec{x} + (v^A_{21_{P_1}} - p_{21_{P_1}}c)\vec{y} + w^A_{21_{P_1}}\vec{z} \end{Bmatrix}_B \\ \{\mathcal{V}_{P_{eq}}(2/1)\} = \begin{Bmatrix} p_{21_{P_2}}\vec{x} + q_{21_{P_2}}\vec{y} + r_{21_{P_2}}\vec{z} \\ v^B_{21_{P_2}}\vec{y} + w^B_{21_{P_2}}\vec{z} \end{Bmatrix}_B \end{cases}$$

Solving this system hence is equivalent to solving the following equation between twists

$$\left\{\begin{array}{c} p_{21_{P_1}}\vec{x}+q_{21_{P_1}}\vec{y}+r_{21_{P_1}}\vec{z} \\ q_{21_{P_1}}c\vec{x}+(v^A_{21_{P_1}}-p_{21_{P_1}}c)\vec{y}+w^A_{21_{P_1}}\vec{z} \end{array}\right\}_B = \left\{\begin{array}{c} p_{21_{P_2}}\vec{x}+q_{21_{P_2}}\vec{y}+r_{21_{P_2}}\vec{z} \\ v^B_{21_{P_2}}\vec{y}+w^B_{21_{P_2}}\vec{z} \end{array}\right\}_B$$

We hence have:

$$\left\{\begin{array}{l} p_{21_{P_1}} = p_{21_{P_2}} = p_{21} \\ q_{21_{P_1}} = q_{21_{P_2}} = q_{21} \\ r_{21_{P_1}} = r_{21_{P_2}} = r_{21} \\ q_{21_{P_1}}c = 0 = u^B_{21} \\ v^A_{21_{P_1}} - p_{21_{P_1}}c = v^B_{21_{P_2}} = v^B_{21} \\ w^A_{21_{P_1}} = w^B_{21_{P_2}} = w^B_{21} \end{array}\right. \Rightarrow \{\mathcal{V}_{P_{\text{eq}}}(2/1)\} = \left\{\begin{array}{c} p_{21}\vec{x}+r_{21}\vec{z} \\ v^B_{21}\vec{y}+w^B_{21}\vec{z} \end{array}\right\}_B$$

The twist $\{\mathcal{V}_{P_{\text{eq}}}(2/1)\}$ is the twist of a line contact pair whose line of contact is $(B,\vec{z})$ and whose tangent plane at the contact is $(B,\vec{y},\vec{z})$. The parallel association of two sphere/plane pairs with coincident planes hence is a line contact pair whose line of contact is the line which goes through the points of contact of both sphere/plane pairs and whose tangent plane at the contact is the common plane of both sphere/plane pairs, as illustrated in Figure 6.15.

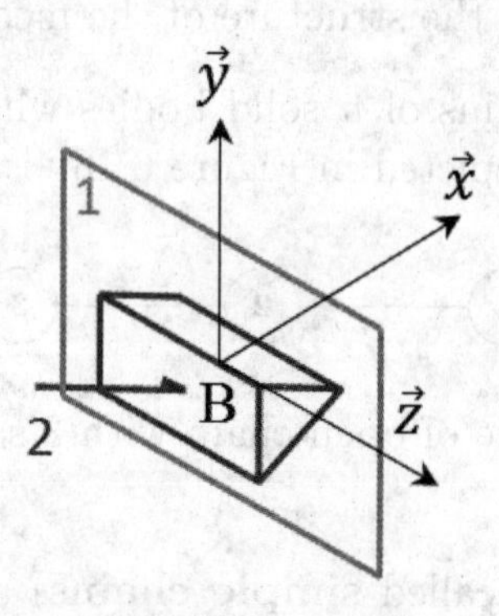

Figure 6.15 kinematic pair which is equivalent to the parallel association of two sphere/plane pairs with coincident planes: a line contact pair

This approach can be generalized to the case of the parallel association of n kinematic pairs.

Property 8 (Parallel Association of n Kinematic Pairs)

The kinematic pair P_{eq} which is equivalent to the parallel association of n kinematic pairs $P_1, \ldots, P_n$ between two solid bodies (1) and (2) can be identified from its twist $\{\mathcal{V}_{P_{\text{eq}}}(2/1)\}$, which is equal to the n twists $\{\mathcal{V}_{P_i}(2/1)\}$ of these n kinematic pairs:

$$\begin{aligned}\{\mathcal{V}_{P_{\text{eq}}}(2/1)\} &= \{\mathcal{V}_{P_1}(2/1)\} = \{\mathcal{V}_{P_2}(2/1)\} = \cdots = \{\mathcal{V}_{P_{n-1}}(2/1)\} \\ &= \{\mathcal{V}_{P_n}(2/1)\} \end{aligned} \tag{6.4}$$

This relation is called the **kinematic compatibility relation**, and it means that a motion is possible if it is allowed by **all** the intermediate liaisons.

6.5 Analysis of Simple Chains

Definition 60 (Chain of Solid Bodies)

A **chain of solid bodies** is a set of rigid solid bodies which are linked together by liaisons.

Different types of chains are considered depending on the number of pieces, the number of liaisons, and the structure of the mechanism:

- **open chains** are chains of n solid bodies which are linked together by $(n-1)$ liaisons, as depicted in Figure 6.16;

Figure 6.16 an example of open chain with 3 solid bodies and 2 liaisons

- **closed chains** (also called **simple chains**) are open chains whose extreme solid bodies are linked by a liaison: they hence are chains of n solid bodies which are linked together by n liaisons and which hence form a loop, as depicted in Figure 6.17;

- **complex chains** are chains of n solid bodies which are linked together by $a > n$ liaisons and hence form many loops.

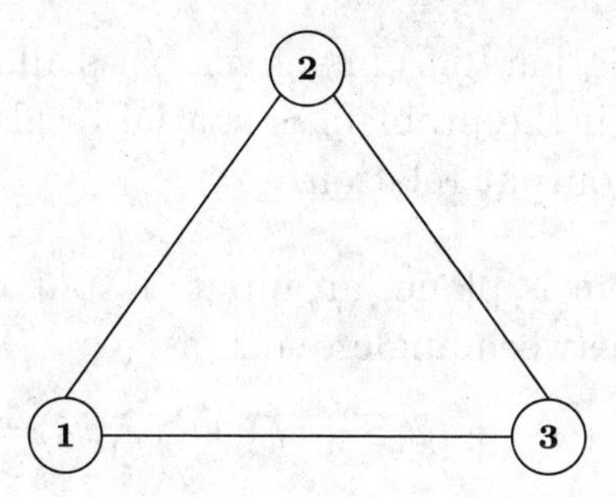

Figure 6.17 an example of closed chain with 3 solid bodies and 3 liaisons

If the structure of a mechanism can be described by a simple chain, then the geometric and kinematic analyses of this simple chain will allow us to determine the relations which exist between the output and input parameters of this mechanism.

6.5.1 Geometric Point of View

The **geometric analysis** of a mechanism allows to identify the *input/output relations* which exist between the output geometric parameters and the input geometric parameters of the mechanism.

Definition 61 (Kinematic Chain)

A **kinematic chain** is the succession of the kinematic pairs which allow to realize a simple chain.

Definition 62 (Input/Output Relation)

For a couple of geometric input/output parameters, the **input/output relation** defines the **geometric transfer function** of the kinematic chain of the system. It results from the following vector relation:

$$\overrightarrow{OA_1} + \sum_{i=1}^{n-1} \overrightarrow{A_i A_{i+1}} + \overrightarrow{A_n O} = \vec{0} \tag{6.5}$$

where the point O is the origin of the immobile frame, and the points A_i are the centers of the kinematic pairs which belong to the kinematic chain.

The projection of the relation (6.5) in a given basis allows to write the 2 (if the problem is plane) or 3 (if the problem is spatial) scalar relations which allow to determine the input/output relation.

If the considered problem is plane, an input/output relation can also be obtained from a relation between angles, such as:

$$(\overrightarrow{x_0},\overrightarrow{x_1})+(\overrightarrow{x_1},\overrightarrow{x_2})+\cdots+(\overrightarrow{x_{n-1}},\overrightarrow{x_n})+(\overrightarrow{x_n},\overrightarrow{x_0})=(\overrightarrow{x_0},\overrightarrow{x_0})=0 \tag{6.6}$$

This relation can be generalized to the case of 3-D problems as follows:

$$P_{b_0}^{b_1}P_{b_1}^{b_2}\dots P_{b_{n-1}}^{b_n}P_{b_n}^{b_0}=I \tag{6.7}$$

where $P_{b_i}^{b_j}$ is the transformation matrix from the basis b_i to the basis b_j and I is the identity matrix.

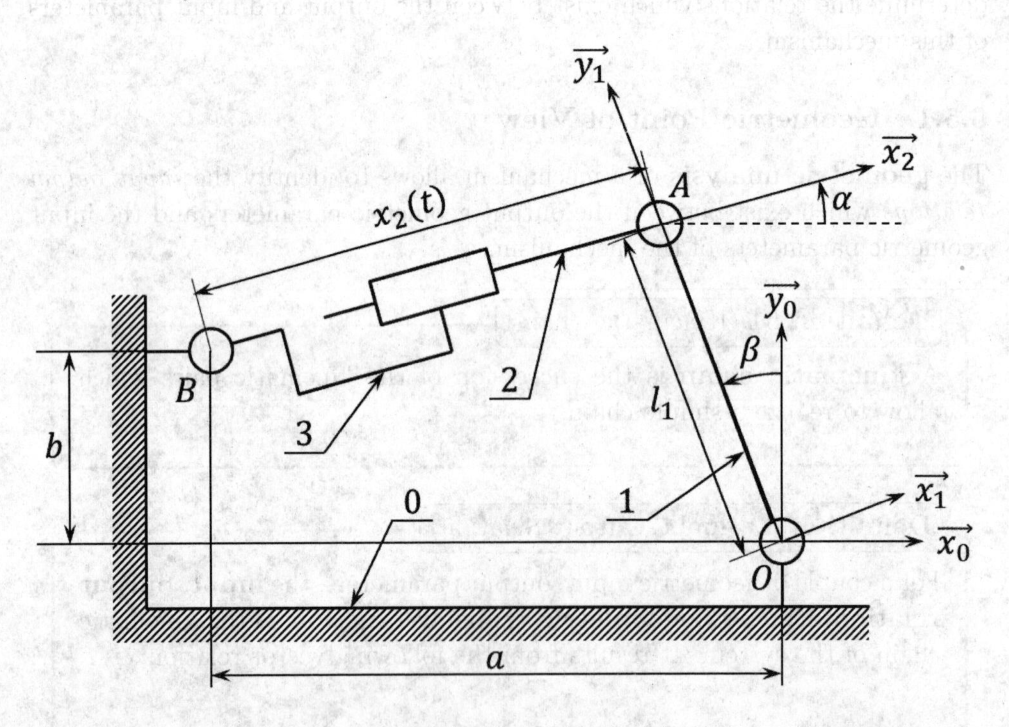

Figure 6.18 parameterized kinematic diagram of the SBA system

In the case of the SBA system, whose parameterized kinematic diagram is shown in Figure 6.18 since the problem is plane, we define the displacement $x_2(t)$ as the input, and the rotation angle β of the rudder arm as the output. The vector relation is

$$\begin{aligned}&\overrightarrow{OB}+\overrightarrow{BA}+\overrightarrow{AO}=\vec{0}\\ \Leftrightarrow &\overrightarrow{BA}-\overrightarrow{OA}-\overrightarrow{BO}=\vec{0}\\ \Leftrightarrow &x_2(t)\overrightarrow{x_2}-l_1\overrightarrow{y_1}-\overrightarrow{BO}=\vec{0}\end{aligned}$$

where

$$\begin{cases}\overrightarrow{BO}=a\overrightarrow{x_0}-b\overrightarrow{y_0}\\ \alpha=(\overrightarrow{x_0},\overrightarrow{x_2})=(\overrightarrow{y_0},\overrightarrow{y_2})\\ \beta=(\overrightarrow{x_0},\overrightarrow{x_1})=(\overrightarrow{y_0},\overrightarrow{y_1})\end{cases}$$

We hence have

$$\begin{aligned}&x_2(t)\overrightarrow{x_2}-l_1\overrightarrow{y_1}-a\overrightarrow{x_0}+b\overrightarrow{y_0}=\vec{0}\\
\Leftrightarrow&\begin{cases}x_2(t)\overrightarrow{x_2}\cdot\overrightarrow{x_0}-l_1\overrightarrow{y_1}\cdot\overrightarrow{x_0}-a\overrightarrow{x_0}\cdot\overrightarrow{x_0}+b\overrightarrow{y_0}\cdot\overrightarrow{x_0}=0\\ x_2(t)\overrightarrow{x_2}\cdot\overrightarrow{y_0}-l_1\overrightarrow{y_1}\cdot\overrightarrow{y_0}-a\overrightarrow{x_0}\cdot\overrightarrow{y_0}+b\overrightarrow{y_0}\cdot\overrightarrow{y_0}=0\end{cases}\\
\Leftrightarrow&\begin{cases}x_2(t)\cos(\overrightarrow{x_0},\overrightarrow{x_2})-l_1\cos(\overrightarrow{x_0},\overrightarrow{y_1})-a=0\\ x_2(t)\cos(\overrightarrow{y_0},\overrightarrow{x_2})-l_1\cos(\overrightarrow{y_0},\overrightarrow{y_1})+b=0\end{cases}\\
\Leftrightarrow&\begin{cases}x_2(t)\cos\alpha-l_1\cos\left(\dfrac{\pi}{2}+\beta\right)-a=0\\ x_2(t)\cos\left(\dfrac{\pi}{2}-\alpha\right)-l_1\cos\beta+b=0\end{cases}\\
\Leftrightarrow&\begin{cases}x_2(t)\cos\alpha+l_1\sin\beta-a=0\\ x_2(t)\sin\alpha-l_1\cos\beta+b=0\end{cases}\\
\Leftrightarrow&\begin{cases}x_2(t)\cos\alpha=a-l_1\sin\beta\\ x_2(t)\sin\alpha=l_1\cos\beta-b\end{cases}\\
\Rightarrow&x_2(t)=\sqrt{(a-l_1\sin\beta)^2+(b-l_1\cos\beta)^2}\end{aligned}\tag{6.8}$$

6.5.2 Kinematic Point of View

The **kinematic analysis** of a mechanism allows to establish the relations which exist between the output kinematic parameters and the input kinematic parameters of the mechanism.

Definition 63 (Kinematic Chain Relation)

The **kinematic chain relation** is the writing of the composition of motions under the form:

$$\{\mathcal{V}(S_n/S_0)\}=\sum_{i=1}^{n}\{\mathcal{V}(S_i/S_{i-1})\}\tag{6.9}$$

This relation needs to be written for **each** loop of a chain of solid bodies. Each such relation between twists is equivalent to 2 vector equations whose projection in a basis provides 6 scalar equations. Some relations between the output kinematic parameters and the input kinematic parameters can be deduced from these equations, which define the **kinematic transfer function** (the input/output relation) of the mechanism.

Let's illustrate the determination of the kinematic chain relation on the SBA system whose parameterized kinematic diagram is shown in Figure 6.18. The kinematic chain relation is

$$\{\mathcal{V}(1/0)\} = \{\mathcal{V}(1/2)\} + \{\mathcal{V}(2/3)\} + \{\mathcal{V}(3/0)\}$$

where

$$\{\mathcal{V}(1/0)\} = \left\{ \begin{array}{c} \dot{\beta}\vec{z} \\ \vec{0} \end{array} \right\}_O, \{\mathcal{V}(1/2)\} = \left\{ \begin{array}{c} (\dot{\beta}-\dot{\alpha})\vec{z} \\ \vec{0} \end{array} \right\}_A,$$
$$\{\mathcal{V}(2/3)\} = \left\{ \begin{array}{c} \vec{0} \\ \dot{x_2}(t)\overrightarrow{x_2} \end{array} \right\}_B, \{\mathcal{V}(3/0)\} = \left\{ \begin{array}{c} \dot{\alpha}\vec{z} \\ \vec{0} \end{array} \right\}_B$$

Let's express these 4 twists at the point A. Thanks to the point change formula, we get

$$\begin{aligned}
\{\mathcal{V}(1/0)\} &= \left\{ \begin{array}{c} \dot{\beta}\vec{z} \\ \vec{0} \end{array} \right\}_O = \left\{ \begin{array}{c} \dot{\beta}\vec{z} \\ \dot{\beta}\vec{z} \times \overrightarrow{OA} \end{array} \right\}_A = \left\{ \begin{array}{c} \dot{\beta}\vec{z} \\ \dot{\beta}\vec{z} \times l_1\overrightarrow{y_1} \end{array} \right\}_A = \left\{ \begin{array}{c} \dot{\beta}\vec{z} \\ -\dot{\beta}l_1\overrightarrow{x_1} \end{array} \right\}_A \\
\{\mathcal{V}(2/3)\} &= \left\{ \begin{array}{c} \vec{0} \\ \dot{x_2}(t)\overrightarrow{x_2} \end{array} \right\}_B = \left\{ \begin{array}{c} \vec{0} \\ \dot{x_2}(t)\overrightarrow{x_2} \end{array} \right\}_A \\
\{\mathcal{V}(3/0)\} &= \left\{ \begin{array}{c} \dot{\alpha}\vec{z} \\ \vec{0} \end{array} \right\}_B = \left\{ \begin{array}{c} \dot{\alpha}\vec{z} \\ \dot{\alpha}\vec{z} \times \overrightarrow{BA} \end{array} \right\}_A = \left\{ \begin{array}{c} \dot{\alpha}\vec{z} \\ \dot{\alpha}\vec{z} \times x_2(t)\overrightarrow{x_2} \end{array} \right\}_A \\
&= \left\{ \begin{array}{c} \dot{\alpha}\vec{z} \\ \dot{\alpha}x_2(t)\overrightarrow{y_2} \end{array} \right\}_A
\end{aligned}$$

We hence get

$$\left\{ \begin{array}{c} \dot{\beta}\vec{z} \\ -\dot{\beta}l_1\overrightarrow{x_1} \end{array} \right\}_A - \left\{ \begin{array}{c} (\dot{\beta}-\dot{\alpha})\vec{z} \\ \vec{0} \end{array} \right\}_A - \left\{ \begin{array}{c} \vec{0} \\ \dot{x_2}(t)\overrightarrow{x_2} \end{array} \right\}_A - \left\{ \begin{array}{c} \dot{\alpha}\vec{z} \\ \dot{\alpha}x_2(t)\overrightarrow{y_2} \end{array} \right\}_A = \left\{ \begin{array}{c} \vec{0} \\ \vec{0} \end{array} \right\}_A$$
$$\Rightarrow -\dot{\beta}l_1\overrightarrow{x_1} - \dot{x_2}(t)\overrightarrow{x_2} - \dot{\alpha}x_2(t)\overrightarrow{y_2} = \vec{0}$$

If we project this relation on the direction $\overrightarrow{x_2}$ to remove $\dot{\alpha}$ from the relation, we get

$$\begin{aligned}
&-\dot{\beta}l_1\overrightarrow{x_1}\cdot\overrightarrow{x_2}-\dot{x_2}(t)\overrightarrow{x_2}\cdot\overrightarrow{x_2}-\dot{\alpha}x_2(t)\overrightarrow{y_2}\cdot\overrightarrow{x_2}=0\\
\Leftrightarrow&-\dot{\beta}l_1\cos(\overrightarrow{x_1},\overrightarrow{x_2})-\dot{x_2}(t)=0\\
\Leftrightarrow&-\dot{\beta}l_1\cos(\beta-\alpha)-\dot{x_2}(t)=0\\
\Leftrightarrow&-\dot{\beta}l_1(\cos\beta\cos\alpha+\sin\beta\sin\alpha)-\dot{x_2}(t)=0
\end{aligned}$$

Some expressions for $\cos\alpha$ and $\sin\alpha$ can then be determined thanks to Figure 6.18. Indeed, we can write the following geometric relations:

$$\left\{\begin{array}{l} x_2(t)\cos\alpha+l_1\sin\beta=a\\ x_2(t)\sin\alpha+b=l_1\cos\beta\end{array}\right.\Leftrightarrow\left\{\begin{array}{l}\cos\alpha=\dfrac{a-l_1\sin\beta}{x_2(t)}\\ \sin\alpha=\dfrac{l_1\cos\beta-b}{x_2(t)}\end{array}\right.$$

We hence have

$$\begin{aligned}
&-\dot{\beta}l_1(\cos\beta\cos\alpha+\sin\beta\sin\alpha)-\dot{x_2}(t)=0\\
\Leftrightarrow&-\dot{\beta}l_1\left(\cos\beta\frac{a-l_1\sin\beta}{x_2(t)}+\sin\beta\frac{l_1\cos\beta-b}{x_2(t)}\right)-\dot{x_2}(t)=0\\
\Leftrightarrow&\,\dot{x_2}(t)x_2(t)=-\dot{\beta}l_1(\cos\beta(a-l_1\sin\beta)+\sin\beta(l_1\cos\beta-b))\\
\Leftrightarrow&\,\dot{x_2}(t)x_2(t)=-\dot{\beta}l_1(a\cos\beta-b\sin\beta)\\
\Leftrightarrow&\,\dot{x_2}(t)x_2(t)=\dot{\beta}l_1(b\sin\beta-a\cos\beta)
\end{aligned}$$

It can be noticed that we can obtain the same relation by differentiating the geometric input/output relation (6.8), which is equivalent to the relation $x_2^2(t)=(a-l_1\sin\beta)^2+(b-l_1\cos\beta)^2$:

$$\begin{aligned}
&\frac{d}{dt}(x_2^2(t))=\frac{d}{dt}((a-l_1\sin\beta)^2+(b-l_1\cos\beta)^2)\\
\Leftrightarrow&\,2\dot{x_2}(t)x_2(t)=-2\dot{\beta}l_1\cos\beta(a-l_1\sin\beta)+2\dot{\beta}l_1\sin\beta(b-l_1\cos\beta)\\
\Leftrightarrow&\,\dot{x_2}(t)x_2(t)=\dot{\beta}l_1(-\cos\beta(a-l_1\sin\beta)+\sin\beta(b-l_1\cos\beta))\\
\Leftrightarrow&\,\dot{x_2}(t)x_2(t)=\dot{\beta}l_1(b\sin\beta-a\cos\beta)
\end{aligned}$$

To conclude, these two approaches (geometric approach and kinematic approach) are perfectly equivalent. The geometric approach remains, most of the time, simpler, even though the kinematic approach is necessary to evaluate the mobilities of a model. This latter point will be tackled in a dedicated chapter in the second volume.

6.6 Conclusions and Perspectives

In this chapter, we saw how to model systems of rigid solid bodies. This modeling is necessary in many fields of modern engineering, since most innovative industrial systems (robots, airplanes, rockets, cars, etc.) contain solid bodies which are linked together. Many works which allow an efficient and pertinent analysis of novel systems are based on this modeling as it allows to predict their mechanical performances. The content of this chapter hence allows the reader to use a rigorous theoretical approach on a problem considered difficult.

However, one must keep in mind that the kinematic pairs presented in this chapter are only a model of the real links which exist between the solid bodies of mechanisms. Indeed, real surfaces have defects, and real links require clearances for the mechanism to work. In order to perform the robust design of innovative technological systems, this study needs to be completed by the isostatic and hyperstatic analyses of structures, which will be presented in the second volume. Only then will the industrial stakes related to the imagination, analysis and design of systems be fully tackled.

Chapter 7

Combinatorial Analysis and Counting

The command system of automated systems often comprises a computer and a Programmable Logic Controller (PLC). Thanks to the reports provided by the sensors or to the instructions provided by a user, it determines the orders which need to be sent to the pre-actuators for them to perform the expected processes correctly with a complete or partial absence of any human being. This behavior is based on the acquisition, treatment and delivering of numerical data which are quantified by means of adapted numbering systems: for instance, the numerical data which are provided by a user on a keyboard are decimal whereas the data which are used by a computer are binary. This chapter presents the mathematical framework which is necessary to such systems.

7.1 Numbering and Coding Systems

The command system of automated systems generally uses variables. In order to study the behavior of such a system according to the variation of the value of these variables, it can be useful to group them into "words", each variable being a "letter" of such a word. These organizations of variables are called **coding systems**, and they are based on **numbering systems**. We are hence going to see the main numbering and coding systems which exist.

Definition 64 (Coding System)

A **coding system** is a bijective correspondence between a set of figures, letters, words, ... and some symbols (which can be of the same type, or of another type). When a coding system is associated with a numbering system, it is called a **weighted coding system**, and classical arithmetic operations can then be applied to the corresponding codes. Otherwise, in the case of **non-weighted coding systems**, such operations cannot be applied to the codes.

It can be noticed that, contrary to a numbering system, which is always weighted, a coding system can be weighted or non-weighted. Besides, a given command system may manipulate data which are expressed in different numbering systems, so we will see how to change the numbering system in which a data is expressed.

7.1.1 Weighted Numbering Systems

In order to manipulate numerical data, it is necessary to **quantify** them, i.e. get an image of them by means of a **numbering system**. The basic element of a numbering system (which allows to differentiate two numbering systems) is its **number base** N, which is a set of numerical values thanks to which it is possible to get any numerical data X.

A number base N is a set of numerical data called **weights**, which are powers of the natural integer N ($N > 1$) which is called the **key** of the number base. A number base N can hence be defined as:

$$\text{number base } N = (\ldots, N^{-n}, N^{-(n-1)}, \ldots, N^{-1}, N^0, N^1, \ldots, N^{p-1}, N^p, \ldots), \quad n, p \in \mathbb{N}^*$$

However, in this chapter, we are going to consider integer data only, so we will restrict number bases to the positive powers of their key. We will hence consider number bases N which are defined as:

$$\text{number base } N \ = (N^0, N^1, \ldots, N^{p-1}, N^p, \ldots), p \in \mathbb{N}^*$$

Any positive integer data X which is expressed in the number base N can hence be expressed under the form:

$$X = (a_p a_{p-1} \ldots a_1 a_0)_N = a_p N^p + a_{p-1} N^{p-1} + \cdots + a_1 N^1 + a_0 N^0 \tag{7.1}$$

where each coefficient a_i is called **digit** of weight i and represents a positive integer data such that $\forall i \in \{0, \ldots, p\}, 0 \leqslant a_i \leqslant (N-1)$.

The decimal numbering system is the most common one. Its origin is most probably related to the usual presence of 10 fingers on the hands of human beings, and it corresponds to the number base 10. When the number base N in which a number is expressed is not indicated, the decimal numbering system will be retained by default. According to the relation (7.1), the number 285 can hence be written under the following form:

$$(285)_{10} = 2.10^2 + 8.10^1 + 5.10^0$$

In this case:

- 10 is the *key* of the number base;
- the *number base* 10 is defined as the base $(10^0, 10^1, \ldots, 10^{p-1}, 10^p, \ldots), p \in \mathbb{N}^*$; and
- the figures 2, 8 and 5 are *digits* which correspond to different *weights*:
  - the figure 5 is the digit of weight 0;
  - the figure 8 is the digit of weight 1;
  - the figure 2 is the digit of weight 2.

Numbering systems are **weighted** systems because the position of each digit a_i is associated with a specific weight.

7.1.2 Binary Numbering System and Binary Coding Systems

Binary Numbering System

The binary numbering system was created during the 17th century by the German mathematician Gottfried Wilhelm Leibniz (1646-1716). It is a weighted numbering system of number base 2 which has two digits: 0 and 1. It is used in every electronic device because it is easy to implement from a technological point of view: indeed, its two digits can be represented by two electrical levels (e.g. 5 V for the digit 1 and 0 V for the digit 0), provided the values of these levels are distinct enough. According to the relation (7.1), the number $(\texttt{1001})_2$ can hence be expressed under the following form:

$$(\texttt{1001})_2 = 1.2^3 + 0.2^2 + 0.2^1 + 1.2^0$$

We hence have $(\texttt{1001})_2 = (9)_{10}$. This numbering system uses a specific vocabulary:

- a digit is called a **bit** (**bi**nary digi**t**) and its symbol is b;
- a **byte** is a word of 8 bits and its symbol is B;
- by extension, we have:
  - 1 kB (kilo byte) $= 2^{10}$ B $= 1,024$ B;
  - 1 MB (mega byte) $= 2^{10}$ kB $= 1,024$ kB $=$ 1,048,576 B;
  - 1 GB (giga byte) $= 2^{10}$ MB $= 1,024$ MB $=$ 1,048,576 kB $=$ 1,073,741,824 B.

It can also be noticed that a variable which can take only two values 0 and 1 is called a logic variable, as we will see in section 7.2.1.

Natural Binary Coding System

The natural binary coding system results from the correspondence between the decimal and binary numbering systems.

> **Definition 65** (Natural Binary Coding System)
>
> The **natural binary coding system** is a coding system which associates words of n logic variables to decimal values by conversion from decimal to binary.

Table 7.1 presents the correspondence between the decimal values from 0 to 15 and the corresponding natural binary code. Let's consider a decimal number N whose natural binary code can be expressed on n bits. We hence have:

$$\begin{aligned}
(N)_{10} &= (a_{n-1}a_{n-2}\dots a_2a_1a_0)_2 \\
&= a_{n-1}.2^{n-1} + a_{n-2}.2^{n-2} + \dots + a_2.2^2 + a_1.2^1 + a_0.2^0 \\
&\leqslant 1.2^{n-1} + 1.2^{n-2} + \dots + 1.2^2 + 1.2^1 + 1.2^0 \\
&\leqslant 1.(1 + 2^1 + 2^2 + \dots + 2^{n-2} + 2^{n-1}) \\
&\leqslant \frac{1-2^n}{1-2} \\
&\leqslant 2^n - 1 \\
&< 2^n
\end{aligned}$$

Since the function $x \longmapsto \ln x$ is increasing on $\mathbb{R}_+^*$, we hence have:

$$\ln N < \ln(2^n)$$
$$\Leftrightarrow \ln N < n \ln 2$$
$$\Leftrightarrow n > \frac{\ln N}{\ln 2}$$

For instance, let us consider the decimal number 145. We can determine that

$$n > \frac{\ln 145}{\ln 2} \approx 7.18$$

logic variables, i.e. 8 logic variables, will be necessary to code the decimal number 145 in natural binary code. We can also notice that n bits allow to code all decimal values from 0 to $(2^n - 1)$.

Table 7.1 correspondence between the decimal values from 0 to 15 and the natural binary code

Decimal value	**Natural binary code** $(a_3a_2a_1a_0)_2 = a_3 2^3 + a_2 2^2 + a_1 2^1 + a_0 2^0$
0	0000
1	0001
2	0010
3	0011
4	0100
5	0101
6	0110
7	0111
8	1000
9	1001
10	1010
11	1011
12	1100
13	1101
14	1110
15	1111

In the case of Table 7.1, 4 bits allow to code all decimal values from 0 to 15, and these 4 bits can be organized in 4 columns. It can also be noticed that:

- for the bit of weight 0, there is a vertical alternance of one '0' and one '1';
- for the bit of weight 1, there is a vertical alternance of two '0' and two '1';

- for the bit of weight 2, there is a vertical alternance of four '0' and four '1'; and finally
- for the bit of weight 3, there is a vertical alternance of eight '0' and eight '1'.

Let's consider the natural binary code for the decimal values from 0 to $(2^n - 1)$. We can hence deduce that, for any bit of weight $p \leqslant n$, there is an alternance of 2^p '0' and 2^p '1' in the natural binary code.

An advantage of the natural binary code is that the conversion from decimal to natural binary is very easy. However, the state of more than one logic variable may change between the codes which correspond to two successive decimal values (for instance, between the two codes `0111` and `1000` which correspond to the decimal values 7 and 8, the state of 4 logic variables simultaneously changes). This simultaneous change of state of many variables can cause technological uncertainties: indeed, it implies that the state of many technological components will simultaneously change, which is impossible whatever the technology retained. Actually, the state of the components will change *successively* in an order which is unknown, and the system will manipulate erroneous data during a period of time which is very short but long enough to be taken into account.

Reflected Binary Coding System (or Gray Coding System)

The **reflected binary coding system**, or Gray coding system (referring to the American engineer Frank Gray who invented it in 1953), is based on the natural binary coding system but includes modifications which suppress the drawback of the natural binary coding system which was presented above: indeed, the specificity of the reflected binary coding system is that the state of only one variable changes between the codes which correspond to two successive decimal values. This specificity is illustrated in Table 7.2, as well as the correspondence between the natural binary code and the reflected binary code, since it is very easy to find the conversion from natural binary to reflected binary, or conversely. This table also emphasizes some symmetries to which this coding system owes its name. These symmetries can be identified on the 2nd, 4th and 8th rows: the first symmetry is related to the 2 bits (1 bit $\times$ 2 rows) of the rightmost column, the second symmetry is related to the 8 bits (2 bits $\times$ 4 rows) of the two rightmost columns, and the third symmetry is related to the 24 bits (3 bits $\times$ 8 rows) of the three rightmost columns.

Table 7.2 correspondence between the decimal values from 0 to 15, the natural binary code and the reflected binary code

Decimal value	Natural binary code	Reflected binary code
0	0000	0000
1	0001	0001
2	0010	0011
3	0011	0010
4	0100	0110
5	0101	0111
6	0110	0101
7	0111	0100
8	1000	1100
9	1001	1101
10	1010	1111
11	1011	1110
12	1100	1010
13	1101	1011
14	1110	1001
15	1111	1000
...	...	...

This specificity of the reflected binary coding system is important, and it is exploited in the implementation of numerical position encoders, since it allows to address the technological issues which are due to the simultaneous change of state of many variables. Besides, as we will see in section 7.2.6, the organization of Karnaugh tables is based on the symmetric disposal of the values of logic variables. However, the coding of a decimal value in reflected binary code is not as easy as its coding in natural binary code, and arithmetic operations cannot be applied to the reflected binary code since it is not weighted.

Binary Coded Decimal (BCD) Coding System

The **BCD coding system** is a binary coding system which associates a binary expression to a decimal number X by concatenating the conversions of each one of its decimal digits from decimal to binary. As each one of these digits is comprised between 0 and 9, the corresponding binary code requires 4 bits. This code is non-weighted, and arithmetic operations can hence not be applied to it. However, it can be associated with very quick coding/decoding algorithms, and one of its main applications is display devices (for instance, in seven-segment displays, as illustrated in Figures 7.1 and 7.2).

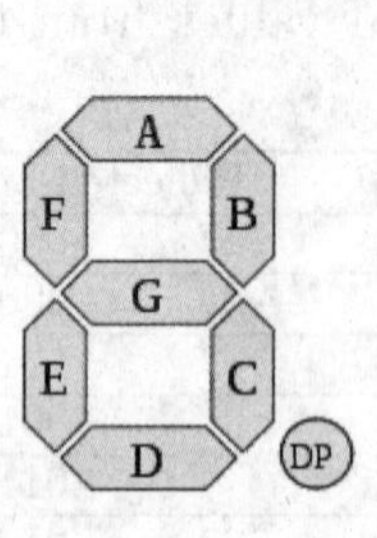

Figure 7.1 a seven-segment display

Figure 7.2 an example of seven-segment display in an elevator

The BCD code for the decimal number 125 can hence be determined as follows:

$$(\underbrace{1}_{0001}\underbrace{2}_{0010}\underbrace{5}_{0101})_{10} = (000100100101)_{\text{BCD}}$$

Conversely, the decimal value which corresponds to a BCD code can be determined by replacing each group of 4 bits from the right to the left by the corresponding decimal digit. For instance, the decimal value which corresponds to the BCD code $(000101100111)_{\text{BCD}}$ can be determined as follows:

$$(\underbrace{0001}_{1}\underbrace{0110}_{6}\underbrace{0111}_{7})_{\text{BCD}} = (167)_{10}$$

The BCD coding system is also called **8421 coding system**: this name refers to the weights of the 4 bits (2^3, 2^2, 2^1 and 2^0) which are needed to code each decimal digit.

7.1.3 Hexadecimal Numbering and Coding Systems

Hexadecimal Numbering System

The hexadecimal numbering system was created to express groups of 4 bits obtained in binary code under a compact form. It is very common in computer science, particularly for memory addresses.

The hexadecimal numbering system is a weighted numbering system of number base 16 ($\varepsilon\xi$, *ex* means "six" in ancient Greek, and *decimus* means "tenth" in

Latin) which has 16 digits: the decimal digits from 0 to 9 for the 10 first digits, and the letters from A to F for the 6 last digits (with the following correspondence: $(A)_{16} = (10)_{10}$, $(B)_{16} = (11)_{10}$, $(C)_{16} = (12)_{10}$, $(D)_{16} = (13)_{10}$, $(E)_{16} = (14)_{10}$, and $(F)_{16} = (15)_{10}$). According to the relation (7.1), the number $(4D5)_{16}$ can hence be expressed under the following form:

$$(4D5)_{16} = 4.16^2 + 13.16^1 + 5.16^0$$

We hence have $(4D5)_{16} = 4 \times 256 + 13 \times 16 + 5 = (1237)_{10}$.

Hexadecimal Coding System

The hexadecimal coding system is associated with the hexadecimal numbering system, and it is much used in Programmable Logic Controllers. It is the "natural" code of the automatician since it allows to group information obtained in binary code in packs of 4 bits. The hexadecimal code is used in operating systems (Linux, Windows, MacOS, etc.) to express in a synthetic way the memory addresses which are assigned to a given peripheral, which can sometimes be coded on 64 bits.

7.1.4 Equivalence between the Decimal, Binary and Hexadecimal Numbering Systems

Table 7.3 presents the equivalence between the three previously presented numbering systems for the decimal values from 0 to 20.

7.1.5 Conversion from a Numbering System to Another One

Six conversions are possible between the 3 numbering systems presented in sections 7.1.1 to 7.1.3. They can be divided into three types of conversion:

- conversions from decimal to binary and from decimal to hexadecimal (and more generally from decimal to a number base N);
- conversions from binary to decimal and from hexadecimal to decimal (and more generally from a number base N to decimal); and
- conversions from binary to hexadecimal and from hexadecimal to binary.

Table 7.3 correspondence between the decimal, binary and hexadecimal values for the decimal values from 0 to 20

Decimal value	Binary value	Hexadecimal value
0	00000	0
1	00001	1
2	00010	2
3	00011	3
4	00100	4
5	00101	5
6	00110	6
7	00111	7
8	01000	8
9	01001	9
10	01010	A
11	01011	B
12	01100	C
13	01101	D
14	01110	E
15	01111	F
16	10000	10
17	10001	11
18	10010	12
19	10011	13
20	10100	14

Conversions from Decimal to Binary and from Decimal to Hexadecimal

Let's first consider the case of the conversion from decimal to a number base N. Let's consider an integer data X whose decimal expression is known. If we want to express X in the number base N, according to the relation (7.1), this expression can be determined under the form:

$$X = a_p N^p + a_{p-1} N^{p-1} + \cdots + a_1 N^1 + a_0 N^0$$

We can hence write:

$$\underbrace{X}_{q_0} = \underbrace{\left(a_p N^{p-1} + a_{p-1} N^{p-2} + \cdots + a_2 N + a_1\right)}_{q_1} N + \underbrace{a_0}_{r_0}$$

This is the well-known expression for the Euclidean division of q_0 by N which provides the quotient q_1 and the remainder $r_0 < N$. The coefficient a_0 hence is the remainder r_0 of the Euclidean division of $q_0 = X$ by N. We can then write:

$$\underbrace{a_pN^{p-1} + a_{p-1}N^{p-2} + \cdots + a_2N + a_1}_{q_1}$$
$$= \underbrace{(a_pN^{p-2} + a_{p-1}N^{p-3} + \cdots + a_3N + a_2)}_{q_2} N + \underbrace{a_1}_{r_1}$$

The coefficient a_1 hence is the remainder r_1 of the Euclidean division of q_1 by N. If we generalize, we can hence write:

$$\underbrace{a_pN^{p-2} + a_{p-1}N^{p-3} + \cdots + a_3N + a_2}_{q_2}$$
$$= \underbrace{(a_pN^{p-3} + a_{p-1}N^{p-4} + \cdots + a_4N + a_3)}_{q_3} N + \underbrace{a_2}_{r_2}$$

$$\underbrace{a_pN^{p-3} + a_{p-1}N^{p-4} + \cdots + a_4N + a_3}_{q_3}$$
$$= \underbrace{(a_pN^{p-4} + a_{p-1}N^{p-5} + \cdots + a_5N + a_4)}_{q_4} N + \underbrace{a_3}_{r_3}$$
$$\cdots$$
$$\underbrace{a_pN^2 + a_{p-1}N + a_{p-2}}_{q_{p-2}} = \underbrace{(a_pN + a_{p-1})}_{q_{p-1}} N + \underbrace{a_{p-2}}_{r_{p-2}}$$
$$\underbrace{a_pN + a_{p-1}}_{q_{p-1}} = \underbrace{a_p}_{q_p} N + \underbrace{a_{p-1}}_{r_{p-1}}$$
$$\underbrace{a_p}_{q_p} = \underbrace{0}_{q_{p+1}} N + \underbrace{a_p}_{r_p}$$

The last relation is implied by the fact that $a_p < N$, since all the digits of a number are always lower than the key of the number base in which it is expressed. We can hence conclude that each coefficient a_i is the remainder of the Euclidean division of q_i by N (where $q_0 = X$ and q_{i+1} is the quotient of the Euclidean division of q_i by N).

If we apply this result to the case of the conversion from decimal to binary, the natural binary code for a decimal number can be obtained by performing successive Euclidean divisions of the decimal number and of the successive quotients by 2 until the obtained quotient is null. The obtained remainders (which can be worth only 0 or 1) allow to get, in reverse order, the searched natural binary code. As illustrated in Figure 7.3, we hence get:

$$(145)_{10} = (10010001)_2$$

The natural binary code for the decimal number 145 requires 8 bits, which is in accordance with the result determined in section 7.1.2.

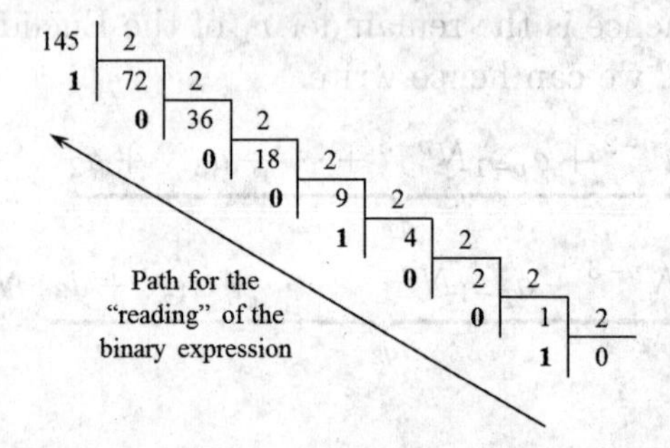

Figure 7.3 example of conversion of the decimal number 145 from decimal to binary

In the same way, if we apply the previously determined result to the case of the conversion from decimal to hexadecimal, the hexadecimal code for a decimal number can be obtained by performing successive Euclidean divisions of the decimal number and of the successive quotients by 16 until the obtained quotient is null. The obtained remainders (which are all comprised between 0 and 15) allow to get, in reverse order, the searched hexadecimal code (by replacing the remainders comprised between 10 and 15 by their corresponding hexadecimal digits A-F). As illustrated in Figure 7.4, we hence get:

$$(6751)_{10} = (1A5F)_{16}$$

since the digits A and F respectively correspond to the decimal numbers 10 and 15.

Conversions from Binary to Decimal and from Hexadecimal to Decimal

The decimal value for a binary number can be obtained directly by using the relation (7.1). We hence get:

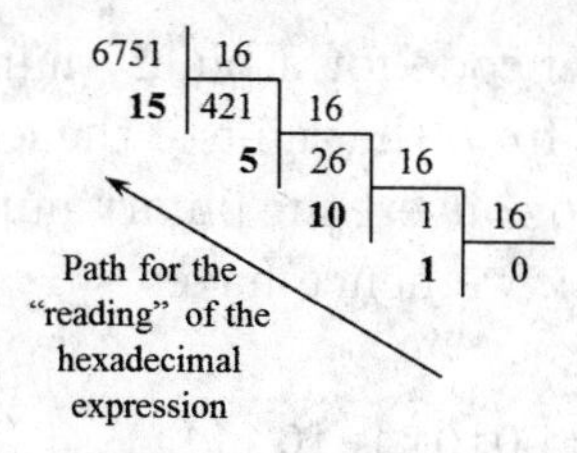

Figure 7.4 example of conversion of the decimal number 6751 from decimal to hexadecimal

$$\begin{aligned}(10010001)_2 &= 1.2^7 + 0.2^6 + 0.2^5 + 1.2^4 + 0.2^3 + 0.2^2 + 0.2^1 + 1.2^0 \\ &= 1\times128+0\times64+0\times32+1\times16+0\times8+0\times4+0\times2+1\times1 \\ &= 128+16+1 = (145)_{10}\end{aligned}$$

In the same way, the decimal value for an hexadecimal number can be obtained directly by using the relation (7.1). We hence get:

$$\begin{aligned}(1A5F)_{16} &= 1.16^3 + 10.16^2 + 5.16^1 + 15.16^0 \\ &= 1\times4096 + 10\times256 + 5\times16 + 15\times1 \\ &= 4096 + 2560 + 80 + 15 = (6751)_{10}\end{aligned}$$

The relation (7.1) obviously allows to do the conversion from any number base N to decimal.

Conversions from Binary to Hexadecimal and from Hexadecimal to Binary

The natural binary code for an hexadecimal number can be obtained by replacing each digit of the hexadecimal number by the equivalent natural binary code on 4 bits. Indeed, each digit of the hexadecimal number is comprised between 0 and F and corresponds to a decimal value comprised between 0 and 15, and it can then be converted into a natural binary code on 4 bits (since natural binary codes on 4 bits correspond to decimal values comprised between 0, for $(0000)_2$, and 15, for $(1111)_2$). If we want to convert the hexadecimal number $(1A5F)_{16}$ from hexadecimal to binary, we hence have:

$$(\underbrace{1}_{0001}\underbrace{A}_{1010}\underbrace{5}_{0101}\underbrace{F}_{1111})_{16} = (0001101001011111)_2$$

Conversely, the hexadecimal code for a binary number can be obtained by replacing each group of 4 bits from the right to the left by the corresponding hexadecimal digit. If we want to convert the binary number $(\texttt{100011010110101111})_2$ from binary to hexadecimal, we hence have:

$$(\underbrace{10}_{2}\,\underbrace{0011}_{3}\,\underbrace{0101}_{5}\,\underbrace{1010}_{A}\,\underbrace{1111}_{F})_2 = (235AF)_{16}$$

7.2 Boolean Logic

Studying logic systems is always equivalent to studying *logic variables.* During the 19th century, the British mathematicians and logicians George Boole (1815-1864) and Augustus De Morgan (1806-1871) wrote the basis of modern mathematic logic which is called the **Boolean algebra** and the logic of classes and relations. The Boolean algebra is dedicated to logic variables and functions of these variables (called logic functions). The specificities of this algebra with respect to traditional algebras are as follows:

- it is a set of two elements, 0 and 1; and
- its operations respect a fundamental property called *idempotence*, which will be detailed later.

It is only during the 20th century, in 1937, that the American physicist Claude Shannon noticed that the mathematic logic proposed by George Boole could be useful for the study of electrical circuits, since the logic states *false* (0) and *true* (1) can respectively be materialized by an open and closed circuit.

7.2.1 Logic Variables

Logic variables are variables which can only take two values 0 and 1. They are also called *binary variables* or *Boolean variables.*

Definition 66 (Equality between two Logic Variables)

Two logic variables a and b are **equal** if $b = 0$ when $a = 0$ and if $b = 1$ when $a = 1$. This is noted $a = b$.

Definition 67 (Complement of a Logic Variable)

Two logic variables a and b are **complementary** if $b = 1$ when $a = 0$ and if $b = 0$ when $a = 1$. b is called the complement of a, which is noted $b = \bar{a}$ (or $a = \bar{b}$). Most of the time, when writing by hand or on a computer, it will be noted $/a$.

7.2.2 Boolean Algebra

Let's consider a set $\mathbb{B} = \{0, 1\}$ and two variables $a, b \in \mathbb{B}$. Some operations can be defined on this set.

Definition 68 (Operation NOT)

The operation NOT (noted $^-$ or $/$) is a unary operation defined as:

$$\begin{aligned} ^-: \mathbb{B} &\longrightarrow \mathbb{B} \\ a &\longmapsto \bar{a} \end{aligned}$$

Definition 69 (Operation OR)

The operation OR (noted $+$) is a binary operation defined as:

$$\begin{aligned} +: \mathbb{B} \times \mathbb{B} &\longrightarrow \mathbb{B} \\ (a, b) &\longmapsto a + b \end{aligned}$$

with:

- $a + b = 1$ if and only if $a = 1$ or $b = 1$ (one variable at the value 1 is enough to get a result equal to 1);
- $a + b = 0$ otherwise (if $a = b = 0$).

Definition 70 (Operation AND)

The operation AND (noted $\cdot$) is a binary operation defined as:

$$\begin{aligned} \cdot : \mathbb{B} \times \mathbb{B} &\longrightarrow \mathbb{B} \\ (a, b) &\longmapsto a \cdot b \end{aligned}$$

with:

- $a \cdot b = 1$ if and only if $a = b = 1$ (both variables must be equal to 1 to get a result equal to 1);
- $a \cdot b = 0$ otherwise (the result is null as soon as one variable is equal to 0).

These operations respect the following laws:

- the operations OR and AND are commutative:

$$\forall a, b \in \mathbb{B}, a + b = b + a \tag{7.2}$$
$$\forall a, b \in \mathbb{B}, a \cdot b = b \cdot a \tag{7.3}$$

- the operations OR and AND are associative:

$$\forall a, b, c \in \mathbb{B}, a + (b + c) = (a + b) + c = a + b + c \tag{7.4}$$
$$\forall a, b, c \in \mathbb{B}, a \cdot (b \cdot c) = (a \cdot b) \cdot c = a \cdot b \cdot c \tag{7.5}$$

- the operations OR and AND have an identity element:

$$\forall a \in \mathbb{B}, a + 0 = a \tag{7.6}$$
$$\forall a \in \mathbb{B}, a \cdot 1 = a \tag{7.7}$$

- the operations OR and AND are distributive over each other:

$$\forall a, b, c \in \mathbb{B}, a + (b \cdot c) = (a + b) \cdot (a + c) \tag{7.8}$$
$$\forall a, b, c \in \mathbb{B}, a \cdot (b + c) = (a \cdot b) + (a \cdot c) \tag{7.9}$$

- the operations OR and AND are idempotent:

$$\forall a \in \mathbb{B}, a + a = a \tag{7.10}$$
$$\forall a \in \mathbb{B}, a \cdot a = a \tag{7.11}$$

- any element $a \in \mathbb{B}$ has a unique complement $\bar{a} \in \mathbb{B}$ which respects the *law of excluded middle* and the *law of contradiction*:

$$\forall a \in \mathbb{B}, a + \bar{a} = 1 \tag{7.12}$$

$$\forall a \in \mathbb{B}, a \cdot \bar{a} = 0 \tag{7.13}$$

Definition 71 (Boolean Algebra)

Because of the laws respected by the operations OR and AND, $(\mathbb{B}, +, \cdot, 0, 1)$ is a **Boolean algebra**.

Proof of the laws:

- the laws (7.2) to (7.7) are implied by the definition of the operations OR and AND;

- the law (7.8) can be proved by means of Table 7.4;

Table 7.4 proof of the law (7.8): $\forall a, b, c \in \mathbb{B}, a + (b \cdot c) = (a + b) \cdot (a + c)$

a	b	c	$b \cdot c$	$a + (b \cdot c)$	$a + b$	$a + c$	$(a + b) \cdot (a + c)$
0	0	0	0	0	0	0	0
0	0	1	0	0	0	1	0
0	1	0	0	0	1	0	0
0	1	1	1	1	1	1	1
1	0	0	0	1	1	1	1
1	0	1	0	1	1	1	1
1	1	0	0	1	1	1	1
1	1	1	1	1	1	1	1

- the law (7.9) can be proved by means of Table 7.5; and

- the laws (7.10) to (7.13) are implied by the definition of the operations NOT, OR and AND.

□

7.2.3 Laws of Boolean Algebras

Boolean algebras respect the following laws:

Table 7.5 proof of the law (7.9): $\forall a, b, c \in \mathbb{B}, a \cdot (b + c) = (a \cdot b) + (a \cdot c)$

a	b	c	$b+c$	$a \cdot (b+c)$	$a \cdot b$	$a \cdot c$	$(a \cdot b) + (a \cdot c)$
0	0	0	0	0	0	0	0
0	0	1	1	0	0	0	0
0	1	0	1	0	0	0	0
0	1	1	1	0	0	0	0
1	0	0	0	0	0	0	0
1	0	1	1	1	0	1	1
1	1	0	1	1	1	0	1
1	1	1	1	1	1	1	1

- the operations OR and AND have an absorbing element:

$$\forall a \in \mathbb{B}, a + 1 = 1 \tag{7.14}$$
$$\forall a \in \mathbb{B}, a \cdot 0 = 0 \tag{7.15}$$

- the following laws hold:

$$\forall a \in \mathbb{B}, \bar{\bar{a}} = a \tag{7.16}$$
$$\forall a, b \in \mathbb{B}, a + (a \cdot b) = a \tag{7.17}$$
$$\forall a, b \in \mathbb{B}, a + (\bar{a} \cdot b) = (\bar{b} \cdot a) + b = a + b \tag{7.18}$$

Proof of the laws:

- law (7.14):

$$a + 1 \overset{(7.12)}{=} a + (a + \bar{a}) \overset{(7.4)}{=} (a + a) + \bar{a} \overset{(7.10)}{=} a + \bar{a} \overset{(7.12)}{=} 1$$

- law (7.15):

$$a \cdot 0 \overset{(7.13)}{=} a \cdot (a \cdot \bar{a}) \overset{(7.5)}{=} (a \cdot a) \cdot \bar{a} \overset{(7.11)}{=} a \cdot \bar{a} \overset{(7.13)}{=} 0$$

- law (7.16): the law (7.16) is implied by the definition of the operation NOT.

- law (7.17):

$$a + (a \cdot b) \overset{(7.7)}{=} (a \cdot 1) + (a \cdot b) \overset{(7.9)}{=} a \cdot (1 + b) \overset{(7.14)}{=} a \cdot 1 \overset{(7.7)}{=} a$$

- law (7.18):

$$a+(\bar{a}\cdot b)\overset{(7.17)}{=}a+(a\cdot b)+(\bar{a}\cdot b)\overset{(7.9)}{=}a+((a+\bar{a})\cdot b)\overset{(7.12)}{=}a+(1\cdot b)\overset{(7.7)}{=}a+b$$

In the proofs above, the symbol $\overset{(7.X)}{=}$ indicates that the expression which follows the symbol is deduced from the expression which precedes the symbol thanks to the law (7.X) of Boolean algebras.

□

7.2.4 De Morgan's Laws

Theorem 9 (De Morgan's Laws)

De Morgan's laws owe their name to the British mathematician Augustus De Morgan (1806-1871). They allow to determine the expression for the complement of a sum or product of terms:

$$\forall a,b\in\mathbb{B}, \overline{a+b}=\bar{a}\cdot\bar{b} \tag{7.19}$$

$$\forall a,b\in\mathbb{B}, \overline{a\cdot b}=\bar{a}+\bar{b} \tag{7.20}$$

Proof: Table 7.6 allows to prove De Morgan's laws. □

Table 7.6 proof of De Morgan's laws

a	b	$\bar{a}$	$\bar{b}$	$a+b$	$\overline{a+b}$	$\bar{a}\cdot\bar{b}$	$a\cdot b$	$\overline{a\cdot b}$	$\bar{a}+\bar{b}$
0	0	1	1	0	1	1	0	1	1
0	1	1	0	1	0	0	0	1	1
1	0	0	1	1	0	0	0	1	1
1	1	0	0	1	0	0	1	0	0

7.2.5 Basic Logic Gates

In the case of real functions, a variable can take an infinity of values, and so do the functions. Then, there exists an infinity of real functions. However, in the case of logic functions, both the variables and the functions of these variables can take only 2 values (0 or 1), and there hence is a finite number of logic functions. It is possible to identify basic logic functions amongst the logic functions of one or two logic variables. These basic logic functions are called *basic logic gates* and they will appear in the expressions for the logic functions of n variables.

Logic Gates with One Input

Let's consider a function $f\colon \mathbb{B} \longrightarrow \mathbb{B}$. The function f can take 2 values for each one of the 2 values that a logic variable a can take, which implies that $2^2 = 4$ different combinations exist for $f(a)$. These combinations are illustrated in Table 7.7.

Table 7.7 the 4 possible logic gates with one input

Variable	Functions			
a	f_1	f_2	f_3	f_4
0	0	0	1	1
1	0	1	0	1

There hence are 4 logic functions of one logic variable:

- the functions f_1 and f_4 have constant values whatever the value of the variable a ($f_1(a) = 0$ and $f_4(a) = 1$): they are **constant functions**;
- the function f_2 is equal to the variable a ($f_2(a) = a$): it is the **identity function**; and
- the function f_3 is equal to the complement of the variable a ($f_3(a) = \bar{a}$): it is the **complement function**.

Logic Gates with Two Inputs

Let's consider a function $f\colon \mathbb{B} \times \mathbb{B} \longrightarrow \mathbb{B}$. The function f can take 2 values for each one of the $2^2 = 4$ values that two logic variables a and b can take, which implies that $2^4 = 16$ different combinations exist for $f(a, b)$. These combinations are illustrated in Table 7.8.
There hence are 16 logic functions of two logic variables:

- 6 of these functions are the same as the functions identified amongst the functions of one logic variable:
  - the functions f_1 and f_{16} have constant values whatever the value of the variables a and b ($f_1(a, b) = 0$ and $f_{16}(a, b) = 1$): they are **constant functions**;

Table 7.8 the 16 logic gates with two inputs

Variables		**Functions**															
a	b	f_1	f_2	f_3	f_4	f_5	f_6	f_7	f_8	f_9	f_{10}	f_{11}	f_{12}	f_{13}	f_{14}	f_{15}	f_{16}
0	0	0	0	0	0	0	0	0	0	1	1	1	1	1	1	1	1
0	1	0	0	0	0	1	1	1	1	0	0	0	0	1	1	1	1
1	0	0	0	1	1	0	0	1	1	0	0	1	1	0	0	1	1
1	1	0	1	0	1	0	1	0	1	0	1	0	1	0	1	0	1

- the function f_4 is equal to the variable a ($f_4(a,b) = a$), and the function f_6 is equal to the variable b ($f_6(a,b) = b$): they respectively are **identity functions** over a and b; and
- the function f_{13} is equal to the complement of the variable a ($f_{13}(a,b) = \bar{a}$), and the function f_{11} is equal to the complement of the variable b ($f_{11}(a,b) = \bar{b}$): they are **complement functions**;

• 2 of these functions correspond to the binary operations of Boolean algebras:

- the function f_8 is the function which corresponds to the **binary operation OR** ($f_8(a,b) = a + b$); and
- the function f_2 is the function which corresponds to the **binary operation AND** ($f_2(a,b) = a \cdot b$);

• amongst the 8 remaining logic functions of two logic variables:

- the function $f_3(a,b) = a \cdot \bar{b}$ is such that the variable b inhibits the variable a (if $b = 1$, then $f_3(a,b) = 0$ whatever the value of a): it is the **inhibition function of** a **by** b, also noted $f_3(a,b) = a \uparrow b$ ($f_3(a,b) = a$ until $b = 1$);
- the function $f_5(a,b) = \bar{a} \cdot b$ is such that the variable a inhibits the variable b (if $a = 1$, then $f_5(a,b) = 0$ whatever the value of b): it is the **inhibition function of** b **by** a, also noted $f_5(a,b) = b \uparrow a$ ($f_5(a,b) = b$ until $a = 1$);
- the function $f_7(a,b) = \bar{a} \cdot b + a \cdot \bar{b}$ is worth 1 if and only if $a = 1$ **or** $b = 1$ (the case where $a = 1$ **and** $b = 1$ being excluded): it is the **exclusive OR (XOR) function**, also noted $f_7(a,b) = a \oplus b$;
- the function $f_9(a,b) = \bar{a} \cdot \bar{b}$ is the complement of the function f_8 ($f_9(a,b) = \overline{f_8(a,b)} = \overline{a+b}$): it is the **NOR function**;

– the function $f_{10}(a,b) = \bar{a}\cdot\bar{b} + a\cdot b$ is worth 1 if and only if $a = b$: it is the **identity function**, also noted $f_{10}(a,b) = a \equiv b$ (it can also be noticed that $f_{10}(a,b) = \overline{f_7(a,b)} = \overline{a \oplus b}$); it is also sometimes noted $f_{10}(a,b) = a \otimes b$ and called the **XAND function**;

– the function $f_{12}(a,b) = \bar{a}\cdot\bar{b}+a\cdot\bar{b}+a\cdot b = a+\bar{a}\cdot\bar{b} = a+\bar{b}$ is worth 1 as long as $b = 0$ (whatever the value of a) and is equal to a as soon as $b = 1$: it is the **implication function of b on a**, also noted $f_{12}(a,b) = b \Rightarrow a$ (it can also be noticed that $f_{12}(a,b) = \overline{f_5(a,b)} = \overline{b \uparrow a}$: the implication function of b on a hence is the complement of the inhibition function of b by a);

– the function $f_{14}(a,b) = \bar{a}\cdot\bar{b}+\bar{a}\cdot b+a\cdot b = \bar{a}\cdot\bar{b}+b = \bar{a}+b$ is worth 1 as long as $a = 0$ (whatever the value of b) and is equal to b as soon as $a = 1$: it is the **implication function of a on b**, also noted $f_{14}(a,b) = a \Rightarrow b$ (it can also be noticed that $f_{14}(a,b) = \overline{f_3(a,b)} = \overline{a \uparrow b}$: the implication function of a on b hence is the complement of the inhibition function of a by b); and finally

– the function $f_{15}(a,b) = \bar{a}\cdot\bar{b}+\bar{a}\cdot b+a\cdot\bar{b} = \bar{a}+\bar{b}$ is the complement of the function f_2 ($f_{15}(a,b) = \overline{f_2(a,b)} = \overline{a\cdot b}$): it is the **NAND function**.

All these results are illustrated in Table 7.9. It can be noticed that the logic functions NOR and NAND are **universal logic functions** (they are also called **complete logic functions**) because these two basic logic functions allow to express any logic function.

Table 7.9 the 16 logic gates with two inputs

Variables		Functions															
		f_1	f_2	f_3	f_4	f_5	f_6	f_7	f_8	f_9	f_{10}	f_{11}	f_{12}	f_{13}	f_{14}	f_{15}	f_{16}
a	b	0	$a\cdot b$	$a\uparrow b$	a	$b\uparrow a$	b	$a\oplus b$	$a+b$	$\overline{a+b}$	$a\equiv b$	$\bar{b}$	$b\Rightarrow a$	$\bar{a}$	$a\Rightarrow b$	$\overline{a\cdot b}$	1
0	0	0	0	0	0	0	0	0	0	1	1	1	1	1	1	1	1
0	1	0	0	0	0	1	1	1	1	0	0	0	0	1	1	1	1
1	0	0	0	1	1	0	0	1	1	0	0	1	1	0	0	1	1
1	1	0	1	0	1	0	1	0	1	0	1	0	1	0	1	0	1

The graphical representation of logic gates is presented in Table 7.10. Two types of symbols are generally used, which correspond to two different standards: the International Electrotechnical Commission (IEC) standard, and the older American ASGS standard. It can be noticed that the logic identity is sometimes named the XAND gate.

Table 7.10 symbols of logic gates in the IEC and ASGS standards

Logic gate	Boolean equation	IEC standard	ASGS standard
YES	$S=a$	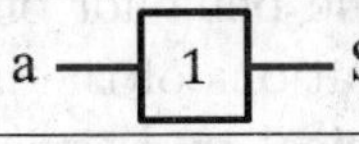 a [1] S	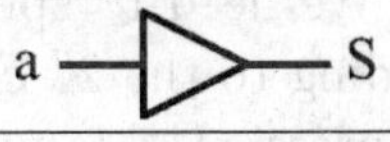 a S
NOT	$S=\bar{a}$	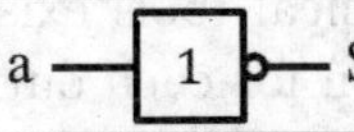 a [1] S	 a S
OR	$S=a+b$	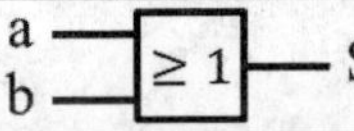 a b [≥ 1] S	 a b S
AND	$S=a\cdot b$	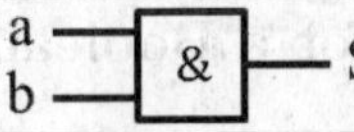 a b [&] S	 a b S
Implication	$S=a\Rightarrow b=\bar{a}+b$	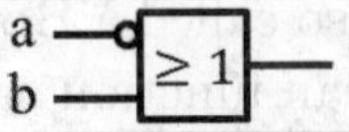 a b [≥ 1] S	 a b S
	$S=b\Rightarrow a=a+\bar{b}$	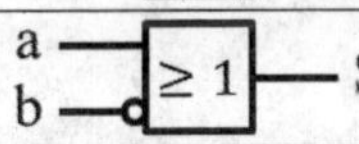 a b [≥ 1] S	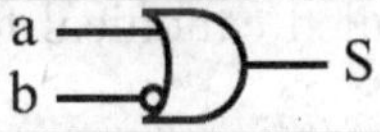 a b S
Inhibition	$S=a\uparrow b=a\cdot\bar{b}$	 a b [&] S	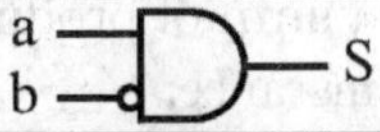 a b S
	$S=b\uparrow a=\bar{a}\cdot b$	 a b [&] S	 a b S
NOR	$S=\overline{a+b}$	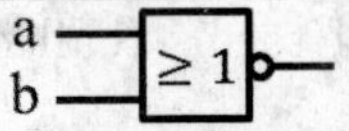 a b [≥ 1] S	 a b S
NAND	$S=\overline{a\cdot b}$	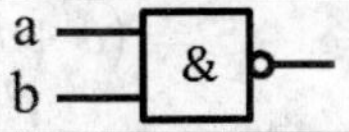 a b [&] S	 a b S
XOR	$S=a\oplus b$	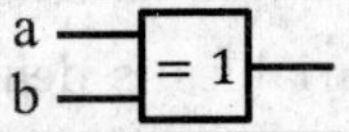 a b [= 1] S	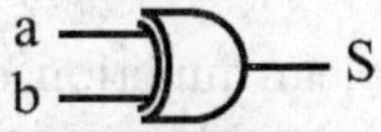 a b S
Logic identity	$S=a\equiv b$ $=\overline{a\oplus b}=a\otimes b$	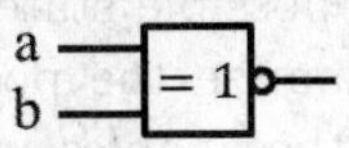 a b [= 1] S	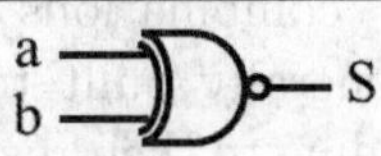 a b S

Logic Gates with n Inputs

Let's consider a function $f\colon\ \mathbb{B}^n \longrightarrow \mathbb{B}$. The function f can take 2 values for each one of the 2^n values that n logic variables $x_1, x_2, \ldots, x_n$ can take, which implies that there are 2^{2^n} different logic functions of n logic variables $f(x_1, x_2, \ldots, x_n)$.

7.2.6 Specification of a Boolean Function, Truth Tables and Karnaugh Tables

Each output variable of a combinatory system can be expressed as a function of the input logic variables of this system, only. Each output variable can hence be expressed as a Boolean function of the n input variables of the system. We need to specify the behavior of this function (i.e. its value 0 or 1) according to the 2^n combinations of the input variables in order to determine its expression. Many graphical tools exist under the form of tables to determine the Boolean expression for each output variable of the system according to its input variables.

Canonical Expressions for a Boolean Function

A logic function (which is also called a **Boolean function**) is a function which has logic variables. A Boolean function can be written under two forms which are called canonical forms:

- a **sum** of products of terms (called the **disjunctive** canonical form), for instance:

$$S = a \cdot b + \bar{a} \cdot \bar{b}$$

- a **product** of sums of terms (called the **conjunctive** canonical form), for instance:

$$S = \left(a + \bar{b}\right) \cdot \left(\bar{a} + b\right)$$

A Boolean function of n variables is defined when its value is known for all the 2^n combinations of the possible states of its n variables. Ordered tables are generally built to summarize the possible combinations of these states. According to their disposal, these tables are called **truth tables** or **Karnaugh tables**.

Truth Tables

The **truth table** of p Boolean functions of n variables is a table with $(n + p)$ columns (n columns for the input logic variables and p columns for the output Boolean functions) and 2^n rows (which correspond to the 2^n possible combinations of the values of the input variables). Each one of these 2^n combinations is written by using one of the codes presented in section 7.1: the natural binary code, the reflected binary code, or the BCD code.

Table 7.11 presents an example of truth table of 2 Boolean functions S_1 and S_2 of three variables a, b, c which uses the natural binary code.

Table 7.11 truth table of 2 Boolean functions of 3 variables in natural binary code

a	b	c	S_1	S_2
0	0	0	1	0
0	0	1	1	0
0	1	0	1	1
0	1	1	1	0
1	0	0	0	0
1	0	1	1	1
1	1	0	0	0
1	1	1	0	1

Truth tables allow to emphasize the combinations of the values of the input variables and the corresponding value of the output functions, and it is possible to establish correspondences between these combinations and other data (such as the corresponding decimal value, for instance). Besides, a truth table can be limited to the interesting combinations only. However, the use of the natural binary code is not optimal, and the size of truth tables can quickly become huge since the height of a truth table doubles each time an input variable is added, which makes the filling of the truth table a hard, painful and fastidious task.

Karnaugh Tables

A **Karnaugh table** is a table which owes its name to the American engineer Maurice Karnaugh (born in 1924). It is a truth table with crossed ordered inputs which gathers, in its 2^n squares (where n is the number of input logic variables), the different states of **a single** output Boolean function. One Karnaugh table hence needs to be built for each Boolean function. This table has 2^r rows and 2^c columns, with $r + c = n$. Two presentations of a Karnaugh table exist for a given function:

- a presentation for which the values of the input variables (0 or 1) are clearly indicated: this presentation should be prefered to avoid confusion regarding the value of the input variables; and
- a presentation for which these values are replaced by lines: a line corresponds to the value 1 whereas the absence of line corresponds to the value 0.

These two presentations are illustrated in Figure 7.5.

c \ ab	00	01	11	10
0	1	1	0	0
1	1	1	0	1

a
b

1	1	0	0
1	1	0	1

c

Figure 7.5 the two possible presentations of a Karnaugh table (on the right, the black lines correspond to the value 1)

A correspondence between a truth table and a Karnaugh table can be easily found by using the reflected binary code and by respecting the disposal of variables, as illustrated in Figure 7.6, since the p-th square of the Karnaugh table corresponds to the p-th row of the truth table.

Karnaugh tables have many advantages:

- their presentation by means of the reflected binary code is adapted to the writing of the expression for functions and to their simplification, as we will see in section 7.2.7; and
- for a given number of variables and a single function, a Karnaugh table is more compact than a truth table (since the Karnaugh table contains 2^n squares whereas the truth table contains $(n+1).2^n$ squares, as illustrated in Figure 7.6).

However, a truth table allows to gather the values of many Boolean functions whereas one Karnaugh table needs to be built for each Boolean function. Besides, the simplification of Boolean expressions that we will see in section 7.2.7 requires much attention when more than 4 logic variables are involved.

7.2.7 Basic simplification Methods: Algebraic Method and Karnaugh's Method

It can be useful to determine a simplified form for a Boolean expression which characterizes the state of an output variable of a combinatory command system according to the state of the n input variables of the system, in order to reduce the number of logic functions which are needed to describe it and hence reduce

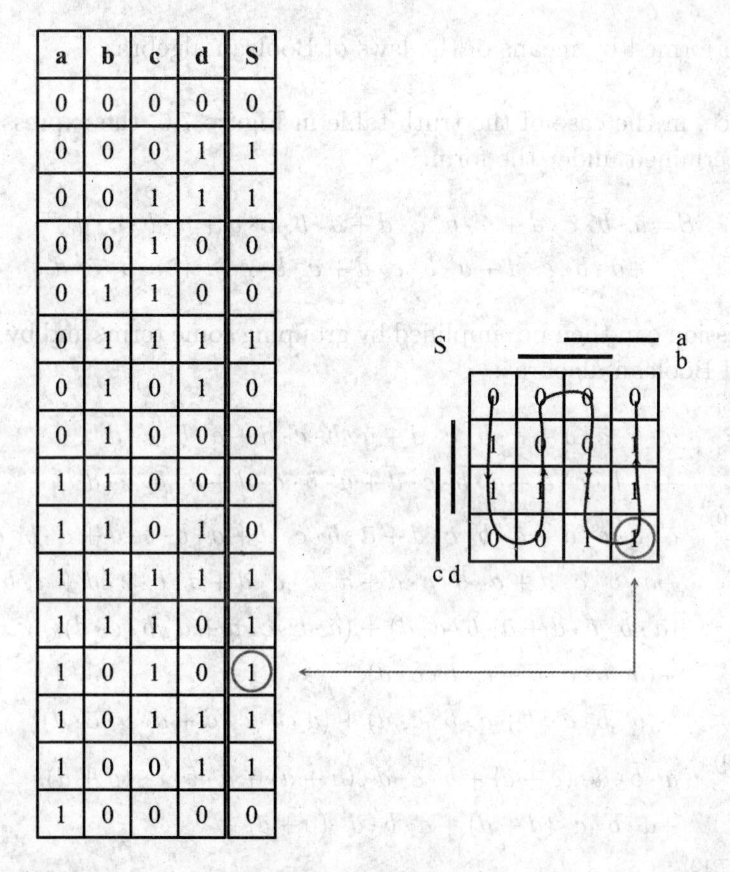

a	b	c	d	S
0	0	0	0	0
0	0	0	1	1
0	0	1	1	1
0	0	1	0	0
0	1	1	0	0
0	1	1	1	1
0	1	0	1	0
0	1	0	0	0
1	1	0	0	0
1	1	0	1	0
1	1	1	1	1
1	1	1	0	1
1	0	1	0	1
1	0	1	1	1
1	0	0	1	1
1	0	0	0	0

Figure 7.6 correspondence between a truth table using the reflected binary code and a Karnaugh table

the number of components that its technological realization requires. Two main basic simplification methods exist.

Algebraic Method

The expression for the function is obtained from the truth table: the rows of the truth table for which the function is equal to 1 are identified, and the corresponding expressions for the input variables for these rows are summed (for each combination of values of the variables, a variable a is indicated in the expression if $a = 1$ in the corresponding row of the truth table, and its complement $\bar{a}$ is indicated if $a = 0$). The expression obtained hence is a disjunctive canonical form. The simplification of the Boolean expression can then be performed by means of the laws of Boolean algebras.

For instance, in the case of the truth table in Figure 7.6, the expression for S can be determined under the form:

$$\begin{aligned}S=&\bar{a}\cdot\bar{b}\cdot\bar{c}\cdot d+\bar{a}\cdot\bar{b}\cdot c\cdot d+\bar{a}\cdot b\cdot c\cdot d+a\cdot b\cdot c\cdot d\\&+a\cdot b\cdot c\cdot\bar{d}+a\cdot\bar{b}\cdot c\cdot\bar{d}+a\cdot\bar{b}\cdot c\cdot d+a\cdot\bar{b}\cdot\bar{c}\cdot d\end{aligned}\tag{7.21}$$

This expression can then be simplified by grouping some terms and by applying the laws of Boolean algebras:

$$\begin{aligned}S&=\bar{a}\cdot\bar{b}\cdot\bar{c}\cdot d+\bar{a}\cdot\bar{b}\cdot c\cdot d+\bar{a}\cdot b\cdot c\cdot d+a\cdot b\cdot c\cdot d\\&\quad+a\cdot b\cdot c\cdot\bar{d}+a\cdot\bar{b}\cdot c\cdot\bar{d}+a\cdot\bar{b}\cdot c\cdot d+a\cdot\bar{b}\cdot\bar{c}\cdot d\\&\overset{(7.10)}{=}\bar{a}\cdot\bar{b}\cdot\bar{c}\cdot d+\bar{a}\cdot\bar{b}\cdot c\cdot d+\bar{a}\cdot b\cdot c\cdot d+a\cdot b\cdot c\cdot d+a\cdot b\cdot c\cdot d\\&\quad+a\cdot b\cdot c\cdot\bar{d}+a\cdot\bar{b}\cdot c\cdot\bar{d}+a\cdot\bar{b}\cdot c\cdot d+a\cdot\bar{b}\cdot c\cdot d+a\cdot\bar{b}\cdot\bar{c}\cdot d\\&=(\bar{a}\cdot\bar{b}\cdot\bar{c}\cdot d+\bar{a}\cdot\bar{b}\cdot c\cdot d)+(\bar{a}\cdot b\cdot c\cdot d+a\cdot b\cdot c\cdot d)\\&\quad+(a\cdot b\cdot c\cdot d+a\cdot b\cdot c\cdot\bar{d})\\&\quad+(a\cdot\bar{b}\cdot c\cdot\bar{d}+a\cdot\bar{b}\cdot c\cdot d)+(a\cdot\bar{b}\cdot c\cdot d+a\cdot\bar{b}\cdot\bar{c}\cdot d)\\&\overset{(7.9)}{=}\bar{a}\cdot\bar{b}\cdot d\cdot(\bar{c}+c)+b\cdot c\cdot d\cdot(\bar{a}+a)+a\cdot b\cdot c\cdot(d+\bar{d})\\&\quad+a\cdot\bar{b}\cdot c\cdot(\bar{d}+d)+a\cdot\bar{b}\cdot d\cdot(c+\bar{c})\\&\overset{(7.7),(7.12)}{=}\bar{a}\cdot\bar{b}\cdot d+b\cdot c\cdot d+a\cdot b\cdot c+a\cdot\bar{b}\cdot c+a\cdot\bar{b}\cdot d\\&=(\bar{a}\cdot\bar{b}\cdot d+a\cdot\bar{b}\cdot d)+b\cdot c\cdot d+(a\cdot b\cdot c+a\cdot\bar{b}\cdot c)\\&\overset{(7.9)}{=}\bar{b}\cdot d\cdot(\bar{a}+a)+b\cdot c\cdot d+a\cdot c\cdot(b+\bar{b})\\&\overset{(7.7),(7.12)}{=}\bar{b}\cdot d+b\cdot c\cdot d+a\cdot c\\&\overset{(7.17)}{=}\bar{b}\cdot d+\bar{b}\cdot c\cdot d+b\cdot c\cdot d+a\cdot c\\&\overset{(7.9)}{=}\bar{b}\cdot d+c\cdot d\cdot(\bar{b}+b)+a\cdot c\\&\overset{(7.7),(7.12)}{=}\bar{b}\cdot d+c\cdot d+a\cdot c\end{aligned}$$

The simplified disjunctive canonical form of the expression for S hence is:

$$S=\bar{b}\cdot d+c\cdot d+a\cdot c\tag{7.22}$$

It is also possible to use a converse method by identifying the rows of the truth table for which the function is equal to 0, by summing the corresponding expressions for the input variables for these rows, and by determining the complement of the disjunctive form obtained. Sometimes, this method allows to obtain the result more quickly, when the function is equal to 1 more often than it is equal to 0. The method chosen hence needs to be adapted to the situation considered.

Karnaugh's Method (or Graphical Method)

The expression for a Boolean function can be obtained in a similar way by identifying the squares of the Karnaugh table for which the function is equal to 1 and by summing the corresponding expressions for the input variables for these squares. However, an interesting specificity of Karnaugh tables – due to their disposal according to the reflected binary code – can be exploited: the state of only one variable changes between two adjacent squares.

We are hence going to identify groups of 2^p adjacent values 1 in order to remove the corresponding p variables whose variation has no impact on the value of the function. These groups must be as big as possible and as few as possible in order to determine an expression which is as simple as possible. For this simplification to be valid, these groups must be disposed symmetrically with respect to the symmetry axes of the table: these groups can hence only be squares or rectangles of 2^p values 1. These symmetry axes correspond to the symmetries which are performed with respect to the reflected binary code. A single square which contains the value 1 can be taken into account in many different groups, but each square which contains the value 1 needs to be taken into account in at least one group.

If two adjacent squares contain the value 1, then it means than there is one input variable whose variation has no impact on the value of the function: this variable can hence be removed from the expression (by implicitly applying the law (7.12)). If four adjacent squares contain the value 1, then it means that there are two input variables whose variation has no impact on the value of the function: these variables can hence be removed from the expression. By generalizing, if 2^p adjacent squares contain the value 1, then it means that there are p input variables whose variation has no impact on the value of the function: these variables can hence be removed from the expression.

It can also be noticed that a Karnaugh table can be considered as a developed cylinder, as illustrated in Figure 7.7, in order to exploit as much as possible the simplification method that it allows. Two squares which are located in a same column but in the upmost and downmost rows hence are adjacent, and so are two squares which are located in a same row but in the leftmost and rightmost columns. The four corners of a Karnaugh table hence are adjacent.

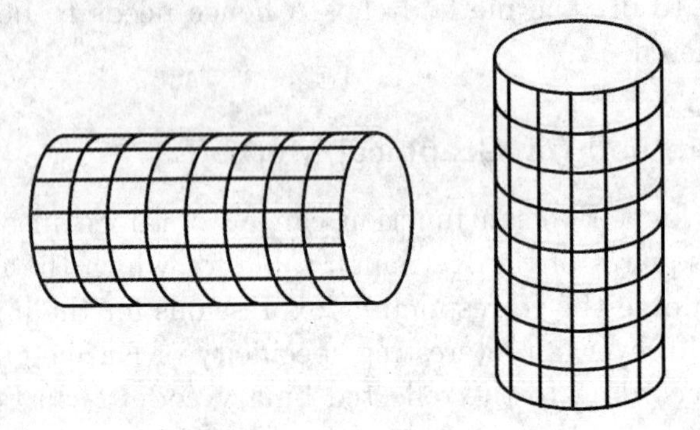

Figure 7.7 Karnaugh tables considered as developed cylinders

This specificity can be exploited to simplify the expression for the previously defined Boolean function S. The Karnaugh table of the function S is recalled in Figure 7.8. Three groups of adjacent squares which contain the value 1 can be identified in this Karnaugh table:

cd \ ab	00	01	11	10
00	0	0	0	0
01	1	0	0	1
11	1	1	1	1
10	0	0	1	1

Figure 7.8 Karnaugh table of the function S

1. on the one hand, a first group of 4 adjacent squares which contain the value 1 can be identified in the third row of the table, as illustrated in Figure 7.9: in these 4 squares, a and b vary whereas c and d do not. a and b can hence be removed since their variation has no impact on the value of S. Since $c = d = 1$, the corresponding expression is $c \cdot d$.

cd \ ab	00	01	11	10
00	0	0	0	0
01	1	0	0	1
11	1	1	1	1
10	0	0	1	1

Figure 7.9 groups of adjacent squares corresponding to the expression $c \cdot d$

2. on the other hand, a second group of 4 adjacent squares which contain the value 1 can be identified in the lower right corner of the table, as illustrated in Figure 7.10: in these 4 squares, b and d vary whereas a and c do not. b and d can hence be removed since their variation has no impact on the value of S. Since $a = c = 1$, the corresponding expression is $a \cdot c$.

cd \ ab	00	01	11	10
00	0	0	0	0
01	1	0	0	1
11	1	1	1	1
10	0	0	1	1

Figure 7.10 groups of adjacent squares corresponding to the expression $a \cdot c$

3. finally, a third group of 4 adjacent squares which contain the value 1 can be identified, as illustrated in Figure 7.11. At first glance, these 4 squares do not look adjacent, but the two groups of 2 squares which contain the value 1 become adjacent if the table is folded so that its leftmost and

rightmost columns are joined. These two groups of 2 squares can hence be fused to obtain a single group of 4 squares. In these 4 squares, a and c vary whereas b and d do not. a and c can hence be removed since their variation has no impact on the value of S. Since $b = 0$ and $d = 1$, the corresponding expression is $\bar{b} \cdot d$.

cd \ ab	00	01	11	10
00	0	0	0	0
01	1	0	0	1
11	1	1	1	1
10	0	0	1	1

Figure 7.11 groups of adjacent squares corresponding to the expression $\bar{b} \cdot d$

cd \ ab	00	01	11	10
00	0	0	0	0
01	1	0	0	1
11	1	1	1	1
10	0	0	1	1

Figure 7.12 groups of adjacent squares which can be identified in the Karnaugh table of the function S

The simplified expression for S can hence be obtained by summing these three expressions:

$$S = \bar{b} \cdot d + c \cdot d + a \cdot c, \tag{7.23}$$

and this expression is the same as the one obtained in (7.22) thanks to the algebraic method, but it was obtained in a more straightforward manner. However, Karnaugh's method is efficient when there are less than 4 or 5 variables.
It can be noticed that:

- in a Karnaugh table with n variables:
  - $2^0 = 1$ square corresponds to the product of n variables or complements of variables;
  - a group of $2^1 = 2$ adjacent squares corresponds to the product of $(n-1)$ variables (since the variations of the n-th variable has no impact on the value of the function);
  - a group of $2^2 = 4$ adjacent squares corresponds to the product of $(n-2)$ variables (since the variations of the 2 additional variables have no impact on the value of the function);
  - ...
  - a group of 2^p adjacent squares corresponds to the product of $(n-p)$ variables (since the variations of the p additional variables have no impact on the value of the function);
- in the same way that it is possible to consider a converse algebraic method, it can be interesting to group the adjacent squares of a Karnaugh table which contain the value 0 if there are more values 1 than there are values 0 in the table: the converse Karnaugh's method hence allows to obtain an expression for $\bar{S}$.

Sometimes, some combinations of values of input variables may not correspond to any physical situation, or the value of the function for these combinations may not matter for the considered problem. The value of the function for these combinations can then be noted x in the Karnaugh table, and these x can then be replaced by 0 or 1 according to the groups of 2^p squares that they may allow to get.

7.3 Application to FPGA and Pneumatic Systems

Boolean functions can be implemented by means of different types of technology. In this book, we will only focus on the electronic and pneumatic technology.

7.3.1 Implementation of Boolean Functions by Means of FPGA

The advent and development of electronics during the second half of the 20th century allowed to design miniaturized electrical circuits called **integrated circuits**. They are made of silicium or germanium and allow to gather, in a really small space, up to tens of thousands of resistors, diods and transistors whose association allows to perform logic gates. Integrated circuits are named according to the number of logic gates that they contain, as illustrated in Table 7.12.

Table 7.12 different types of integrated circuits

Name	Number of logic gates	Applications
Small Scale Integration (SSI)	Less than 10	Basic gates, flip-flops
Middle Scale Integration (MSI)	Between 10 and 99	Counters, registers
Large Scale Integration (LSI)	Between 100 and 9,999	Memories,
Very Large Scale Integration (VLSI)	More than 10,000	microprocessors

A Field-Programmable Gate Array (FPGA) is a type of Programmable Logic Device (PLD) (and of VLSI integrated circuit) designed to be configured after manufacturing, which is the reason why it is called "field-programmable". An example of FPGA is depicted in Figure 7.13, and the nano board to which it belongs is depicted in Figure 7.14.

Figure 7.13 a Field-Programmable Gate Array with 25,000 gates

Figure 7.14 nano board of the FPGA in Figure 7.13

Such development boards are much used for the implementation of acquisition and treatment interfaces. The most common FPGA architecture consists of an array of programmable logic components called **logic blocks**, input/output (I/O) cells, and interconnect resources, as illustrated in Figure 7.15.

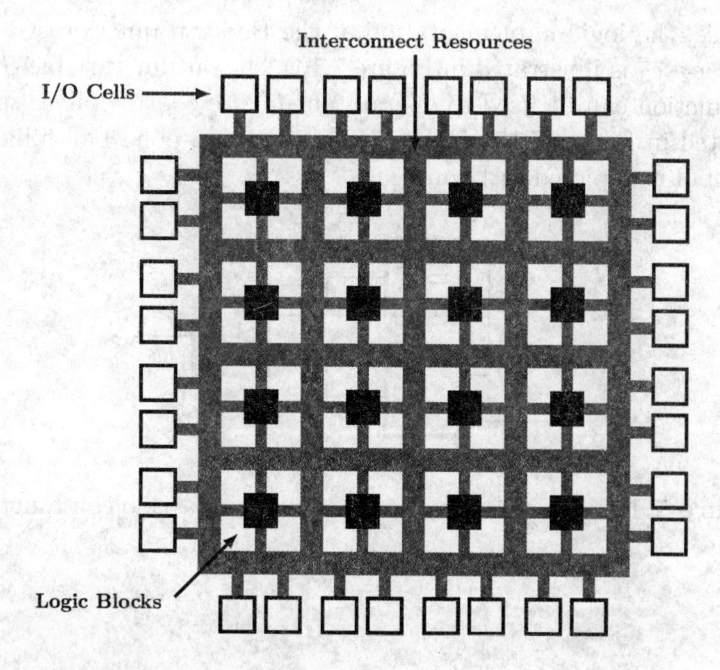

Figure 7.15 the internal structure of a FPGA

Each logic block is made of a Look-Up Table (LUT) and of a flip-flop. The LUT allows to implement logic equations which generally have $n \in \{4, 5, 6\}$

input variables and one output variable: a single LUT hence allows to perform any of the 2^{2^n} functions of n logic variables. These logic blocks, which can be between a few thousands and a few millions on a single FPGA, are connected by means of configurable interconnect resources to perform simple logic gates or complex combinatorial functions.

7.3.2 Implementation of Boolean Functions in Pneumatic Systems

Instead of electricity, oil or water (which require hydraulic elements) or even air (which requires pneumatic elements) can be used to perform logic gates. Logic gates which are based on fluid energies are more expensive, and they require more space than electronic elements. However, they can be really useful in agressive (e.g. salty environments) or highly constrained environments (mines, powder mills, chemical industries, etc.). These elements may also allow to use the same type of energy for both the command system and the operative system, which may make the design of the system easier.

For instance, the logic implementation of the Boolean function $S = e_1 \cdot \overline{e_2} + \overline{e_1} \cdot e_2 + e_1 \cdot e_2 \cdot \overline{e_3}$ is illustrated in Figure 7.16. The output variable S of such a Boolean function can be used to operate pneumatic control valves such as the ones depicted in Figure 7.17, which belong to a ping-pong ball filling system, whose symbol is depicted in Figure 7.18.

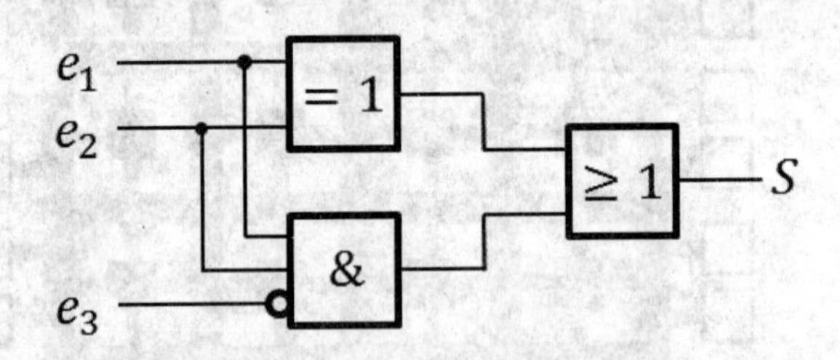

Figure 7.16 pneumatic implementation of a Boolean function

Figure 7.17 an example of logically operated control valves

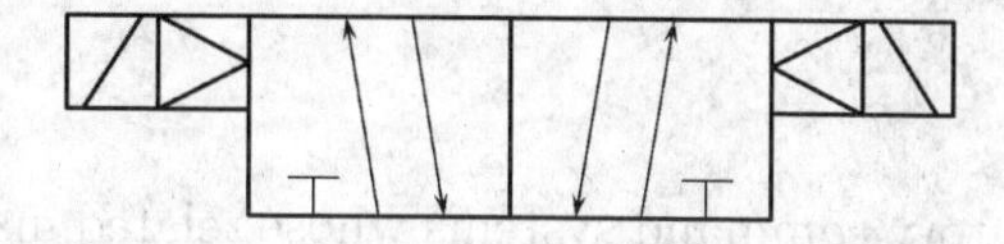

Figure 7.18 symbol of one of the pneumatic control valves in Figure 7.17

Chapter 8

Modeling of the Sequencing of Operations

Even if the combinatory command systems whose related mathematical framework was described in chapter 7 allow to control simple processes, the control of more complex processes requires memory capabilities in order to take into account previously occurred events in the behavior of the system. *Sequential systems* offer such capabilities.

In this chapter, after defining what a sequential evolution is, we will present some tools which allow to describe such sequential evolutions and give some practical applications of these tools.

8.1 Principle of a Sequential Evolution

8.1.1 Limits of Combinatorial Systems

Let's suppose that we want to design the command system of an electrical motor whose behavior is as follows:

- two monostable push buttons are used to give the orders ON and OFF to the motor;
- pushing the button ON causes the rotation of the motor if it was not working already;
- pushing the button OFF makes the motor stop if it was working; and
- pushing the button ON has the priority over pushing the button OFF.

This system is depicted in Figure 8.1.

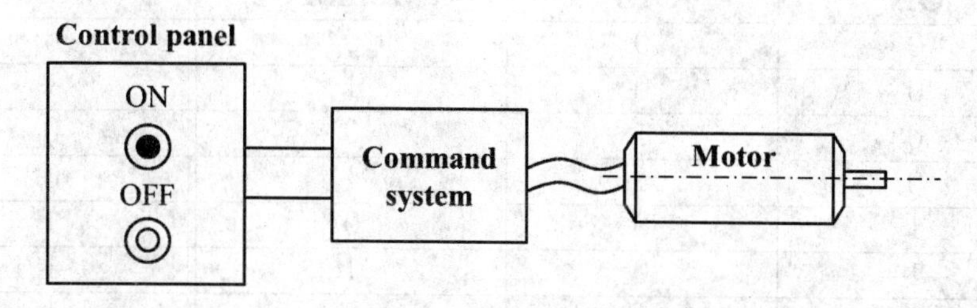

Figure 8.1 command system of an electrical motor

This expected behavior allows us to begin filling the truth table of the command function of the motor, which is presented in Table 8.1, where the values 0 and 1 of the logic variable *motor* respectively correspond to the orders OFF and ON. However, it can be noticed that, when no button is pushed, the output order is variable: if the motor is not working, we want it to stay offline, whereas if it is working, we want it to go on working. It is hence necessary to know the state (working or not working) of the motor to know the state of the output order when both input variables are equal to 0.

Table 8.1 incomplete truth table of the command function of the motor

ON	OFF	Motor
0	0	?
0	1	0
1	0	1
1	1	1

To address this problem, it is possible to introduce an additional variable which indicates the state of the motor at a given instant t. The output variable of the system hence represents the state of the motor at the following instant $t + dt$. If this variable is called *memory* (since it allows to memorize the previous state of the motor), the complete truth table of the command function of the motor can be determined: it is presented in Table 8.2. As illustrated by the two first rows of the truth table, if both input variables are equal to 0 (if no button is pushed), then the output variable keeps the same value as the one memorized at the previous instant ($motor = memory$), whereas the memorized value has no impact on the value of the output variable in all the other cases. The memory indicates if the motor was in motion ($memory = 1$) or not ($memory = 0$) before the push of the button ON or OFF.

Table 8.2 complete truth table of the command function of the motor

ON	OFF	Memory	Motor
0	0	0	0
0	0	1	1
0	1	0	0
0	1	1	0
1	0	0	1
1	0	1	1
1	1	0	1
1	1	1	1

The logic expression for the output variable can be deduced from this truth table as follows:

$$\begin{aligned}\text{Motor} &= \text{ON} + \overline{\text{ON}} \cdot \overline{\text{OFF}} \cdot \text{Memory} \\ &= \text{ON} + \text{ON} \cdot \overline{\text{OFF}} \cdot \text{Memory} + \overline{\text{ON}} \cdot \overline{\text{OFF}} \cdot \text{Memory} \\ &= \text{ON} + \underbrace{(\text{ON} + \overline{\text{ON}})}_{1} \cdot \overline{\text{OFF}} \cdot \text{Memory} \\ &= \text{ON} + \overline{\text{OFF}} \cdot \text{Memory}\end{aligned}$$

A combinatory command system **does not allow** to control this motor, since the additional knowledge of the previous state of the motor is necessary to determine the value of the output variable. A sequential command system hence is necessary, such as the memory operator depicted in Figure 8.2.

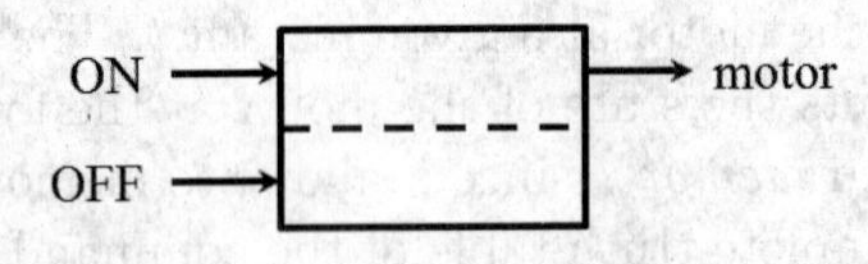

Figure 8.2 memory operator for the command of the electrical motor

Practically, all automated systems are sequential. Only some parts of these systems are purely combinatory, mainly because their transient evolution is very short in comparison with the transient evolution of the other elements of the studied system.

8.1.2 Structure of a Sequential System

As said in section 8.1.1, a **sequential system** is a system whose evolutions depend on its logic input variables (like a combinatory system) **and** on its previous state. This state is characterized by internal variables whose evolution depends on their own memorized value and on the input variables of the system.

The structure of a sequential system hence comprises two combinatory functional blocks, as illustrated in Figure 8.3. The memorization of the internal variables Y_i is either performed by means of flip-flops[1] or taken into account in the construction of the system itself. It can be noticed that either the value 1 or 0 can be memorized.

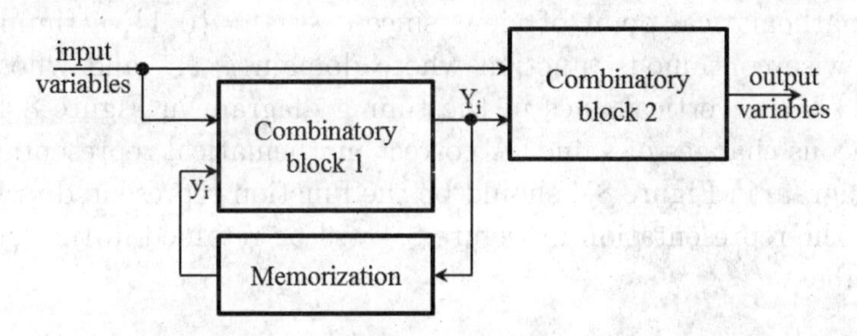

Figure 8.3 structure of a sequential system

Contrary to a **combinatory evolution**, which only depends on the input variables of the system, a **sequential evolution** depends on the input variables of the system **and** on the previous evolutions of this system.

8.2 Description Tools

8.2.1 Timing Diagrams

Definition

Definition 72 (Timing Diagram)

The **timing diagram** of a Boolean variable a is the temporal representation of the evolution of its state.

[1] A flip-flop, or latch, is a circuit that has two stable states and which can be used to store state information. It is the basic storage element in sequential logic.

This definition is illustrated in Figure 8.4. One should not mistake a timing diagram with a Gantt chart, though, since a Gantt chart is a type of bar chart that illustrates a project schedule.

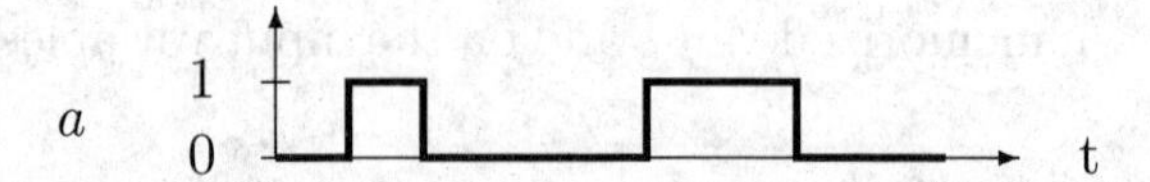

Figure 8.4 timing diagram of a Boolean variable a

From a mathematical point of view, since $a \in \mathbb{B} = \{0, 1\}$, a timing diagram is a piecewise continuous function whose domain is $\mathbb{R}^+$ and whose range is $\mathbb{B} = \{0, 1\}$. The vertical lines in the timing diagram in Figure 8.4 represent instantaneous changes of value. A correct mathematical representation of the timing diagram in Figure 8.4 should be the function represented in Figure 8.5. However, the representation in Figure 8.4 will be retained in the remainder of this chapter.

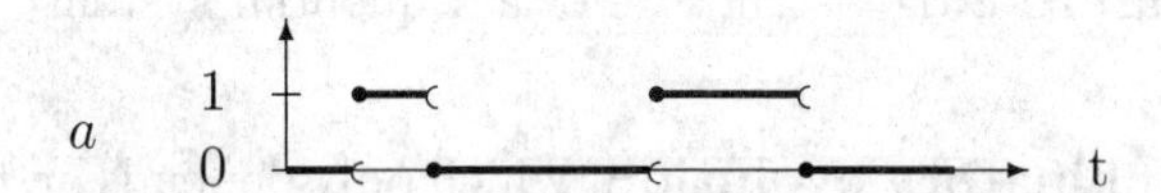

Figure 8.5 mathematical representation of the timing diagram in Figure 8.4

It can be noticed that, practically, the changes of value of variables will not be instantaneous but will be delayed by a period of time which corresponds to the **transmission time** of the related component.

Timing diagrams are of much interest if they are represented with the same time scale one under the other, as illustrated in Figure 8.6. It is thus possible to represent, one under the other, the timing diagrams of the evolution of the input variables of a system and the timing diagram of the evolution of the output variable of this system according to the evolution of its input variables, in order to graphically emphasize the behavior of the system. A set of timing diagrams is often called a timing diagram too.

Rising Edge and Falling Edge

Definition 73 (Rising Edge, Falling Edge)

Two specific variables can be associated with each Boolean variable a:

- the variable $\uparrow a$, which is called the **rising edge** of a and is defined as:

$$\begin{cases} \uparrow a = 1 \text{ when the value of } a \text{ changes from 0 to 1} \\ \uparrow a = 0 \text{ otherwise} \end{cases}$$

- the variable $\downarrow a$, which is called the **falling edge** of a and is defined as:

$$\begin{cases} \downarrow a = 1 \text{ when the value of } a \text{ changes from 1 to 0} \\ \downarrow a = 0 \text{ otherwise} \end{cases}$$

It can be noticed that the notation $\uparrow a$ has no relation with the notation used for the inhibition functions presented in section 7.2.5.

It is possible to illustrate this definition on the timing diagram of the variable a in Figure 8.4. The timing diagrams of the variables $\uparrow a$ and $\downarrow a$ are depicted in Figure 8.6.

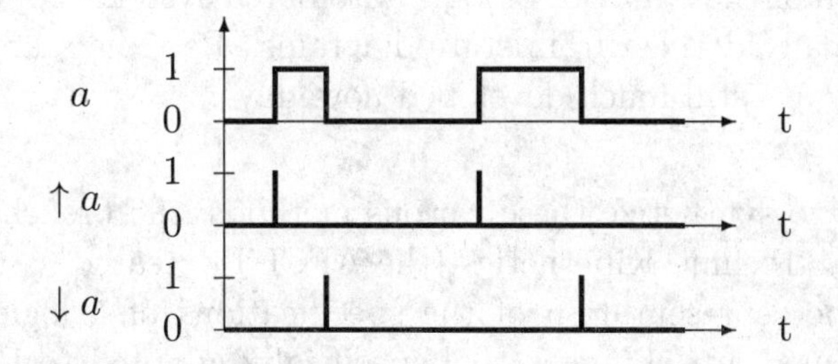

Figure 8.6 rising and falling edges of a Boolean variable a

Once again, a correct mathematical representation of the timing diagrams in Figure 8.6 should be under the form of the functions in Figure 8.7. However, the representation in Figure 8.6 will be retained in the remainder of this chapter.

It can be noticed that a logic variable is true for a given period of time, whereas

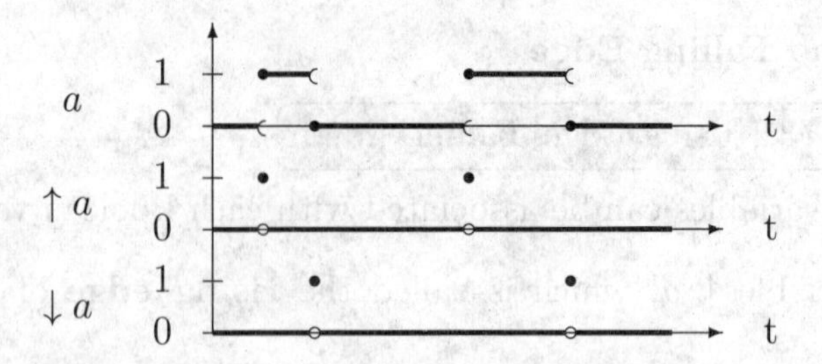

Figure 8.7 mathematical representation of the timing diagrams in Figure 8.6

a rising or falling edge variable is true only at some instants. The change of value of a logic variable is called an **event** related to this logic variable.

8.2.2 GRAFCET

History of GRAFCET

During the 60s and the 70s, the generalization of more and more complex automated production systems led to the emergence of methods aiming at standardizing the design of their command system. Various modeling formalisms for the description of the behavior of logic automated systems appeared, which were morphologically derived from state diagrams. Petri nets are such a formalism, and they are still much developed nowadays.

In 1975, in order to synthetize these various methods, the French Association for Economic and Technic Kibernetics (the AFCET) created a standardizing committee for the representation of the specifications of a logic automated system, which aimed at defining a tool specifically adapted to the definition of the specifications of logic command systems. In 1977, this committee defined a graphical formalism called **GRAFCET** ("Graphe Fonctionnel de Commande par Étapes et Transitions", which is the French acronym for "functional graph for control by means of steps and transitions"). This model, which was completed by the French national Agency for the Development of Automated Production (ADEPA), gave birth to a French standard in 1982 (which was reedited in 1995) and to an international standard in 1988 (which was reedited in 2002).

Definition 74 (Grafcet chart)

A **grafcet chart** is a functional diagram which uses the GRAFCET language.

Another standard exists which defines the characteristics of the functional diagram under the form of sequences or **SFC** (**S**equential **F**unction **C**hart). It is a programming formalism which is mainly dedicated to Programmable Logic Controllers (PLC) and which uses most of the graphical principles of the GRAFCET language.

However, one should not mistake the SFC and the GRAFCET languages. The GRAFCET language is a **specification** language for logic command systems whereas the SFC language is a **programming** language for the PLCs which must control such systems.

Domain of Validity of the GRAFCET Language

The main objective of the GRAFCET language is to allow the modeling of the relation which exists between the input variables (the instructions given by the user and the reports provided by the operative system) and the output variables (the orders sent to the operative system and the information sent to the user) of the command system when:

- these input/output variables are **Boolean**; and
- the relation which exists between them is **sequential**.

Besides, the command systems for which the GRAFCET language was designed must respect two additional properties because of their high safety constraints:

- **reactivity**: a system is **reactive** if it is sensitive to any variation of its input variables as soon as it occurs, and if it **instantaneously** delivers the corresponding value of its output variables.
- **determinism**: a system is **deterministic** if a sequence of variations of its input variables **always** produces the same sequence of variations of its output variables.

The GRAFCET language hence is specifically adapted to the modeling of the behavior of command systems which are:

- logic;
- sequential;
- reactive; and
- deterministic.

Basic Elements of the GRAFCET Language

A simple grafcet chart example is depicted in Figure 8.8.

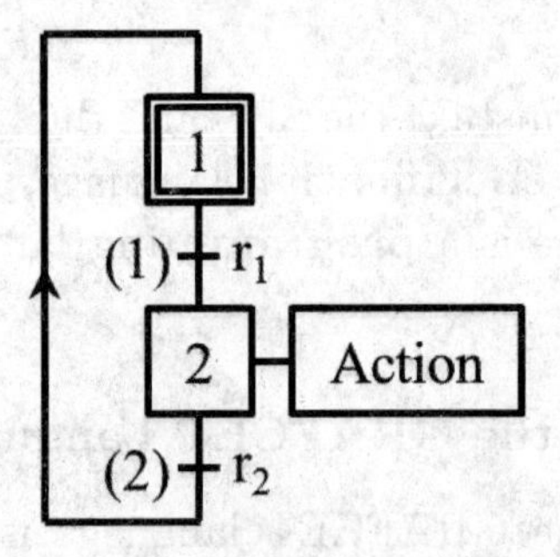

Figure 8.8 a simple grafcet chart example

A grafcet chart comprises two types of symbols: **steps** and **transitions**. A grafcet chart is **always** an alternance of steps and transitions which are connected by means of directed **links**. These links are implicitly from top to bottom, which is the reason why an arrow will always indicate links from bottom to top (like the link on the left-hand side of the grafcet chart in Figure 8.8). The steps and transitions of a grafcet chart have the following characteristics:

- the **steps** of a grafcet chart represent the **states of the system** whose behavior is modeled by the grafcet chart.

  A step is represented by a square and identified by a label. Each step can be **active** or **inactive**. The steps which are active at the initial instant are called **initial steps**, and they are represented by means of a double square: this is the case of the step 1 in the grafcet chart in Figure 8.8.

  A **step variable** is associated with each step i of a grafcet chart: it is noted X_i and characterizes the state of the step. If the step i is active, then $X_i = 1$; if the step i is inactive, then $X_i = 0$. The state of a sequential system at a given instant can hence be characterized by the set of the steps of the corresponding grafcet chart which are active at that instant. The set of these active steps is called the **situation** of the grafcet chart.

The actions which are executed by a system are related to the states of this system, which is the reason why they are associated with the steps of the grafcet chart. They correspond to the output variables of the command system. An action associated with a step is represented in a rectangle linked to the step by a line and represented on the right of the step, like the action associated with the step 2 in Figure 8.8. An action is executed if the corresponding step is active, and it is stopped as soon as the related step becomes inactive.

- the **transitions** of a grafcet chart indicate a possible evolution of the activity between two or more steps.

  Each transition is represented by an horizontal line located between one or many preceding steps and one or many succeeding steps. The transition designation is indicated on its left, between parentheses. However, even if the specification of the label of a step is mandatory, the specification of the transition designation is optional. The clearing of a transition represents a change of state of the system (and hence a change of step of the corresponding grafcet chart). A transition represents a single possible evolution of the system.

  A transition is **enabled** if all the preceding steps are active. The logic expression which conditions a transition is called the **transition-condition** associated with the transition, and it is indicated on its right. This transition-condition is a logic combinatory function of the input variables of the system and sometimes of the internal state of the system (by means of step variables) and of time. A transition-condition can be **true** or **false**. The always true transition-condition is noted $\underline{1}$ (it was formerly noted "$= 1$").

  Finally, if a transition is enabled and if the related transition-condition is true, then this transition can (and must, as we will see later) be cleared.

  The simple example of the grafcet chart part in Figure 8.9 allows to illustrate these notions:

  - the transition 2 is *enabled* if the step 2 is *active*;
  - the related transition-condition is *true* if the input variables a and b are true ($a \cdot b = 1$) or if the step 10 is active ($X_{10} = 1$);

– the transition 2 can hence be cleared if it is *enabled* and if the related transition-condition is *true*.

(1) $a + b$

2

(2) $a \cdot b + X_{10}$

Figure 8.9 grafcet chart example illustrating the vocabulary related to transitions

Basic Structures of a Grafcet Chart

The basic structures of a grafcet chart are configurations which are related to the basic concepts of logic systems. They allow to express successions of states, selections of sequences, and parallel sequences.

These basic structures are as follows:

- a **sequence** is a succession of steps in which each step (except the last one) has only one succeeding transition and each step (except the first one) has only one preceding transition enabled by a single step of the sequence. If a grafcet chart comprises only one sequence, then it is called a **linear grafcet chart**. The grafcet chart in Figure 8.8 is such a linear grafcet chart.
- a **steps skip** allows to skip one or many steps when the related actions do not need to be executed. If we consider the grafcet chart in Figure 8.10, it is possible to go from the step 2 to the step 3, or to skip the step 3, depending on the value of the variable a. If a is false, then the step 3 becomes active, but if a is true, then the step 3 can be skipped.
- a **sequence backward skip** allows a sequence to be repeated as long as a condition does not hold. If we consider the grafcet chart in Figure 8.11, the sequence $\{3, 4\}$ is repeated as long as c is false, and the step 5 becomes active only when c becomes true.

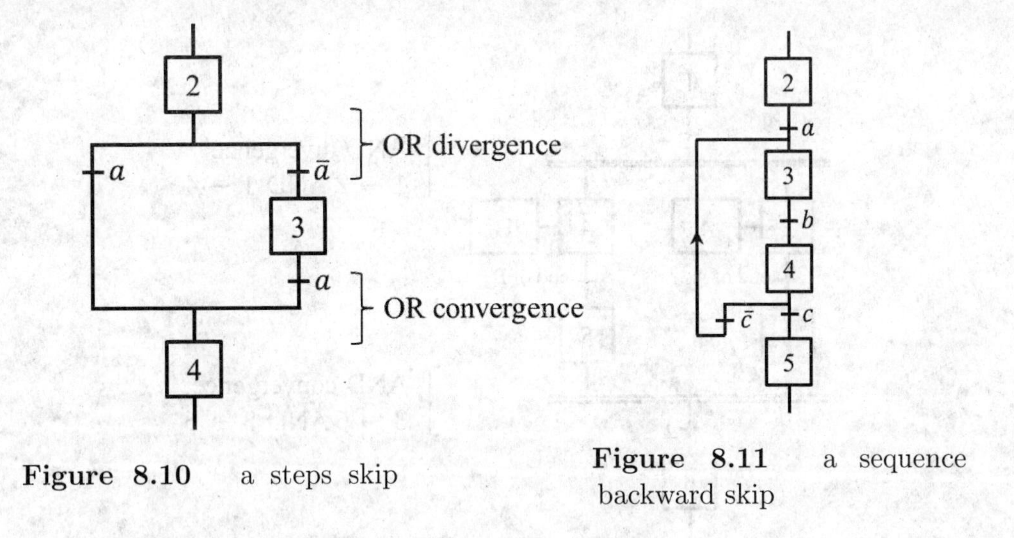

Figure 8.10 a steps skip

Figure 8.11 a sequence backward skip

- a **selection of sequences** represents different possible evolutions to many steps. The transition-conditions related to the transitions of a selection of sequences must be **exclusive** so that only one of them is true. An example of a selection of sequences is depicted in Figure 8.12.

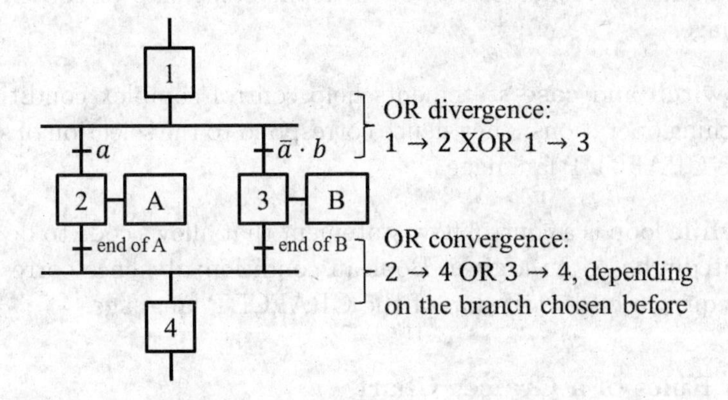

Figure 8.12 an example of a selection of sequences

- **parallel sequences** represent a set of sequences which can evolve independently after the clearing of a transition which simultaneously activates many steps. A parallel sequences example is depicted in Figure 8.13.

It can be noticed that some of these basic structures also naturally exist in some programming languages. For instance, in the C language:

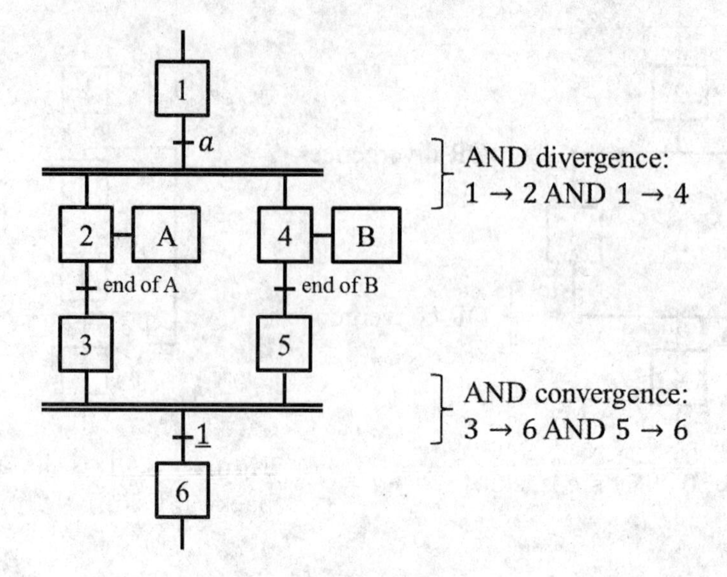

Figure 8.13 a parallel sequences example

- the **goto** statement performs a one-way transfer of control to another line of code: it hence corresponds to the steps skip of the GRAFCET language.
- the **switch** and **case** statements help control complex conditional and branching operations: they hence correspond to the selection of sequences of the GRAFCET language.
- the **while** loop is a control flow statement that allows code to be executed repeatedly based on a given Boolean condition: it hence corresponds to the sequence backward skip of the GRAFCET language.

Evolution Rules of a Grafcet Chart

The evolutions of a grafcet chart must respect 5 fundamental rules. These rules will be illustrated on the grafcet chart example in Figure 8.14. These 5 rules are as follows:

1. **Initial situation**. The initial situation of a grafcet chart characterizes the initial behavior of the command system with respect to the operative system and/or to the elements of the environment. It corresponds to the steps of the grafcet chart which are active at the initial time, and hence

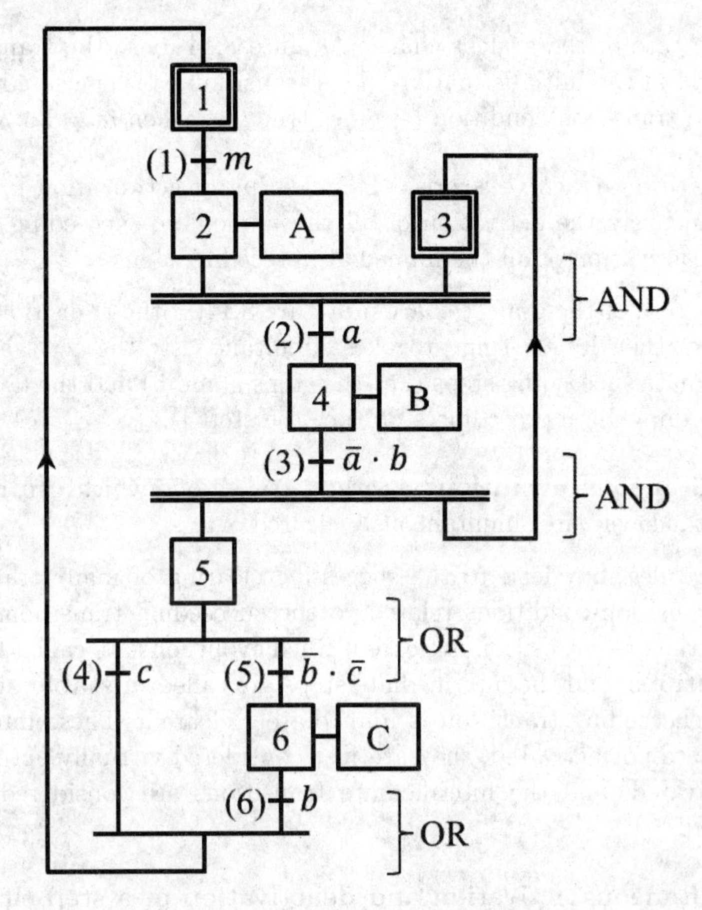

Figure 8.14 a grafcet chart example

to the set of the initial steps. A grafcet chart necessarily has one initial situation.

In the case of the grafcet chart in Figure 8.14, since the steps 1 and 3 are initial steps, the initial situation is the situation $S_0 = \{1, 3\}$.

2. **Clearing of a transition**. A transition can and **must** be cleared if the two following conditions hold:

   - the transition is enabled (i.e. all the preceding steps are active); and

   - the related transition-condition is true.

In the case of the grafcet chart in Figure 8.14, only the transition 1 is enabled in the initial situation. This transition can be cleared when the related transition-condition becomes true, i.e. when $m = 1$.

3. **Evolution of active steps**. The clearing of a transition provokes simultaneously the activation of all the immediate succeeding steps and the deactivation of all the immediate preceding steps.

   In the case of the grafcet chart in Figure 8.14, if the transition 3 can be cleared, then its clearing provokes simultaneously the activation of the immediate succeeding steps (i.e. the steps 3 and 5) and the deactivation of the immediate preceding step (i.e. the step 4).

4. **Simultaneous evolutions**. Several transitions which can be cleared simultaneously are simultaneously cleared.

   These rules may lead to the successive clearing of many transitions if the transition-conditions related to the succeeding transitions are true when the first transition is cleared. Such evolutions are called **transient evolutions**, and the intermediate steps are called **unstable steps** since their succeeding transition is immediately cleared. These intermediate steps are not activated, they are just considered virtually activated and deactivated, and the intermediate transitions are considered virtually cleared.

5. **Simultaneous activation and deactivation of a step**. If an active step is simultaneously activated and deactivated, then it remains active. This case is illustrated in Figure 8.15: the deactivation of the step 0 is effective when $a = 1$, which activates the step 1 and reactivates the step 0.

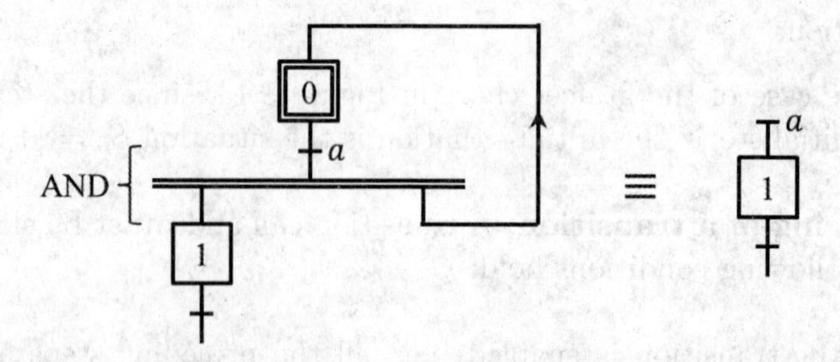

Figure 8.15 simultaneous activation and deactivation of the step 0

Different Points of View

The grafcet chart of a system can be determined according to different points of view:

- the **"system" point of view**, according to which the grafcet chart describes the external behavior of the system without taking into account the technologies used. A grafcet chart according to the system point of view is simple and quick to determine, and perfectly adapted to a coordinated description of the macro-evolutions of the system. However, it does not take into account the velocity of the components, and it is difficult to optimize.

- the **"command system" point of view**, according to which the grafcet chart describes the orders which are given by the command system and their consequences. A grafcet chart according to the command system point of view is close from the behavior of a PLC, and easy to implement. However, it is difficult to optimize when the technologies used are not taken into account.

- the **"operative system" point of view**, according to which the grafcet chart describes the power distribution (from the pre-actuators point of view), the power conversion (from the actuators point of view), or the power transmission (from the effectors point of view) while taking into account the technologies used. A grafcet chart according to the operative system point of view is close from the real behavior of the system. However, it is dependent on the technologies used, and difficult to optimize structurally.

The analysis of a system hence depends on the retained point of view, and so do the consequences of this analysis on the evolutions or the optimizations of the system. This is the reason why a "divide to conquer" approach is generally used, each module of a system being modeled by one grafcet chart, all these grafcet charts being coordinated by a global grafcet chart.

8.3 Practical Applications

8.3.1 Asynchronous Counter

Counters represent a fundamental application of flip-flops. They are used in systems dedicated to quantification, such as chronometers, numeric clocks, ...

They allow to convert a sequence of logic information into a piece of information which is either logic (e.g. a Boolean result) or numeric.

The flip-flops which are used to perform counters are JK flip-flops such as the one depicted in Figure 8.16.

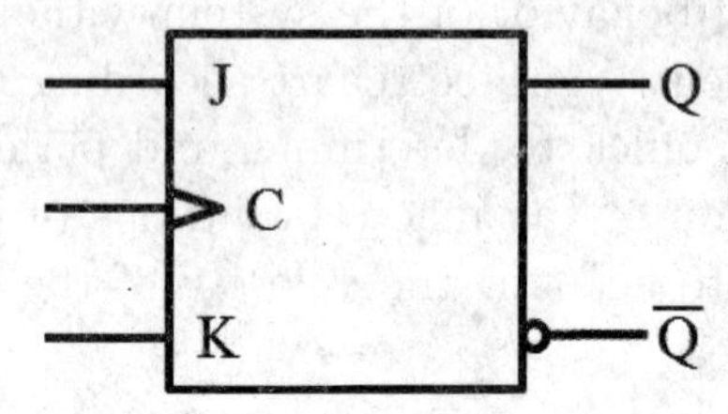

Figure 8.16 symbol of a JK flip-flop

A JK flip-flop has 3 input variables J, K, C and two output variables Q and $\bar{Q}$. The two input variables J and K are taken into account only on a rising edge of the input variable C; otherwise, the output variables remain the same. The input variable J is the memorization input variable, and the input variable K is the erasement input variable. Let's note Q_{n+1} the current state of the output variable and Q_n the internal variable which characterizes the previous state of the output variable. The specificity of this flip-flop is that, when $\uparrow C = 1$ and $J = K = 1$, the value of the state Q_{n+1} of the output variable Q is inverted with respect to the value of the previous state Q_n ($Q_{n+1} = \overline{Q_n}$): this corresponds to a **toggle** command. The truth table of a JK flip-flop is presented in Table 8.3.

Table 8.3 truth table of a JK flip-flop

$\uparrow C$	J	K	Q_{n+1}	$\overline{Q_{n+1}}$
0	X	X	Q_n	$\overline{Q_n}$
1	0	0	Q_n	$\overline{Q_n}$
	0	1	0	1
	1	0	1	0
	1	1	$\overline{Q_n}$	Q_n

The JK flip-flop owes its name neither to its inventor (who was named Eldred Nelson), nor to any scientific personality ... but to playing cards! Indeed, by combining the names of its two input variables J and K and the name of its output variable Q, we get a series of letters (J,Q,K) which correspond to the initials of three playing cards: the Jack, the Queen and the King. The input variable C is used as a Clock to synchronize the output variables, hence its name. This is the reason why the JK flip-flop is called a synchronous flip-flop. Counters can be performed by associating n JK flip-flops in toggle mode. The complement $\overline{Q_i}$ of the output of each flip-flop i is used as the clock for the $(i+1)$-th flip-flop. The output variables Q_i provide the natural binary code for the value Q of the counter:

$$(Q)_{10} = (Q_{n-1}Q_{n-2}\cdots Q_2Q_1Q_0)_2$$

The association of n JK flip-flops in toggle mode hence allows to perform a n-bit counter. For instance, the behavior of the association of 3 JK flip-flops illustrated in Figure 8.17 can be described by means of the timing diagrams in Figure 8.18. These timing diagrams consider that the changes of state of the flip-flops of the counter are simultaneous. However, practically, the electrical signals have a finite transmission speed, the changes of state of the flip-flops hence cannot be simultaneous, and there is a delay δt between each cause and its consequence which is not represented in the timing diagrams in Figure 8.18. This delay corresponds to the **transmission time** of the flip-flops. As a n-bit counter is performed by means of the association of n flip-flops, and since the clock input variable of each flip-flop is an output variable of the previous flip-flop, these transmission times hence cumulate and may cause counting errors if the cumulated delay becomes greater than the period of the clock signal. This is the reason why such a counter is called an asynchronous counter.

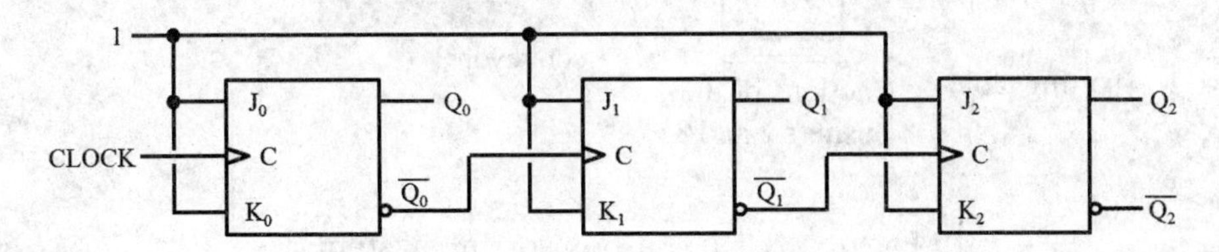

Figure 8.17 a 3-bit counter

8.3.2 Box Sorting System

Let's consider the box sorting system depicted in Figure 8.19. This system aims at sorting small and large boxes which are brought by the conveyor 1 by moving some of them to the conveyor 2 and the other ones to the conveyor 3. To do so, a sensor located near the pusher 1 detects the size (small or large) of the box in front of it. If the box is small, then the pusher 1 pushes it in front

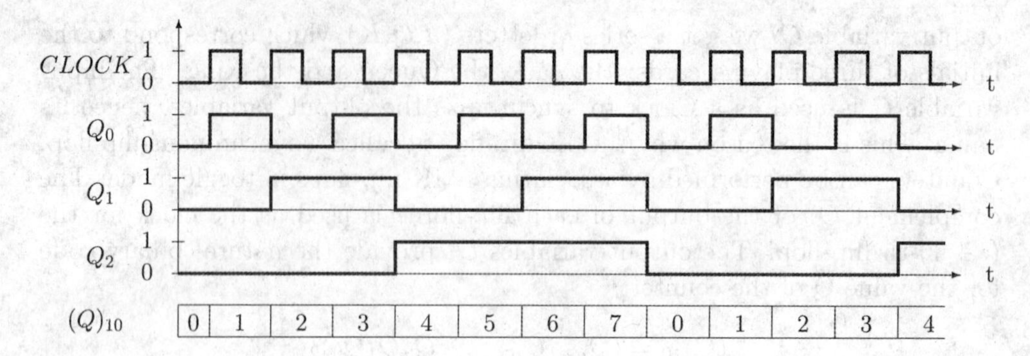

Figure 8.18 timing diagrams of a 3-bit counter using 3 JK flip-flops in toggle mode

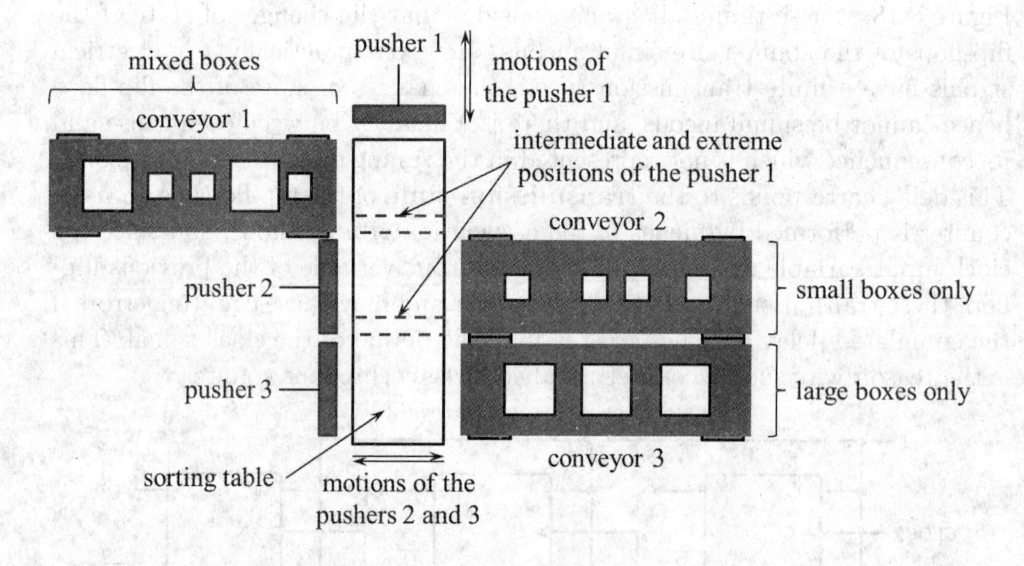

Figure 8.19 a box sorting system

of the pusher 2, which pushes it on the conveyor 2; if the box is large, then the pusher 1 pushes it in front of the pusher 3, which pushes it on the conveyor 3. The conveyors 2 and 3 are supposed to be continuously powered whereas the conveyor 1 needs to stop when a new box is inserted in the system: the conveyor 1 needs to stop when a box is detected by the sensor (i.e. when a box is in front of the pusher 1) until the box has been evacuated on the conveyor 2 or 3 and the associated pushers (pusher 1 and pusher 2 or 3) have come back to their initial position. We are going to determine the grafcet chart which describes the behavior of this system.

First of all, we need to define the actions which have to be executed as well as the information which have to be provided by the sensors to the system. The actions are as follows:

- powering of the conveyor 1 (that we will call PC1), which is supposed to be continuous (i.e. the conveyor is powered as long as PC1 is active; otherwise, it is stopped);
- forward motion of the pusher 1 to the position 2, FM P1/2 (which is the intermediate position right before the pusher 2);
- forward motion of the pusher 1 to the position 3, FM P1/3 (which is the extreme position right before the pusher 3);
- backward motion of the pusher 1, BM P1 (the pusher 1 comes back to its initial position);
- forward motion of the pusher 2, FM P2 (the pusher 2 pushes the box on the conveyor 2);
- backward motion of the pusher 2, BM P2 (the pusher 2 comes back to its initial position);
- forward motion of the pusher 3, FM P3 (the pusher 3 pushes the box on the conveyor 3); and
- backward motion of the pusher 3, BM P3 (the pusher 3 comes back to its initial position).

Some other notations could be used as well, e.g. M+ and M− for a forward and backward motion.

A sensor must be located near the pusher 1 to detect the size of the boxes. We will suppose that this sensor provides two logic variables lb (Large Box) and sb (Small Box) which are equal to 1 when the corresponding box is in front of the pusher 1. Besides, for the motions of the pushers to be correctly controlled, some sensors should allow to detect their position (in or out, and even out in an intermediate position for the pusher 1). The information provided by the sensors will hence be as follows:

- detection of a large box: lb;

- detection of a small box: sb;
- detection of the IN position of the pusher 1: p10;
- detection of the intermediate OUT position (position 2) of the pusher 1: p12;
- detection of the extreme OUT position (position 3) of the pusher 1: p13;
- detection of the IN position of the pusher 2: p20;
- detection of the OUT position of the pusher 2: p21;
- detection of the IN position of the pusher 3: p30; and
- detection of the OUT position of the pusher 3: p31.

All these actions will be associated with steps of the grafcet chart whereas all these detection information will be associated with transitions of the grafcet chart. Let's suppose that, initially, all the pushers are in their IN position, that the conveyor 1 is powered, and that boxes are transported by the conveyor 1 to the sorting system:

- the order PC1 hence is the only order executed at the initial step of the grafcet chart.
- the first event which will deactivate the initial step is the arrival of a box (lb or sb is true). A selection of sequences must hence be inserted below the initial step to take into account both cases. The grafcet chart is divided into two branches, each branch corresponding to a specific size of box. These two branches will have a similar structure:
  - on the one hand, if the box detected is large, then the pusher 1 must be moved to the extreme position 3 thanks to the action FM P1/3. When the box is in front of the pusher 3 (p13 is true), the pusher 1 comes back to its initial position and the pusher 3 moves forward, this forward motion being immediately followed by the backward motion of the pusher 3 as soon as its OUT position is detected (which means that the box is on the conveyor 3). Parallel sequences after the transition-condition p13 allow these operations, and the cycle is considered over when both the pushers 1 and 3 are back to their initial position.

– on the other hand, if the box detected is small, then the pusher 1 must be moved to the intermediate position 2 thanks to the action FM P1/2. When the box is in front of the pusher 2 (p12 is true), the pusher 1 comes back to its initial position and the pusher 2 moves forward, this forward motion being immediately followed by the backward motion of the pusher 2 as soon as its OUT position is detected (which means that the box is on the conveyor 2). Parallel sequences after the transition-condition p12 allow these operations, and the cycle is considered over when both the pushers 1 and 2 are back to their initial position.

The corresponding grafcet chart is depicted in Figure 8.20. Another structure based on many grafcet charts can also be used: such a structure will be presented in the second volume.

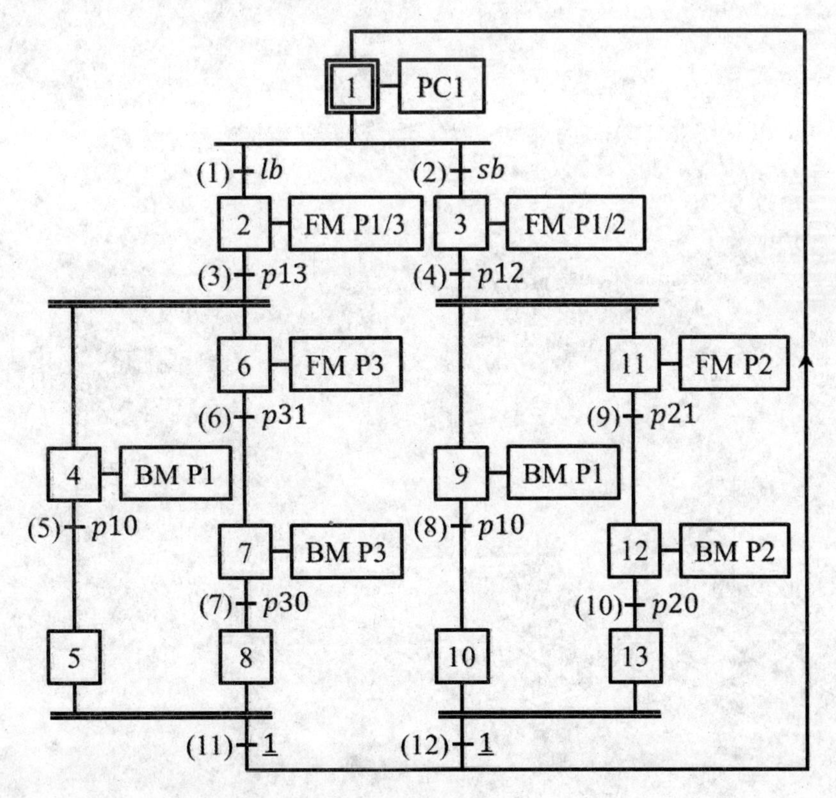

Figure 8.20 grafcet chart describing the behavior of the box sorting system in Figure 8.19